GNSS Atmospheric Seismology

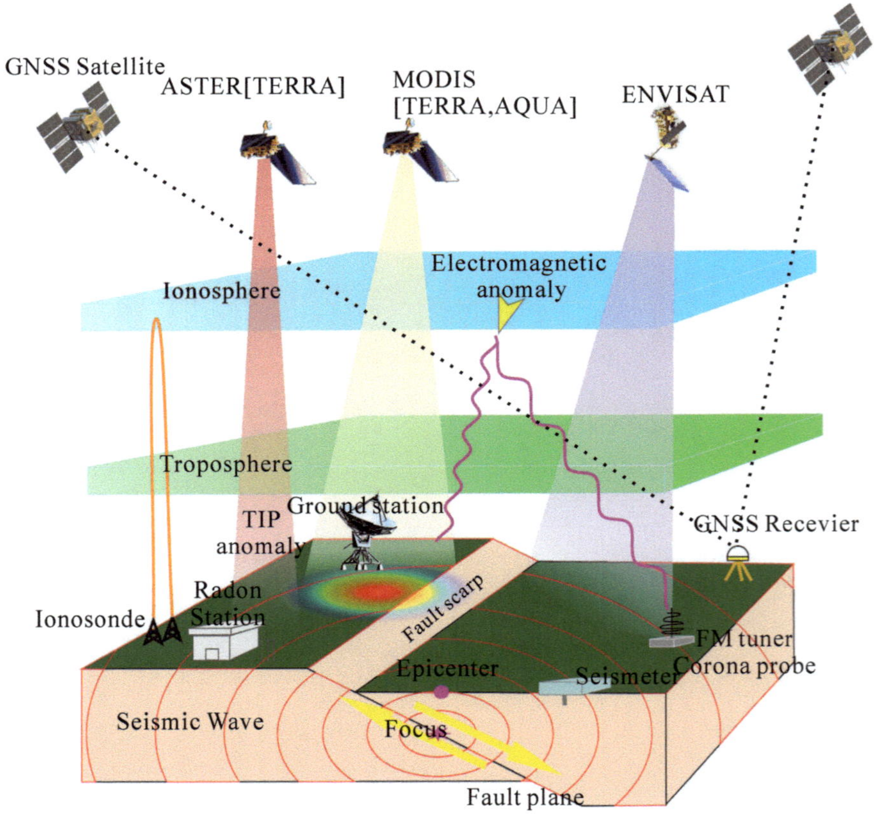

GNSS Satellite
ASTER[TERRA]
MODIS
[TERRA,AQUA]
ENVISAT
Electromagnetic
anomaly
Ionosphere
Troposphere
Ground station
TIP
anomaly
GNSS Recevier
Radon
Station
Ionosonde
Fault scarp
FM tuner
Corona probe
Epicenter
Seismeter
Seismic Wave
Focus
Fault plane

Shuanggen Jin · R. Jin · X. Liu

GNSS Atmospheric Seismology

Theory, Observations and Modeling

Springer

Shuanggen Jin
Shanghai Astronomical Observatory
Chinese Academy of Sciences
Shanghai, China

Nanjing University of Information Science and Technology
Nanjing, China

R. Jin
Shanghai Astronomical Observatory
Chinese Academy of Sciences
Shanghai, China

X. Liu
Shanghai Astronomical Observatory
Chinese Academy of Sciences
Shanghai, China

ISBN 978-981-10-3176-2 ISBN 978-981-10-3178-6 (eBook)
https://doi.org/10.1007/978-981-10-3178-6

This Springer imprint is published by the registered company Springer Nature Singapore Pte Ltd.
The registered company address is: 152 Beach Road, #21-01/04 Gateway East, Singapore 189721, Singapore

Preface

The earthquake is a very complex and broad topic, which is related to various scales motions of the Earth's surface mass and interior as well as the microscopic processes, such as the generation of electric charge and chemical reactions. Also, earthquakes often occur and result in civilian casualties and huge damages as well as secondary disasters, such as tsunami and landslide. Therefore, monitoring and understanding of earthquakes are still a major objective for many countries. The worldwide seismometers could estimate rapidly the location, the magnitude, and the focal mechanism of earthquakes, but the detailed rupture and pre-seismic anomalies are still not clear due to the lack of dense near-field observations and limited observations. Global Navigation Satellite Systems (GNSS), Interferometric Synthetic Aperture (InSAR), seismometer, and gravity measurements could provide unique insights on the size and rupture of the earthquake. However, understanding and prediction of earthquakes are still challenging or difficultly confirmed from traditional technique observations. Nowadays, seismic atmospheric disturbance observations may help us to get a more comprehensive and profound knowledge on the Earth's atmospheric response to the earthquake, volcano, tsunami, and solid Earth/ionosphere coupling, e.g., ionospheric or electromagnetic observations.

GNSS is a powerful tool not only for the crust deformation but also the seismic atmospheric–ionospheric variations, especially for regions with dense GNSS continuous operating stations. The neutral atmospheric parameters and ionospheric Total Electron Content (TEC) can be precisely estimated from ground-based GNSS and spaceborne GNSS Radio Occultation, which can be used to investigate the seismic atmospheric–ionospheric disturbances and may provide insights on the earthquake. Recently, several significant seismic ionospheric disturbances were observed from continuous GPS measurements with acoustic waves, Rayleigh wave, and gravity waves, e.g., the 2011 Mw = 9.1 Japan earthquake. The seismic ionospheric disturbances are probably driven by the ground-coupled airwaves from ground vertical motion of seismic waves propagating. Some mechanism on atmospheric/ionospheric anomalies and coupling processes between the atmosphere and solid Earth were also discussed. Furthermore, some pre-seismic atmospheric and ionospheric anomalies with several hours to several days before the main shock onset were also observed in the temperature, TEC, and NmF2 time series. Therefore, GNSS may provide us a new perspective to monitor and understand the earthquake with seismic atmospheric–ionospheric disturbances.

However, the detailed pattern, evolution, and mechanism of the seismic ionospheric disturbance are not clear together with earthquake sources. The relationship between the earthquake and the atmospheric–ionospheric variation is not understood comprehensively up to now. The earthquake threshold and pre-seismic ionospheric disturbances following earthquakes with different magnitudes, focal mechanism, depth, and external conditions are still challenging as well as the mechanism of solid Earth/ocean–atmosphere–ionosphere coupling, which needs to be improved for more cases and be verified with more GNSS measurements. This book presents GNSS atmospheric seismology with some recent progresses.

GNSS Atmospheric Seismology: Methods, Observations and Modelling, has been written as a monograph and textbook that guides the reader through the theory and practice of seismic atmospheric disturbances sounding as well as possible applications. This book includes Chap. 1: Introduction, Chap. 2: Atmospheric Changes and Observations, Chap. 3: GNSS Tropospheric Sounding, Chap. 4: GNSS Ionospheric Sounding, Chap. 5: Detection Methods for Ionospheric Disturbances, Chap. 6: Seismic Lower Atmospheric Anomalies, Chap. 7: Pre-seismic Ionospheric Anomalies, Chap. 8: Co-/Post-seismic Ionospheric Disturbances, Chap. 9: Two-Mode Seismo-ionospheric Disturbances, Chap. 10: Seismo-ionospheric Rayleigh Waves, Chap. 11: Epicenter from Ionospheric Disturbances, Chap. 12: Tsunami Ionospheric Disturbances, Chap. 13: Volcano Atmospheric Disturbances, Chap. 14: Volcanic Plumes Detection from GNSS SNR, and Chap. 15: Summary and Prospective. Chapters 1–4, 6–10 and 12–15 were contributed from Prof. Shuanggen Jin, Chaps. 5 and 11 and part of Chaps. 8 and 10 were contributed from R. Jin, Chaps. 12 and 13 were contributed from X. Liu, Chap. 14 was contributed from Qinyun Zhang, and part of Chap. 10 was contribute from Yuhan Liu as well as part of Chaps. 6 and 7 from Munawar Shah.

This book presents the theory, methods, results, and modeling of GNSS atmospheric seismology for scientists and users who have basic background and experiences in GNSS and seismology. Furthermore, it is also useful for the increasing number of next generation multi-GNSS scientists, engineers, and users' community as well as hazards mitigation and reduction. We would like to thank Assistant Editor's help and Springer-Verlag for their cordial collaboration and help during the process of publishing this book.

Shanghai/Nanjing, China — Shuanggen Jin
Shanghai, China — R. Jin
Shanghai, China — X. Liu

Contents

Part II Seismic Atmospheric Disturbances

Abbreviations

ABL	Atmospheric Boundary Layer
AGW	Acoustic Gravity Wave
ART	Algebraic Reconstruction Technique
ASTER	Advanced Spaceborne Thermal Emission and Reflection Radiometer
AVHRR	Advanced Very High Resolution Radiometer
AVO	Alaska Volcano Observatory
BDS	BeiDou Navigation Satellite System
BPR	Bottom Pressure Records
C/A	Coarse/Acquisition
CAS	Chinese Academy of Sciences
CDMA	Code Division Multiple Access
CEA	China Earthquake Administration
CHAMP	Challenging Mini-satellite Payload
CID	Co-seismic Ionospheric Disturbances
CIT	Computerized Ionospheric Tomography
CMONOC	Crustal Movement Observation Network of China
CMT	Centroid Moment Tensor
CODE	Center for Orbit Determination in Europe
COSMIC	Constellation Observing System for Meteorology, Ionosphere, and Climate
CYGNSS	CYclon GNSS
DART	Deep-ocean Assessment and Reporting of Tsunamis
DCB	Differential Code Biases
DD	Double Difference
DEMETER	Detection of Electro-Magnetic Emissions Transmitted from Earthquake Regions
DGPS	Differential GPS
DOT	Department of Transportation
DOY	Day of Year

Dst	Disturbance storm time
ECMWF	European Centre for Medium-Range Weather Forecasts
EIA	Equatorial Ionization Anomaly
EISCAT	European Incoherent Scatter Scientific Association
ENVISAT	ENVIronmental SATellite
ESA	European Space Agency
EST	Equivalent Slab Thickness
EU	European Union
FDMA	Frequency Division Multiple Access
FDSN	International Federation of Digital Seismograph Networks
FDTD	Finite Difference Time Domain
FFT	Fast Fourier Transform
GAGAN	GPS-Aided Geo Augmented Navigation
Galileo	Galileo Navigation Satellite System
GAMIT	GPS analysis software by Massachusetts Institute of Technology
GBAS	Ground-Based Augmentation Systems
GDOP	Geometric Dilution of Precision
GEO	Geostationary Earth Orbit
GEONET	GPS Earth Observation Network in Japan
GFZ	GeoForschungsZentrum Potsdam
GIM	Global Ionospheric Map
GIPSY-OASIS	GNSS-Inferred Positioning System and Orbit Analysis
GLONASS	GLOnass NAvigation Satellite System of Russia
GMF	Global Mapping Function
GNSS	Global Navigation Satellite Systems
GNSS-R	GNSS-Reflectometry
GO	Geometric Optics
GOCE	Gravity Field and Steady-State Ocean Circulation Explorer
GPS	Global Positioning System
GPS/MET	GPS/Meteorology
GRACE	Gravity Recovery and Climate Experiment
GSI	Geographical Survey Institute in Japan
GSN	Global Seismographic Network
GTS	Global Telecommunication System
HFI	Hardy Function Interpolation
IAG	International Association of Geodesy
IERS	International Earth Rotation and Reference System Service
IGS	International GNSS Service
IGW	Internal Gravity Wave
InSAR	Interferometric Synthetic Aperture Radar
IPP	Ionosphere Pierce Point
IRI	International Reference Ionospheric
IRIS	Incorporated Research Institutions for Seismology
IRNSS	India's Regional Navigation Satellite Systems
ISR	Ionospheric Scatter Radar

ITRF	IERS Terrestrial Reference Frame
JMA	Japan Meteorological Agency
JPL	Jet Propulsion Laboratory
LAAS	Local Area Augmentation System
LC	Linear Combination
LEO	Low Earth Orbit (satellite)
LiDAR	Light Detection And Ranging
LOS	Line of Sight
LS	Least Squares (adjustment)
LST	Land surface temperature
LT	Local Time
MART	Multiplicative Algebraic Reconstruction Technique
MDA	Maximum Disturbance Azimuth
MDD	Maximum Disturbance Distance
MDP	Maximum Disturbance Point
MEO	Medium Earth Orbit
mHz	milli-Hertz
MIT	Massachusetts Institute of Technology
MODIS	MODerate-resolution Imaging Spectroradiometer
MSS	Mean of the Square Slopes
NASA	National Aeronautics and Space Administration
NCAR	National Center for Atmospheric Research
NGDC	US National Geophysical Data Center
NMF	Niell Mapping Function
NOAA	National Oceanic and Atmospheric Administration
NWP	Numerical Weather Prediction
OBSIP	Ocean Bottom Seismograph Instrument Pool
OLR	Outgoing Long-wave Radiation
OMI	Aura/Ozone Monitoring Instrument
PASSCAL	Portable Array Seismic Studies of the Continental Lithosphere
PBO	Plate Boundary Observatory
PNT	Positioning, Navigation and Timing
POD	Precise Orbit Determination
PPP	Precise Point Positioning
PRN	Pseudo-Random Noise
PWV	Precipitable Water Vapor
QOA	Quasi-Optimum Algorithm
QZSS	Quasi-Zenith Satellite System
RINEX	Receiver Independent Exchange
RMS	Root Mean Square
RO	Radio Occultation
RTK	Real-Time Kinematic
SAC-C	Satellite de Aplicaciones Cientificas-C
SD	Single Difference
SHAO	Shanghai Astronomical Observatory

SID	Seismic Ionospheric Disturbance
SIP	Sub-ionospheric Pierce Points
SLM	Single Layer Model
SNR	Signal-to-Noise Ratio
SPIDR	Space Physics Interactive Data Resource
STD	Slant Tropospheric Delay
STEC	Slant Total Electron Content
SVD	Singular Value Decomposition
TEC	Total Electron Content
TECU	Total Electron Content Unit
TID	Traveling Ionospheric Disturbances
TIE-GCM	Thermosphere–Ionosphere–Electrodynamics General Circulation Model
UHF	Ultra High Frequency
UNAVCO	University NAVSTAR Consortium
USGS	US Geological Survey
UT	Universal Time
UTC	Coordinated Universal Time
VEI	Volcanic Explosivity Index
VHF	Very High Frequency
VMF	Vienna Mapping Functions
WGS	World Geodetic System
WVR	Water Vapor Radiometry
ZHD	Zenith Hydrostatic Delay
ZTD	Zenith Tropospheric Delay
ZWD	Zenith Wet Delay

Part I
Background and Observations

Chapter 1
Introduction

1.1 History

Earthquakes often occurred with rapid release of energy from the Earth's crust and resulted in huge casualties and damages. Furthermore, secondary disasters are sometimes caused by earthquakes, such as tsunamis, landslides, and quake lakes. According to the US Geological Survey recorded (USGS), annual earthquakes of Mw > 6.5 are nearly 60 times, particularly recent 2004 Mw = 9.1 Sumatra, 2008 Mw = 8.0 Wenchuan, 2010 Mw = 8.8 Chile 2011 Mw = 9.0 Tohoku, and 2015 Mw = 7.8 Nepal earthquakes (Fig. 1.1). Therefore, it has been a hot topic to monitor and predict the earthquake. The earthquake is a very complex and broad topic, which is related to various scales motions of the Earth's surface mass and interior as well as the microscopic processes, such as the generation of electric charge and chemical reactions (Pulinets and Boyarchuk 2005). A number of techniques and observations have been developed to monitor and understand earthquakes. Seismometers around the globe could rapidly estimate the magnitude, the location and the focal mechanism of earthquakes, but the source mechanism and the extent of the rupture still have large uncertainty due to the lack of near-field observations and the method of earthquake source inversion. The horizontal extent of the source plays a fundamental role in the estimation of the earthquake risk for warning system. Global Navigation Satellite Systems (GNSS), Interferometric Synthetic Aperture Radar (InSAR), gravimeter, and strong motion measurements could provide unique insights on the size and kinematic rupture of the earthquake (Jin et al. 2006, 2007, 2010; Cambiotti et al. 2011). However, it is still challenging and difficult to fully monitor and predict earthquakes based on traditional measurements (Burchfiel et al. 2008; Jin et al. 2013).

Earthquakes are normally caused by the collision of tectonic plates. The accumulated huge energy release and generated waves from the Earth's interior may propagate into the atmosphere through weak zones on the Earth's crust, which will cause anomalies in the temperature, the pressure and electromagnetic fields in the atmosphere. The first distinctive signal in the air pressure was detected before the 1964 Alaskan earthquake (Bolt 1964), which was the generation of strong electric

S. Jin et al., *GNSS Atmospheric Seismology*,
https://doi.org/10.1007/978-981-10-3178-6_1

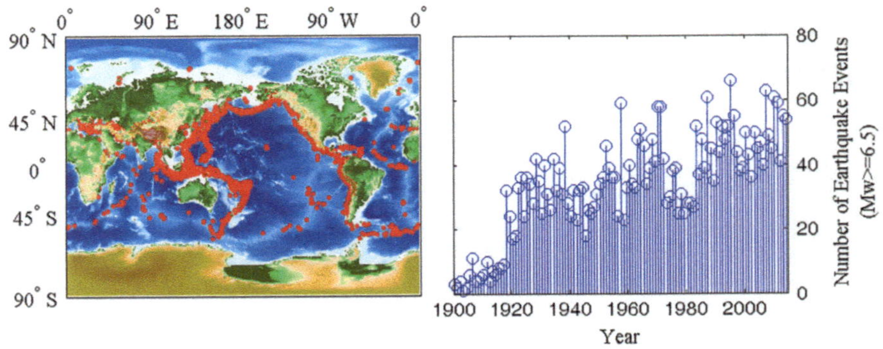

Fig. 1.1 The distribution of historical earthquakes from 1990 to 2018

field around the epicenter and as a result of this electric field plasma drift from earthquake epicenter excited free ions motion in the atmosphere. This finding was later paid more attention for special interests and initiatives to monitor and understand earthquakes from the atmospheric observations, including pre-, co-, and post-seismic atmospheric anomalies.

1.2 Seismo-Atmospheric Anomalies

Thermal anomalies were found near the earthquake preparation zone due to the release of uncontrolled energy and stress (Freund 2000), which spread out like electric dynamo from the Earth's interior. Dobrovolsky et al. (1979) have estimated thermal anomalies along geographic latitude and longitude around earthquake preparation zones. Later, more surface temperature anomalies following big earthquakes were found from measurements. A number of significant anomalies following big earthquakes were detected recently by ground and satellite observations. For example, Blackett et al. (2011) found anomalous temperature patterns before the 2001 Gujarat earthquake from 6 years of MODIS Land surface temperature (LST) data. Spiky peaks have been observed near the earthquake as thermal anomalies. It was practically proved in laboratory that the porosity of the rocks near the epicenter is truncated. It could be possible that the variations in these gaps might be due to enormous energy discharges at some dynamic zone while the remaining non-active zones might not suffer large energy release (Freund et al. 1993).

Electromagnetic anomalies have been also recorded before or after earthquakes, which attracted more scientists to study such kind of fluctuations near the epicenter. Cole and Pierce (1965) investigated the vertical electron profile anomalies before the earthquake and the threshold height variations of the atmosphere that suffer from electromagnetic waves prior to earthquakes, including the retroacted wave path of VLF and ULF signals in the ionosphere and changes in the motion of infrasonic waves

(Fraser-Smith et al. 1994). The electric current from epicenter has been overlapped by electron tides at the equatorial region, which was seen in the form of equatorial ionization anomalies (Tzur and Roble 1985). Anomalies of the electromagnetic field around the earthquake preparation zone were also observed by the low frequency electromagnetic waves (Fraser-Smith et al. 1990; Hayakawa and Fujinawa 1994). Major interdisciplinary signatures of the shearing crustal rocks have been recorded by radio waves and other low frequency waves in the form of Acoustic-Gravity Waves (AGWs) that highlighted the changes in the ionosphere preceded the earthquakes (Molchanov et al. 1995, 1998). Furthermore, a developed comprehensive model of the seismogenic electromagnetic waves prior to earthquake was shown the lithosphere and ionosphere couplings. In addition, Sharma et al. (2010) and Kuo et al. (2011) showed that electric field around the earthquake preparation zone was responsible to drift the SIP eastward, and found a spiky peak of EIA that bulges out eastward prior to earthquakes.

In addition, the upper atmospheric perturbation related to seismic events was firstly detected in 1964 mainly based on the analyses of Doppler records and vertical sounder ionograms after the 1964 Mw = 9.2 Alaska earthquake (Davies and Baker 1965; Leonard and Barnes 1965). The acoustic-gravity wave associated to tectonic deformation following the earthquake was found at the F region (Row et al. 1966, 1967). These preliminary results implicated that the energy released during the rupture of the fault was not only propagating in the solid Earth but also leaking out to the Earth's atmosphere and even to the ionosphere, resulting in strong perturbations of plasma density and plasma speed. Many others Doppler sounder observations related to earthquake events with magnitude of larger than 6.5 followed the Alaska event showed that the Doppler sounders could be considered as ionospheric seismometers. Particular emphasis was related to the Doppler sounder detection of the ionospheric signature induced by Rayleigh wave propagation.

Furthermore, some seismo-atmospheric anomalies were observed in the pressure, temperature, electromagnetic field, and the ionosphere in the past time, which may provide new insights on earthquakes. However, the lithosphere–atmosphere–ionosphere coupling is very complex or controversial (Pulinets and Ouzounov 2011). Kamogawa (2006) proposed the possible mechanisms for energy transport from the lithosphere to the atmosphere–ionosphere in Fig. 1.2. Generally, the atmospheric electric field generated on or near the ground surface during the seismic period is assumed to cause the atmospheric anomalies.

1.3 Seismo-Ionospheric Disturbances

Since the first ionospheric perturbation was detected after the 1964 Alaska earthquake by Doppler records and vertical sounder ionograms in 1965 (Davies et al. 1965), later many observation results were reported by various techniques, including Doppler sounder, ionosonde, Global Positioning System (GPS) (Jin et al. 2011a, 2015). The vertical displacement, at teleseismic distance, induced by Rayleigh waves

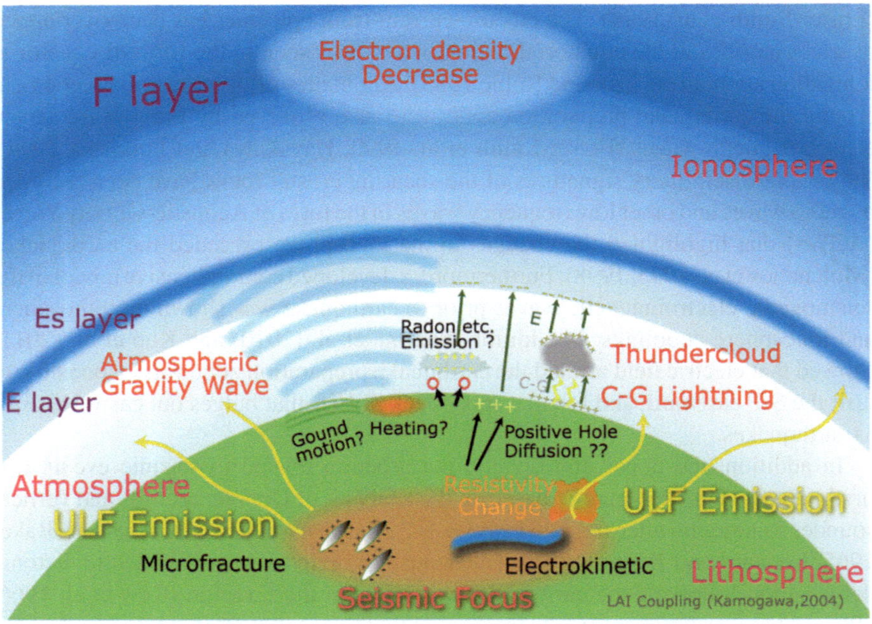

Fig. 1.2 Lithospheric–atmospheric–ionospheric disturbances and coupling (Kamogawa 2006)

produces an acoustic wave that propagates vertically until the ionosphere by dynamic coupling. During the upward propagation, the double effect of the conservation of kinetic energy and the exponential decreasing of the atmospheric density, strongly amplify the acoustic wave to reach the ionosphere and create strong variations in the plasma density and plasma velocity (detected by Doppler sounder). Consequently, the propagation proprieties of Rayleigh waves could be estimated by observing the ionopshere (Najita and Yuen 1979; Tanaka et al. 1984; Artru et al. 2004). Recently, more observation evidences on seismo-ionospheric anomalies were reported from different measurements (Jin et al. 2010, 2011b, 2014; Occhipinti et al. 2010, 2013).

Many satellites are continuously orbiting around the Earth in far and near orbits to detect magnetic field variations as well as changes in natural constituent of the Earth, which suffered changes related to earthquakes. There are some refined missions to monitor continuous and transient variations in the atmosphere related to earthquakes. All the satellite missions have a vital role in specifying variations in the ionosphere related to earthquake but the French micro-satellite of Detection of Electro-Magnetic Emissions Transmitted from Earthquake Regions (DEMETER) was the most promising one. The DEMETER mission was launched on 29 June, 2004 with providing electromagnetic fluctuation data related to earthquakes. DEMETER data have been examined in various ways and methods to define the perturbations of the ionosphere caused by acoustic gravity waves from earthquakes (Venkatraman and Heelis 2000; Pisa et al. 2011; Li and Parrot 2013; Ryu et al. 2014). In addition, spaceborne GPS Radio Occultation provides high-spatial resolution coverage of global ionospheric density profiles to investigate ionospheric disturbances related to earthquakes, particularly Constellation Observing System for Meteorology, Ionosphere, and Climate (COSMIC) with six microsatellites moving in the sun-synchronous orbits with hourly occultation of ionospheric profiles.

1.3.1 Pre-seismic Ionospheric Anomalies

Several pre-seismic ionospheric anomalies were recently observed in the TEC, NmF2 (F2 peak electron density) or hmF2 (F2 peak height) time series (Liu et al. 2006; Heki 2011), however, the mechanism is still unclear and the debate is still open. The pre-seismic ionospheric disturbances are found significantly because of the probability of a new earthquake precursor, while the lithosphere–atmosphere–ionosphere coupling process is still unknown. Ionospheric anomalies appearing several hours to several days before the main shock onset were reported by lots of experts (e.g., Liu et al. 2004; Heki and Enomoto 2013). Generally, the atmospheric electric field generated on or near the ground surface during the seismic period is assumed to cause the ionospheric anomalies. However, whether these pre-seismic ionospheric disturbances are related to the earthquake or not is still controversial. It is difficult to extract the clear background variation of ionospheric TEC. Up to now, there is no explicit definition for pre-seismic ionospheric disturbances. The effect of TEC filtering or de-trending is negligible especially for considering the disturbance amplitude. Furthermore, the pre-seismic ionospheric anomaly is probably fake signal induced by the data processing but not earthquake precursor with limited observations (Kamogawa and Kakinami 2013; Jin et al. 2015).

Denser GPS TEC has been widely used to investigate ionospheric anomalies prior to earthquake. Statistical analysis of GPS TEC/foF2 prior to earthquakes has been reported since 1970. For example, Liu et al. (2006) showed increase/decrease of GPS TEC anomalies 5 days prior to $M > 5.0$ earthquakes. They further described the phenomena of $E \times B$ enhancement for Taiwan, which was major reason of bulky anomalous TEC. Also, seismo-ionospheric anomalies have been statistically proved that it is depended on magnitude as well as focal depth of earthquakes (Le et al. 2011). Large magnitude earthquakes with a shallow focal depth are significant to cause huge variations in the top atmospheric layers as compared to small magnitude earthquakes. It has been observed that the plasma fountain in the form of bubbles raised from epicenter and inducted in the ionosphere at specific latitudes raises the temperature of F2 layer (Pulinets et al. 2003). These plasma bubbles moved upward in the atmosphere at specific altitude and caused panic in the free ions, which was seen in the form seismo-ionospheric anomalies prior to earthquakes. Acoustic-gravity waves from epicenter prior to earthquakes bulge into the troposphere, which is further indented in the ionosphere to disturb the normal motion of free ions and particle in the top layer atmosphere for many different earthquakes (Rolland et al. 2010; Saito et al. 2011). The pattern and regularity of energy that is responsible to create panic in the top layer atmosphere has been clarified by Huba et al. (2009a, b, c).

In addition, pre-seismic crustal motion and their effects on behavior of VHF, UHF as well as GPS TEC signals were analyzed with a mathematical model in a precise way for the 2011 massive Tohuku earthquake (Heki 2011; Kamogawa and Kakinami 2013). To find the intensity of ground shaking along the fault lines, various GPS observations were deployed to monitor the short-term fluctuation in the Earth crust. Seismo-ionospheric anomalies were observed before ground shaking from GPS observations by analyzing their data both mathematically and statistically (Rishbeth 2007). Apart from this, vertical height (hmF2) of the ionosphere F2 layer, frequency of F2 layer (foF2) were determined with ionosonde data to show the drastic changes in vertical profile of ionospheric free electrons (Liu et al. 2006). Variability of F2 layer has links with upcoming earthquake because of unexpected random motion of vertical electrons in utmost layer of atmosphere, which has been shown by IRI 2007, Nequick and TIE-GCM models (Roble et al. 1988; Wanliss and Showalter 2006). These models solved a major issue in scanning the atmosphere beyond 100 km for searching precursors of earthquake.

Electrons cloud in the ionosphere F2 layer might suffer ascending and descending due to enormous amount of energetic gases evolved from epicenter to the atmosphere along different latitudes (Rishbeth 2006, 2009; Rishbeth and Mendillo 2001). However, it must be distinguished that these sudden variations in F2 layer might not be due to foreshocks of earthquakes. Possible mechanism of enhancement in TEC before the 2008 Wenchuan earthquake with simulation models and spatial pattern was discussed due to contemporary earthquake and other cause of TEC enhancement has been rejected with possible arguments (Zhao et al. 2008). Some noteworthy research by observing the spatial and temporal analysis of GPS TEC showed abrupt enhancement/depletion related to earthquakes (Liu et al. 2001; Riousset et al. 2010). Many statistical and mathematical models have been established for disclosing the seismo-ionospheric anomalies in the confined region for earthquakes and Tsunami (Tsuji et al. 1995). However, all these pre-earthquake anomalies are still not convinced yet and need more studies.

1.3.2 Co-/Post-seismic Ionospheric Anomalies

Since the first time the Total Electron Content(TEC) anomalous signal was observed by Calais, and Minster (1995) after the Northridge earthquake in California on January 17, 1994, numerous studies have been intended to study co-/post-seismic ionospheric disturbances related to earthquakes (Liu et al. 2004; Afraimovich et al. 2010; Heki and Enomoto 2013). For instance, Ducic et al. (2003) detected TEC perturbations related to the acoustic waves associated with the Rayleigh waves measured by the dense Californian GPS network following the Denali Earthquake. The measurement of the horizontal speed of the ionospheric perturbations roughly agrees with the speed of Rayleigh waves propagating on the Earth surface. Heki and Ping (2005) revealed the apparent velocity and directivity based on the traveling time diagram for the 2003 Tokachi-oki earthquake (M 8.0) from GEONET GPS observations, and highlighted a north–south asymmetry of the seismic ionospheric effects related to the geomagnetic field inclination. Co-seismic ionospheric disturbances detected by GNSS near the epicenter were thought to be related to the acoustic-gravity waves excited sequentially along the rupture of the fault (Heki et al. 2006). Recently, a more comprehensive understanding of seismic ionospheric effects for the 2011 Tohoku earthquake (M 9.0) was revealed by abundant GNSS observations collected by GEONET near the epicenter (Rolland et al. 2011; Liu et al. 2011; Tsugawa et al. 2011; Occhipinti et al. 2013). The extremely dense TEC observation performed by the GEONET network clearly highlighted acoustic-gravity waves generated at the epicenter, acoustic waves coupled with Rayleigh wave, and gravity waves coupled with the propagation of the following tsunami (Occhipinti et al. 2013). As clearly showed by Shinagawa et al. (2013) and Rolland et al. (2011), the TEC observations supplied by the dense GEONET network during the 11 March 2011 Tohoku earthquake and the 25 September 2003 Tokachi-Oki earthquake, provide a unique chance to investigate the seismic ionospheric disturbances related to the fault rupture in vicinity of the epicenter, as well as TEC perturbation related to the propagation of Rayleigh waves and tsunami in far-field. Indeed, the Japanese GNSS network (GEONET) is the densest of the world, with over 1200 continuous GNSS stations operated by Geographical Survey Institute (GSI) Japan. Astafyeva et al. (2013) showed the possibilities to estimate the parameters of seismic source by observing the first post-seismic signal appearing in the ionosphere 8 min after the rupture. As a result of the integrated nature of the TEC, the sensitivity of GPS measurement to the ionospheric perturbation related to the seismic activity is mainly limited to larger events.

The seismic ionospheric disturbance amplitude and propagation characteristics are still not understood comprehensively following earthquakes with different magnitudes, focal mechanisms, and space weather. The mechanism of solid Earth/ocean–atmosphere–ionosphere coupling still needs to be improved for different cases and be verified with more GNSS measurements. Perevalova et al. (2014) examined the earthquakes with magnitudes from 4.1 to 8.8 in different regions (Baikal region, Kuril Islands, Japan, Greece, Indonesia, China, New Zealand, Salvador, and Chile) and found that there was a threshold magnitude (near $Mw = 6.5$), below which there were no pronounced earthquake-induced TEC disturbances. The seismic ionospheric disturbance amplitude increases proportionally with earthquake magnitudes generally. Comparing to strike-slip earthquakes, normal/reverse fault earthquakes are usually followed with much more intense seismic ionospheric disturbances. It seems to show that the vertical displacement in the earthquake zone plays a more important role in the formation of seismic ionospheric disturbances. In addition, the geomagnetic condition also has a significant effect on the amplitude of seismic ionospheric disturbance. The disturbance amplitude is highly relative in disturbed geomagnetic days, which show that the filtered TEC series is affected by not only earthquake generated disturbances but also the geomagnetic variations. As we are known, the variation of Earth's ionosphere is complex because of its high temporal dynamic and spatial discrepancy under the influence of various factors such as space weather, natural event, anthropogenic activities, etc. It is still a challenge to extract clear and exact earthquake generated ionospheric disturbances for the worldwide earthquakes happened at different time with different external conditions (space weather, time of day, location et al.). The relationship between the earthquake and the ionospheric variation is not clear or understood comprehensively up to now. The earthquake threshold, propagation characteristics, and coupling process of seismic ionospheric disturbances following earthquakes with different magnitudes, focal mechanisms, depths, and external conditions will be still a hot topic in next decades with more observation data.

1.4 Summary

Since the first ionospheric perturbation related to seismic events was detected in 1965, more evidences are found later. Earthquakes will induce atmospheric and ionospheric disturbances, which may provide new insights on the details of fault rupture with high-spatial-temporal resolution. With development of multi-frequency and multi-systems of Global Navigation Satellite Systems (GNSS), including US's Global Positioning System (GPS), Russia's GLONASS, the EU's Galileo and China's Bei-Dou, it can monitor not only the crust deformation but also the seismic ionospheric variations in details, especially for regions with dense GNSS continuous operating stations. The tropospheric and ionospheric delays derived from ground-based GNSS networks and spaceborne GNSS Radio Occultation provide a good chance to monitor and understand seismic atmospheric disturbances, which may give us a new perspective to observe and forecast the earthquake in the future.

This book will present GNSS atmospheric seismology with the theory, observations, and modeling, including techniques, methods, results, characteristics, and coupling between solid Earth, the atmosphere and the ionosphere. Case studies and modeling on earthquakes, volcanoes, and tsunamis are presented as well as global statistical analysis of seismic ionospheric disturbances related to the global earthquakes of Mw $\geq$ 5.0 from GNSS observations.

References

Afraimovich, E. L., Ding, F., Kiryushkin, V., Astafyeva, E., Jin, S. G., & Sankov, V. (2010). TEC response to the 2008 Wenchuan earthquake in comparison with other strong earthquakes. *International Journal of Remote Sensing, 31*(13), 3601–3613.

Artru, J., Farges, T., & Lognonné, P. (2004). Acoustic waves generated from seismic surface waves: propagation properties determined from Doppler sounding observations and normal-mode modelling. *Geophysical Journal International, 158*(3), 1067–1077.

Astafyeva, E., Rolland, L., Lognonné, P., Khelfi, K., & Yahagi, T. (2013). Parameters of seismic source as deduced from 1 Hz ionospheric GPS data: Case study of the 2011 Tohoku-oki event. *Journal of Geophysical Research: Space Physics, 118*(9), 5942–5950.

Blackett, M., Wooster, M. J., & Malamud, B. D. (2011). Exploring land surface temperature earthquake precursors: A focus on the Gujarat (India) earthquake of 2001. *Geophysical Research Letters, 38,* L15303. https://doi.org/10.1029/2011GL048282.

Bolt, B. A. (1964). Seismic air waves from the great 1964 Alaskan earthquake. *Nature, 202,* 1094–1095.

Burchfiel, B. C., Royden, L. H., van der Hilst, R. D., et al. (2008). A geological and geophysical context for the Wenchuan earthquake of 12 May 2008, Sichuan, People's Republic of China. *GSA Today, 18*(7), 4–11.

Calais, E., & Minster, J. B. (1995). GPS detection of ionospheric perturbations following the January 17, 1994, Northridge earthquake. *Geophysical Research Letters, 22*(9), 1045–1048.

Cambiotti, G., Bordoni, A., Sabadini, R., & Colli, L. (2011). GRACE gravity data help constraining seismic models of the 2004 Sumatran earthquake. *Journal of Geophysical Research: Solid Earth, 116*(B10), B10403. https://doi.org/10.1029/2010JB007848.

Cole, R. K., Jr., & Pierce, E. T. (1965). Electrification in the Earth's atmosphere for altitudes between 0 and 100 kilometers. *Journal of Geophysical Research, 70,* 2735–2749. https://doi.org/10.1029/JZ070i012p02735.

Davies, K., & Baker, D. M. (1965). Ionospheric effects observed around the time of the Alaskan earthquake of March 28, 1964. *Journal of Geophysical Research, 70,* 2251–2253.

Dobrovolsky, I. P., Zubkov, S. I., & Miachkin, V. I. (1979). Estimation of the size of earthquake preparation zones. *Pure and Applied Geophysics, 117,* 1025–1044. https://doi.org/10.1007/BF00876083.

Ducic, V., Artru, J., & Lognonné, P. (2003). Ionospheric remote sensing of the Denali Earthquake Rayleigh surface waves. *Geophysical Research Letters, 30*(18), 1951. https://doi.org/10.1029/2003GL017812.

Fraser-Smith, A. C., Bernardi, A., McGill, P. R., Ladd, M. E., Helliwell, R. A., & Villard, O. G., Jr. (1990). Low-frequency magnetic field measurements near the epicenter of the Ms 7.1 Loma Prieta earthquake. *Geophysical Research Letters, 17,* 1465–1468.

Fraser-Smith, A. C., McGill, P. R., Helliwell, R. A., & Villard, O. G., Jr. (1994). Ultra-low frequency magnetic field measurements in southern California during the Northridge earthquake of 17 January, 1994. *Geophysical Research Letters, 21,* 2195–2198.

Freund, F. (2000). Time-resolved study of charge generation and propagation in igneous rocks. *Journal of Geophysical Research, 105,* 11001–11019. https://doi.org/10.1029/1999JB900423.

Freund, F., Freund, M. M., & Batllo, F. (1993). Critical review of electrical conductivity measurements and charge distribution analysis of magnesium oxide. *Journal of Geophysical Research, 98,* 22209–22229.

Hayakawa, M., & Fujinawa, Y. (Eds.) (1994). *Electromagnetic phenomena related to earthquake prediction*, Terra Scientific Pub. Comp., Tokyo, pp. 677.

Heki, K. (2011). Ionospheric electron enhancement preceding the 2011 Tohoku-Oki earthquake. *Geophysical Research Letters, 38,* L17312. https://doi.org/10.1029/2011GL047908.

Heki, K., & Enomoto, Y. (2013). Preseismic ionospheric electron enhancements revisited. *Journal of Geophysical Research: Space Physics, 118,* 6618–6626. https://doi.org/10.1002/jgra.50578.

Heki, K., & Ping, J. (2005). Directivity and apparent velocity of the coseismic ionospheric disturbances observed with a dense GPS array. *Earth and Planetary Science Letters, 236*(3), 845–855.

Heki, K., Otsuka, Y., Choosakul, N., Hemmakorn, N., Komolmis, T., & Maruyama, T. (2006). Detection of ruptures of Andaman fault segments in the 2004 great Sumatra earthquake with coseismic ionospheric disturbances. *Journal of Geophysical Research: Solid Earth,* 111(B9), B09313. https://doi.org/10.1029/2005jb004202.

Huba, J. D., Joyce, G., Krall, J., & Fedder, J. (2009a). Ion and electron temperature evolution during equatorial spread F. *Geophysical Research Letters, 36,* L15102. https://doi.org/10.1029/2009GL038872.

Huba, J. D., Krall, J., & Joyce, G. (2009b). Atomic and molecular ion dynamics during equatorial spread F. *Geophysical Research Letters, 36,* L10106. https://doi.org/10.1029/2009GL037675.

Huba, J. D., Ossakow, S. L., Joyce, G., Krall, J., & England, S. L. (2009c). Three-dimensional equatorial spread F modeling: Zonal neutral wind effects. *Geophysical Research Letters, 36,* L19106. https://doi.org/10.1029/2009GL040284.

Jin, S. G., Li, Z. C., & Park, P. H. (2006). Seismicity and GPS constraints on crustal deformation in the southern part of the Korean Peninsula. *Geosciences Journal, 10*(4), 491–497. https://doi.org/10.1007/BF02910442.

Jin, S. G., Park, P. H., & Zhu, W. Y. (2007). Micro-plate tectonics and kinematics in Northeast Asia inferred from a dense set of GPS observations. *Earth and Planetary Science Letters, 257*(3–4), 486–496. https://doi.org/10.1016/j.epsl.2007.03.011.

Jin, S. G., Zhu, W., & Afraimovich, E. (2010). Co-seismic ionospheric and deformation signals on the 2008 magnitude 8.0 Wenchuan Earthquake from GPS observations. *International Journal of Remote Sensing, 31*(13), 3535–3543.

Jin, S. G., Feng, G. P., & Gleason, S. (2011a). Remote sensing using GNSS signals: current status and future directions. *Advances in Space Research, 47*(10), 1645–1653. https://doi.org/10.1016/j.asr.2011.01.036.

Jin, S. G., Han, L., & Cho, J. (2011b). Lower atmospheric anomalies following the 2008 Wenchuan Earthquake observed by GPS measurements. *Journal of Atmospheric and Solar-Terrestrial Physics, 73*(7), 810–814.

Jin, S. G., van Dam, T., & Wdowinski, S. (2013). Observing and understanding the Earth system variations from space geodesy. *Journal of Geodynamics, 72,* 1–10. https://doi.org/10.1016/j.jog.2013.08.001.

Jin, S. G., Jin, R., & Li, J. H. (2014). Pattern and evolution of seismo-ionospheric disturbances following the 2011 Tohoku earthquakes from GPS observations. *Journal of Geophysical Research: Space Physics, 119*(9), 7914–7927.

Jin, S. G., Occhipinti, G., & Jin, R. (2015). GNSS ionospheric seismology: Recent observation evidences and characteristics. *Earth Science Reviews, 147,* 54–64. https://doi.org/10.1016/j.earscirev.2015.05.003.

Kamogawa, M. (2006). Preseismic lithosphere-atmosphere-ionosphere coupling. *EOS, Transactions of the American Geophysical Union, 87*(40), 417–424.

Kamogawa, M., & Kakinami, Y. (2013). Is an ionospheric electron enhancement preceding the 2011 Tohoku-Oki earthquake a precursor? *Journal of Geophysical Research: Space Physics, 118,* 1751–1754. https://doi.org/10.1002/jgra.50118.

Kuo, C. L., Huba, J. D., Joyce, G., & Lee, L. C. (2011). Ionosphere plasma bubbles and density variations induced by pre-earthquake rock currents and associated surface charges. *Journal of Geophysical Research, 116*(A10), A10317. https://doi.org/10.1029/2011JA016628.

Le, H., Liu, J. Y., & Liu, L. (2011). A statistical analysis of ionospheric anomalies before 736 M6.0 + earthquakes during 2002–2010. *Journal of Geophysical Research, 116,* A02303. https://doi.org/10.1029/2010ja015781.

Leonard, R. S., & Barnes, R. A., Jr. (1965). Observation of ionospheric disturbances following the Alaska earthquake. *Journal of Geophysical Research, 70,* 1250–1253.

Li, M., & Parrot, M. (2013). Statistical analysis of an ionospheric parameter as a base for earthquake prediction. *Journal of Geophysical Research: Space Physics, 118,* 3731–3739. https://doi.org/10.1002/jgra.50313.

Liu, J. Y., Chen, Y. I., Chuo, Y. J., & Tsai, H. F. (2001). Variations of ionospheric total electron content during the Chi-Chi earthquake. *Geophysical Research Letters, 28,* 1383–1386. https://doi.org/10.1029/2000GL012511.

Liu, J.-Y., Chuo, Y. J., Shan, S. J., Tsai, Y. B., Chen, Y. I., Pulinets, S. A., et al. (2004). Pre-earthquake ionospheric anomalies registered by continuous GPS TEC measurements. *Annales Geophysicae, 22*(5), 1585–1593.

Liu, J. Y., Chen, Y. I., Chuo, Y. J., & Chen, C. S. (2006). A statistical investigation of pre earthquake ionospheric anomaly. *Journal of Geophysical Research, 111,* A05304.

Liu, J.-Y., Chen, C.-H., Lin, C.-H., Tsai, H.-F., Chen, C.-H., & Kamogawa, M. (2011). Ionospheric disturbances triggered by the 11 March 2011 M9. 0 Tohoku earthquake. *Journal of*

Geophysical Research: Space Physics (1978–2012), *116*(A6), A06319. https://doi.org/10.1029/2011ja016761.
Molchanov, O. A., & Hayakawa, M. (1998). Subionospheric VLF signal perturbations possibly related to earthquakes. *Journal of Geophysical Research, 103*, 17,489–17,504.
Molchanov, O. A., Hayakawa, M., & Rafalsky, V. A. (1995). Penetration characteristics of electromagnetic emissions from underground seismic source into the atmosphere, ionosphere and magnetosphere. *Journal of Geophysical Research, 100,* 1691–1712.
Najita, K., & Yuen, P. C. (1979). Long-period oceanic rayleigh wave group velocity dispersion curve from HF Doppler sounding of the ionosphere. *Journal of Geophysical Research: Space Physics, 84*(A4), 1253–1260.
Occhipinti, G., Dorey, P., Farges, T., & Lognonné, P. (2010). Nostradamus: The radar that wanted to be a seismometer. *Geophysical Research Letters, 37,* L18104. https://doi.org/10.1029/2010GL044009.
Occhipinti, G., Rolland, L., Lognonné, P., & Watada, S. (2013). From Sumatra 2004 to Tohoku-Oki 2011: The systematic GPS detection of the ionospheric signature induced by tsunamigenic earthquakes. *Journal of Geophysical Research: Space Physics, 118*(6), 3626–3636.
Perevalova, N. P., Sankov, V. A., Astafyeva, E. I., & Zhupityaeva, A. S. (2014). Threshold magnitude for ionospheric TEC response to earthquakes. *Journal of Atmospheric and Solar-Terrestrial Physics, 108,* 77–90.
Pisa, D., Parrot, M., & Santolík, O. (2011). Ionospheric density variations recorded before the 2010 Mw8.8 earthquake in Chile. *Journal of Geophysical Research, 116*, A08309. https://doi.org/10.1029/2011ja016611.
Pulinets, S., & Boyarchuk, K. (2005). *Ionospheric precursors of earthquakes*. Springer Science & Business Media.
Pulinets, S., & Ouzounov, D. (2011). Lithosphere–Atmosphere–Ionosphere Coupling (LAIC) model—An unified concept for earthquake precursors validation. *Journal of Asian Earth Sciences, 41*(4), 371–382.
Pulinets, S. A., Legen'ka, A. D., Gaivoronskaya, T. V., Depuev, V. K. (2003). Main phenomenological features of ionospheric precursors of strong earthquakes. *Journal of Atmospheric and Solar-Terrestrial Physics, 65*, 1337–1347. https://doi.org/10.1016/j.jastp.2003.07.011.
Riousset, J. A., Pasko, V. P., & Bourdon, A. (2010). Air-density-dependent model for analysis of air heating associated with streamers, leaders, and transient luminous events. *Journal of Geophysical Research, 115,* A12321. https://doi.org/10.1029/2010JA015918.
Rishbeth, H. (2006). F-region links with the lower atmosphere? *Journal of Atmospheric and Solar-Terrestrial Physics, 68,* 469–478.
Rishbeth, H. (2007). Do earthquake precursors really exist? *EOS Transactions, AGU, 88*(29), 296. https://doi.org/10.1029/2007EO290008.
Rishbeth, H., & Mendillo, M. (2001). Patterns of F2-layer variability. *Journal of Atmospheric and Solar-Terrestrial Physics, 63,* 1661–1680.
Rishbeth, H., Mendillo, M., Wroten, J., & Roble, R. G. (2009). Day-by-day modelling of the ionospheric F2-layer for year 2002. *Journal of Atmospheric and Terrestrial Physics, 71,* 484–856.
Roble, R. G., Ridley, E. C., Richmond, A. D., & Dickson, R. E. (1988). A coupled thermosphere/ionosphere general circulation model. *Geophysical Research Letters, 15,* 1325–1328.
Rolland, L. M., Occhipinti, G., Lognonné, P., & Loevenbruck, A. (2010). Ionospheric gravity waves detected offshore Hawaii after tsunamis. *Geophysical Research Letters, 37,* L17101. https://doi.org/10.1029/2010GL044479.
Rolland, L. M., Lognonné, P., Astafyeva, E., Kherani, E. A., Kobayashi, N., Mann, M., et al. (2011). The resonant response of the ionosphere imaged after the 2011 off the Pacific coast of Tohoku Earthquake. *Earth, Planets and Space, 63*(7), 853–857.
Row, R. V. (1966). Evidence of long-period acoustic-gravity waves launched into the F-region by the Alaskan earthquake of March 28 1964. *Journal of Geophysical Research, 71,* 343–345.
Row, R. V. (1967). Acoustic-gravity waves in the upper atmosphere due to a nuclear detonation and an earthquake. *Journal of Geophysical Research, 72,* 1599–1610.

Ryu, K., Lee, E., Chae, J. S., Parrot, M., & Pulinets, S. (2014). Seismo-ionospheric coupling appearing as equatorial electron density enhancements observed via DEMETER electron density measurements. *Journal of Geophysical Research: Space Physics, 119,* 8524–8542. https://doi.org/10.1002/2014JA020284.

Saito, A., Tsugawa, T., Otsuka, Y., Nishioka, M., Iyemori, T., Matsumura, M., et al. (2011). Acoustic resonance and plasma depletion detected by GPS total electron content observation after the 2011 off the Pacific coast of Tohoku Earthquake. *Earth Planets Space, 63,* 863–867.

Sharma, K., Dabas, R. S., Sarkar, S. K., Das, R. M., Ravindran, S., & Gwal, A. K. (2010). Anomalous enhancement of ionospheric F2 layer critical frequency and total electron content over low latitudes before three recent major earthquakes in China. *Journal of Geophysical Research, 115,* A1131. https://doi.org/10.1029/2009JA014842.

Shinagawa, H., Tsugawa, T., Matsumura, M., Iyemori, T., Saito, A., Maruyama, T., et al. (2013). Two-dimensional simulation of ionospheric variations in the vicinity of the epicenter of the Tohoku-oki earthquake on 11 March 2011. *Geophysical Research Letters, 40,* 5009–5013. https://doi.org/10.1002/2013GL057627.

Tanaka, T., Ichinose, T., Okuzawa, T., Shibata, T., Sato, Y., Nagasawa, C., & Ogawa, T. (1984). HF-Doppler observations of acoustic waves excited by the Urakawa-Oki earthquake on 21 March 1982. *Journal of Atmospheric and Terrestrial Physics*, *46*(3), 233–245.

Tsugawa, T., Saito, A., Otsuka, Y., Nishioka, M., Maruyama, T., Kato, H., et al. (2011). Ionospheric disturbances detected by GPS total electron content observation after the 2011 off the Pacific coast of Tohoku Earthquake. *Earth, Planets and Space, 63*(7), 875–879.

Tsuji, H., Hatanaka, Y., Sagiya, T., & Hashimoto, M. (1995). Coseismic crustal deformation from the 1994 Hokkaido-Toho-Oki earthquake monitored by a nationwide continuous GPS array in Japan. *Geophysical Research Letters, 22,* 1669–1672. https://doi.org/10.1029/95GL01659.

Tzur, I., & Roble, R. G. (1985). The interaction of a dipolar thunderstorm with its global electrical environment. *Journal of Geophysical Research, 90,* 5989–5999. https://doi.org/10.1029/JD090iD04p05989.

Venkatraman, S., & Heelis, R. (2000). Interhemispheric plasma flows in the equatorial topside ionosphere. *Journal of Geophysical Research, 105,* 18,457–18,464. https://doi.org/10.1029/2000ja000012.

Wanliss, J. A., & Showalter, K. M. (2006). High-resolution global storm index: Dst versus SYM-H. *Journal of Geophysical Research, 111,* A02202.

Zhao, B., Yu, T., Wang, M., Wan, W., Lei, J., Liu, L., et al. (2008). Is an unusual large enhancement of ionospheric electron density linked with the 2008 great Wenchuan earthquake? *Journal of Geophysical Research, 113,* A11304. https://doi.org/10.1029/2008JA013613.

Chapter 2
Atmospheric Changes and Observations

2.1 Atmospheric Structure

Earth's atmosphere is usually air layer of the solid Earth outer, which is an important part of the Earth's natural environment. It is the condition indispensable to the survival of life on Earth and also protective for life on Earth. With developments and advances in space technologies, human activities are involving into the upper atmosphere and space, and therefore monitoring and study on Earth's atmosphere and space environments have become an international hot topic. To know compositions and structure variations of Earth's atmosphere as well as various processes and phenomena, it is important to precisely monitor Earth's atmospheric parameters and its variations in details. According to its ionized characteristics, the atmosphere of the Earth can be divided into two major parts, i.e., the ionosphere and neutral atmosphere. The ionosphere is the upper atmosphere of the Earth ionized by solar radiation with altitude from about 60 km to over 2000 km. The neutral atmosphere, in which the atmospheric components are electrically neutral, is the lower atmosphere below the ionosphere ranging from the surface to about 60 km, including the troposphere and stratosphere.

According to the temperature profiles, the Earth's atmospheric structure is shown in Fig. 2.1. The lower layer is the troposphere where we live and the second one is the stratosphere. The separating boundary between the troposphere and the stratosphere is the most important, called the tropopause. The tropopause with the lowest temperature near the Earth's surface is very sensitive to atmospheric changes. The troposphere has abundant of water vapor and all weather activities take place in this layer. The tropopause affects the water vapor transportation into the stratosphere as well as the Earth's surface temperature and climate changes. The stratosphere layer can be divided into the lower and upper stratosphere but the lower stratosphere is more important that can affect the troposphere. Normally the stratosphere height starts from above 17 km and reaches 50 km of height. The next layer mesosphere height starts from 50 to 80 km and then thermosphere height starts about from 80 km. The top of the mesosphere is known as the mesopause that is the coldest layer in

S. Jin et al., *GNSS Atmospheric Seismology*,
https://doi.org/10.1007/978-981-10-3178-6_2

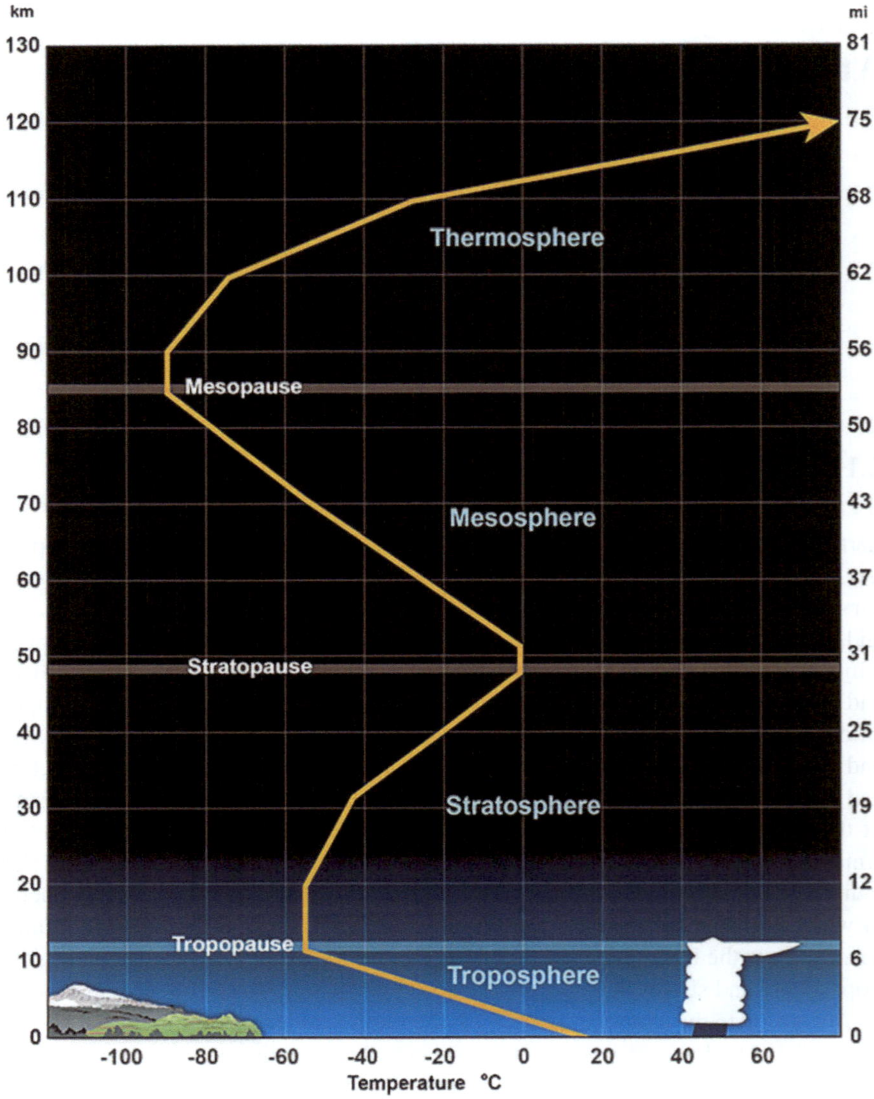

Fig. 2.1 Atmospheric structure (Courtesy by NOAA)

the atmospheric layers. After mesopause, the next layer is the thermosphere whose height reaches 700 km, and its temperature is increasing with the height and has a higher temperature than the lower atmospheric layers. Exosphere is the outermost layer of the Earth's atmosphere and the gases which are directly affected by the solar radiations (Khan 2016).

Due to the energy exchange and solar activity, the Earth's atmosphere activities are very complex, which directly affect the human environment and climate change. Therefore, monitoring the atmospheric changes, motion, circulation, and composition helps to understand the source of atmospheric change and the resulting impact on climate change in the neutral atmosphere and ionosphere, including human activities and natural disasters, such as the effects of earthquakes and volcanoes.

2.2 Atmospheric Changes

2.2.1 Neutral Atmospheric Changes

The Earth's atmosphere derived from the exchange of solar heat, the interaction of solar radiation, underlying surface forcing and atmospheric circulation causes the neutral atmosphere change. Changing activities of air motion and pressure system make the energy and matter exchange between ocean and land, south and north, ground and upper air, form complex changes of weather and climate. The Earth's atmosphere mainly includes nitrogen, oxygen, argon, carbon dioxide, nitrous oxide, two nitrogen, water vapor, carbon monoxide, sulfur dioxide, ozone, aerosol particles, etc. The main parameters of neutral atmosphere include atmospheric pressure, temperature, humidity and moisture content, etc.

Direct measurement is to put the detector in the atmosphere through the direct effect of the atmosphere on the detector and to obtain the parameters feature of the atmosphere. Indirect measurement usually receives radio signals through the target medium which is to be measured by the detector, and analyzes the change characteristics of the signal to obtain the atmospheric parameters. Human's monitoring of the atmospheric parameters started in the use of traditional measuring instruments on the Earth surface, including temperature, pressure, wind, and humidity. Due to the advancement of these detecting instruments, the atmospheric sounding has stepped into the stage of rational and quantificational analysis. The gradual establishment of the weather station networks is the first stage of the atmospheric sounding. The determination of basic parameters such as temperature and pressure has profound significance to the development of the atmospheric science theory. However, all these measurements are confined to the surface. Until 1930s, people began to work in the area of the atmospheric vertical structure, with the widespread use of radiosonde. After 1950s, with the development of aerospace technology, meteorological satellites had been launched with carrying various payloads such as Langmuir probe, mass spectrometry, electric field measuring instrument, etc., as well as remote sensing such as meteorological rocket, laser, microwave, and infrared sensors had been used. All these technologies greatly enhance the detecting ability of the atmosphere and broaden the understanding of the horizontal and vertical atmospheric variations.

2.2.2 *Ionospheric Changes*

Usually, the change of the ionosphere is characterized by the parameters, such as total electron content, electron density, temperature, ion density, and ion temperature. Observing these parameters is the basis of the study on the ionospheric formation process, the shape mode, the energy transport, etc. Due to the ionization properties of the ionosphere, the electron density is the most basic parameter. Direct measurements of the ionosphere include rocket, mass spectrometer, Langmuir probe, electric meter, and magnetometer, etc. These measurements mainly detect the ionospheric ion composition, density, temperature, high altitude magnetic field, etc. Based on the refraction and scattering by radio signals in the ionospheric plasma, indirect method measures radio delay or ionospheric variations to determine ionospheric parameters. The major methods of detecting ionospheric electron density include ionosonde (vertical ionosonde and slant ionosonde), scattering radar measurement, measurement of Faraday rotation, and electronic content based on radio delay (Jin 2016).

Due to the refraction effect of the ionosphere, the radio signals have a phase delay when passing through the ionosphere. The ionosphere is the dispersion medium, and thus the electronic content of the ionosphere can be obtained by using the double-frequency or multifrequency Global Navigation Satellite System (GNSS) observations, which has gradually become the main method of the ionospheric electron density detection. GNSS can monitor the ionospheric variation with real time and global/regional advantages, because of its low cost, high precision, all-weather, high coverage, high temporal and spatial resolution, etc. (Jin et al. 2004).

The ionospheric variations are very complex in the temporal and spatial structure. Unlike the periodic variation of the ionosphere, the formation mechanism of the ionospheric anomalies is more varying and complicated. The anomalous ionospheric disturbances may be generated by the climate changing in the upper ionosphere, and may also be caused by the bottom of the ionosphere, which is the upward propagation of ground disturbance sources. From the origin of the abnormal ionospheric disturbance, it can be divided into two categories. One is the Sun's activity caused by the sudden disturbance of ionosphere anomalies, while the other one is caused by other man-made or natural events in the temporal and spatial distribution of ionospheric disturbances. The sudden disturbance of solar activity is mainly the flare eruption, coronal mass ejections, etc. The solar flare can cause solar radiation to increase rapidly in a short time, and then causes the abnormal disturbance of the ionospheric electron content (Tsurutani et al. 2009). The ejection of high-energy ions after a few hours of moving to the Earth and the interacting with geomagnetism causes severe global geomagnetic disturbances, namely the magnetic storm. The violent changes of the geomagnetic elements can cause large and complicated global ionospheric disturbances, and the disturbances will last for several hours to several days (Mendillo 2006; Danilov 2013). Compared with the ionospheric anomalies caused by solar activity disturbances, the ionospheric disturbances caused by the incident of the ionosphere bottom have a smaller scale, and the duration time lasts for a few minutes to several hours and the special scale is about hundreds to thousands kilometers. The ionospheric disturbances have complexity and diversity caused by the different forms of disturbance sources, e.g., regional and global changes. In essence, the ionospheric anomalies caused by the local ionospheric events such as earthquake, explosion, volcano eruption, and rocket explosion mainly are a solid Earth—atmosphere–ionosphere coupling (Calais et al. 1998; Heki et al. 2004; Jin et al. 2014, 2015). The study of seismic ionospheric anomalies is a great complement to investigate the ionospheric variations and has great significance for the further understanding of the irregular variation characteristics of the ionosphere.

2.3 Observation Techniques

2.3.1 Radiosonde

Radiosonde is a battery-powered telemetry instrument, which is carried into the atmosphere by a balloon and measures various atmospheric parameters, e.g., altitude, relative humidity, pressure, temperature, wind, etc. Radiosondes are operated at a radio frequency of 403 MHz or 1680 MHz. A radiosonde that is dropped from an airplane and falls is called a dropsonde. Currently, there are about 1300 radiosonde launched stations worldwide and most data are open and sharing through international agreements. Normally all routine radiosonde launches occur 45 min before the official observation time of 00:00 UTC and 12:00 UTC, twice daily, which is especially important for numerical modeling. Therefore, radiosondes are an essential source for meteorological and climate study all over the world daily.

2.3.2 Ionosonde

Ionosonde by emitting electromagnetic wave with different frequencies and receiving the echo signal, obtaining ionogram, echo intensity, ionospheric delay, and Doppler shift determines the ionosphere electron density at different altitudes. So far the ionosonde technology is still the basic measurement of the ionosphere. The ionosonde stations located across the globe play an irreplaceable role in precisely measuring the vertical section of electron density. But ionosonde can only detect the ionosphere from ground to about 500 km in altitude, so combining other techniques is needed to detect the top of the ionosphere. The distribution of global ionosonde stations can be found at http://spidr.ionosonde.net/spidr.

2.3.3 Ionospheric Scatter Radar

Ionospheric Scatter Radar (ISR) detection is mainly divided into high-frequency coherent scattering radar detection and noncoherent scattering radar detection. The coherent scattering radar by measuring the power of the scattered signal, the angle of arrival, and the Doppler shift generated by the nonhomogeneity moving of the electron density calculates the electron density, the drift velocity, the size of the heterogeneous body, etc. Incoherent scattering detection calculates ionospheric parameters such as electron density, electron temperature, ion temperature, through scattering signal caused by random thermal fluctuations of quasi-equilibrium electron density. Radar scattering parameters with high temporal and spatial resolution can not only obtain the ionospheric electron density, electron/ion temperature, drift velocity but also indirectly obtain the ionospheric conductivity, electric field, thermal wind, ion collision frequency. The EISCAT (European Incoherent Scatter Scientific Association) 3-D project led by EISCAT represents the recent progress of global scattering radar (Ding et al. 2014). New type of coherent scattering radar based on phased-array is of flexible control, high-frequency resolution, multi-beam synthetic imaging, interference measurement capability, and has important significance to obtain the ionospheric temporal features completely. However, the structure of the scattered radar is very complicated, and the global scattering radar station is fairly limited at present. Although scattering radar is an effective method for the regional ionospheric detection, it is difficult to apply this technique to analyze the temporal and spatial characteristics of the global ionosphere.

2.3.4 Faraday Rotation Detection

Faraday rotation detection is the measurement of the rotation between polarization surface generated by the linear polarization electromagnetic wave through the electromagnetic field and incident wave under the effect of electromagnetic field. This rotation effect is proportional to the electron density, which is often used in the measurement of the ionospheric total electron content (TEC) (Garriott et al. 1970; Ganguly et al. 2001). Comparing with ionospheric detection based on radio delay, Faraday rotation detection is more sensitive to the structure of geomagnetic field. When the radio signal is perpendicular to the magnetic field line, Faraday rotation is not affected by the plasma. Based on this foundation, joint measurements can be used in the correlation research of plasma.

2.3.5 MODIS and OLR Observations

The Moderate resolution Imaging Spectroradiometer (MODIS) satellite was designed in the mid-1995 and launched in 1999 by the National Aeronautics and Space Administration (NASA), which is dedicated to monitoring the Earth's atmosphere. The MODIS satellite has two nodes: descending node (Terra) and ascending node (Aqua), which are continuously moving in Sun-synchronous and polar orbits at 705 km above mean sea level. It surveys the land surface in 36 spectral bands in a range from 0.4 μm to 14.4 μm with 250 m, 500 m, and 1 km spatial resolutions. The footprint of MODIS is 2330 km on land surface by 10 km beneath the nadir (along track at nadir). The data are combined with available daytime and nighttime images. The MODIS provided the daily Land Surface Temperature (LST) and Emissivity daily data, which are retrieved at 1 km pixels by the generalized split-window algorithm and at 6 km grids by the day/night algorithm. In the day/night algorithm, daytime and nighttime LSTs and surface emissivities are retrieved from pairs of day and night MODIS observations in seven TIR bands. MODIS (LST) data (1×1 km^2) are routinely published at http://modis.gsfc.nasa.gov/data.

Outgoing Long-wave Radiation (OLR) is the measurement of the amount of energy emitted to space by Earth's surface, oceans, and atmosphere, and therefore, it is a critical component of the Earth's radiation budget. The wavelength range of more than 99% of OLR radiation is between 4 and 100 μm, which is termed as thermal infrared part of the electromagnetic spectrum. OLR observations are made via the Advanced Very High Resolution Radiometer (AVHRR) instrument aboard the NOAA polar orbiting spacecraft. The raw ascending and descending swath data have been spatially and temporarily interpolated onto grids across equatorial areas from 160°E to 160°W longitude. The spatial resolution is $2.5° \times 2.5°$ while the temporal resolution varies on a daily basis. More information can be found from the official website of NOAA Climate Prediction Center and NCEP data center.

Some anomalies in Land Surface Temperature (LST) and energy are probably associated with earthquakes (Shah 2017). For example, Pulinets and Ouzounov (2011) observed a 5 °C increase in land surface temperature when the amount of greenhouse gases doubles, which may be due to the contribution of the earthquake to the atmosphere. Some anomalous land surface temperature and outgoing long-wave radiation observations were found prior to earthquakes in India, Romania, and Haiti (Rawat et al. 2011; Xiong et al. 2010).

2.3.6 Electromagnetic Observations

DEMETER (Detection of Electromagnetic Emissions Transmitted from Earthquake Regions) is a French satellite, primarily launched for detecting the ionospheric vulnerabilities related to earthquakes. It was dedicated to measure the variations in plasma density over the most severe seismogenic zones on Earth. The mission was launched on June 29, 2004 with very few payloads. After a coupling of months of successful flight, all the scientific payloads were started to active. The mission was officially terminated on December 9, 2009, with ending of all the scientific observations. In the first observation phase, DEMETER satellite measured electromagnetic variations at altitude of 710 km. DEMETER performed all the observations in Sun-synchronous orbits. The flight of DEMETER was then lowered to 660 km in December 2005 and the direction of ascending node remains the same as altitude of 710 km. DEMETER observations were all made at 10:30 LT (daytime) and 22:30 LT (nighttime).

The scientific payloads on DEMETER satellite are the Langmuir probe (ISL: Instrument Sonde de Langmuir) and an ion analyzer (IAP). Among other payloads, the ICE and IMSC instruments are responsible to determine the electric and magnetic field variations related to earthquakes. DEMETER has two modes of scientific data (survey and burst mode), based on the size of data around the whole Earth. Both the modes have the same data but different sizes. DEMETER covers the whole Earth from $+65^\circ$ to -65° along latitude with ascending and descending modes. It covers the whole Earth in 28 flights (14 ascending and 14 descending). The revisited period of the ascending as well as descending orbits of DEMETER is 16 days over a single zenith angle. It is a microsatellite with a weight of 130 kg at the altitude of 710 km and later 600 km while the inclination angle is 98.3°. The Electron density (N_e) and Temperature (T_e) from ISL and IAP payload are widely used in several works for determination of possible ionosphere anomalies related to earthquakes.

2.3.7 GNSS Observations

Nowadays, Global Navigation Satellite Systems (GNSS), including US's Global Positioning System (GPS), Russia's GLONASS, the EU's Galileo and China's BeiDou (Fig. 2.2), have been widely used in navigation, positioning, timing, geodesy, and surveying. GNSS is characterized as a highly precise, continuous, all-weather, and near-real-time microwave (L-band) technique with signals through the Earth's atmosphere. When the GPS signal propagates through the Earth's atmosphere, it is delayed by the atmospheric refractivity, which results in lengthening of the geometric path of the ray. In contrast, the tropospheric and ionospheric:delay delays can be extracted from ground and space-based GNSS observations as well as atmospheric parameters, e.g., atmospheric refractive index, pressure, density profiles in the troposphere, temperature with up to the lower stratosphere (35–40 km), and the ionospheric total electron content (TEC) as well as electron density profiles (Rocken 1997; Hajj and Romans 1998), which can be used to improve weather analysis and forecasting, and monitor ionospheric events.

GPS has been widely used for Positioning, Navigation, and Timing (PNT) globally to all users at all times since 1994. Since GPS is highly precise, continuous, all-weather, and real time with signals that travel through the Earth's atmosphere, additional scientific applications of GPS were being explored by scientists and engineers.

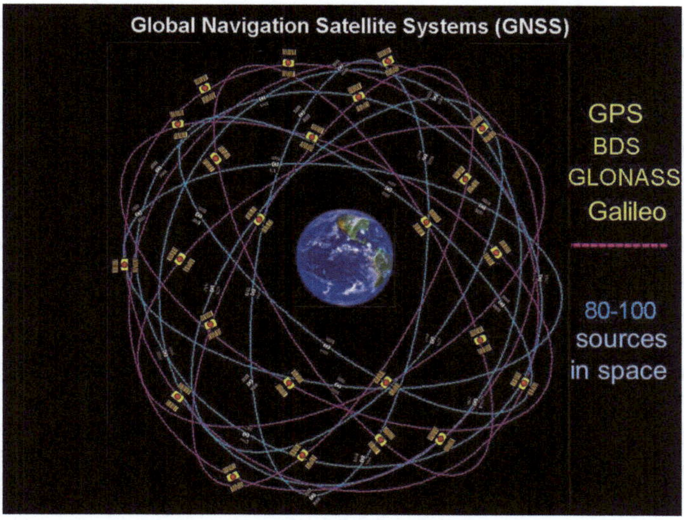

Fig. 2.2 GNSS constellations

Notably, Ware (1992) suggested limb sounding the Earth's atmosphere and ionosphere using GPS atmospheric refracted delay signals. Next, the GPS/Meteorology Mission (GPS/MET) using the GPS radio occultation technique was launched and successfully provided accurate, all-weather, global refractive index, pressure, temperature, and the ionospheric Total Electron Content (TEC) and electron density profiles, which have been widely applied in the atmosphere and ionosphere (Rocken 1997; Hajj and Romans 1998).

GLOnass NAvigation Satellite System (GLONASS) by the former Soviet Union was fully operational since 1995 but suspended and partially operated after the collapse of USSR and now has got the latest up-gradation and is fully operational since 2010 by Russia. The GLONASS (GLOnass NAvigation Satellite System) is the Russian-owned constellation of positioning satellites). It reached full operational capability in 1995, after the efforts made by the collapsed Soviet Union were continued by the Russian Federation. Unfortunately, due to the inability to maintain the system, the constellation slowly reduced, reaching a minimum of eight operational satellites in 2002. The turning of the century saw a renewed interested in restoring the full capability by the Russian government (Dvorkin and Karutin 2006), and the system is currently (2018) formed by 26 operational satellites. The GLONASS space segment is composed of 24 satellites distributed over three orbital planes (separated by 120 right ascension of the ascending node) with radius equal to 25,500 km and 65-inclined. GLONASS satellites broadcast signals in L1 and L2 bands, using Frequency Division Multiple Access (FDMA) technique as channel access method technique. The GLONASS ground control center (in Moscow) and the telemetry and tracking stations are all within the Russian Federation borders.

The Galileo Navigation Satellite System, named after the Italian astronomer and physicist Galileo Galilei, was started following political reasoning that wanted European countries to be independent of satellite navigation, from the US's GPS and Russian GLONASS. The Galileo system was intended to provide a more precise navigation service than the one provided by either GPS or GLONASS (Mendizabal et al. 2008). In March 2002, The European Union (EU) and European Space Agency (ESA) agreed to develop the Galileo positioning system, which is expected to be fully functional by 2020. Galileo is currently under development to be completed within the decade 2011–2020, which is expected to be compatible with the modernized GPS system. Compared to US's GPS and Russian GLONASS, Galileo is designed specifically for civilian and commercial purposes. Currently, a number of countries participated in the Galileo project. Currently, more than four Galileo satellites were launched and operating.

The BeiDou navigation system (BDS) has been developed since the early 1990s. Similar to GPS, different services will be offered: a standard signal for civilian users and a more precise (encrypted) signal for reserved uses. A first experimental satellite, named BeiDou M-1, was launched in 2007 and became operational in 2011. Its orbit has a low eccentricity, an altitude of approximately 21,150 km and an inclination of 55°. The satellite transmits two signals (I and Q, in quadrature) in three bands: B1, B2, and B3. Nowadays, the BeiDou Navigation Satellite System (BDS) provided the global services since December 27, 2018. The constellation of BDS-3 will consist of 3 GEO, 3 IGSO, and 24 MEO satellites. The main function of the BDS is the positioning, velocity measurement, one-way and two-way timing, and short message communications (http://www.beidou.gov.cn).

The BDS signals are QPSK-modulated (Quadrature Phase Shift Keying), broadcasting by CDMA (Code Division Multiple Access) techniques (BeiDou Navigation Satellite System Signal In Space Interface Control Document, http://www.beidou.gov.cn). GPS satellites continuously broadcast right-hand circularly polarized signals with both the navigation message and ranging codes modulated on two L-band microwave carrier frequencies, i.e., the L1 ($f1 = 1.57542$ GHz, $\lambda 1 \approx 19.0$ cm) and L2 ($f2 = 1.2276$ GHz, $\lambda 2 \approx 24.4$ cm). Starting from 2006, as part of the GPS modernization effort, two new civilian-use signals L2C (L2 frequency) and L5 ($fL5 = 1.17645$ GHz, $\lambda L5 \approx 25.5$ cm) are broadcasting on the new generation GPS satellites. Also, another civilian-use frequency L1C will be available in the near future. The BDS signal frequencies are similar or close to the GPS and all GNSS use CDMA mode except GLONASS which uses FDMA (Frequency Division Multiple Access) at the same time. Since the GNSS systems are being updated and under construction, signal frequencies and modulation modes may be changed for improving resistance against interference, higher transfer rate, and system compatibility.

Because of the unpredictable of the occurrence location and time of the earthquake, people cannot make a targeted observation for a seismic event. Large-area coverage from GNSS tropospheric and ionospheric signals may provide other observations for seismic ionospheric disturbances. The GNSS atmospheric sounding is mainly divided into the space-based radio occultation and the ground-based observations, including the tropospheric delay, precipitable water vapor, ionospheric electron density, and Total Electron Content (TEC). Since 1990s, the GPS-MET succeed, and many countries had carried out their respective plans of GNSS radio occultation, e.g., CHAMP and TerraSAR-X (TSX) of Germany, COSMIC of Taiwan/USA, GPS-MET, GRACE, CNOFS of USA, METOP of European Union, SAC-C of Argentina, etc. After nearly 20 years of accumulation, space-based GNSS radio occultation provides an abundant of atmospheric/ionospheric products. Although the temporal and spatial resolution of GNSS radio occultation is not currently able to detect the small disturbance, because of its high vertical resolution, space-based GNSS observations play an important role in study of global vertical atmospheric disturbances from a time scale.

With the increase of the navigation constellation and the ground-based GNSS continuous tracking station, the temporal and spatial resolution has been greatly improved. The IGS ionosphere Working Group was established in 1998 that was supported by CODE, ESA, JPL, UPC, and other agencies, and since then the Global Ionospheric Map (GIM) of 2.5 by 5° has become the important information of temporal and spatial changes of global ionospheric TEC. With the gradual increase of the station used in the global ionospheric calculation of each analysis center and the optimization of the station distribution, the accuracy and resolution are gradually increasing. On the one hand, the global ionospheric map can accurately depict the global ionospheric electron content time-space structure, which provides a basis for scientific research of ionospheric variations. On the other hand, it is an important product for ionospheric correction of satellite navigation, communications, and other radio signals. However, seismic ionospheric disturbances are mostly regional effects. Although the global ionospheric map can accurately characterize the ionospheric temporal and spatial variation at large scale, for regional features the spatial resolution of global ionospheric map is difficult to meet the requirements. The dense GNSS continuous observing station network provides the possibility to detect the ionospheric disturbance in details, such as CMONOC (crustal movement observation network of China), PBO (Plate Boundary Observatory) network, GEONET (GNSS Earth Observation Network in Japan). Figure 2.3 is the distribution of these GNSS observation network stations. CMONOC is managed by the National Development and Reform Commission, led by the China Earthquake Administration, in order to build a high precision and resolution observation network to monitor crustal movement, gravity field, water vapor, ionospheric electron content, etc. Since 2012, CMONOC has operated the GNSS continuous observations with about 260 stations. The PBO Observation network, located along the Western Pacific coast of the United States, was built by the end of 2008 and was a part of the United States EarthScope program. It has more than 1100 GPS continuous observation stations which can provide high-resolution real-time data. Currently, GEONET of Japan is the most intensive GNSS observation network all over the world. The average horizontal distance of more than 1300 stations is about 20 km, which provide high-resolution and high precision observations of regional crustal deformation as well as the possibility for detecting seismic ionospheric disturbances in smaller scale. Therefore, these GNSS observations provided by the intensive network can be used to observe earthquake ionospheric disturbances.

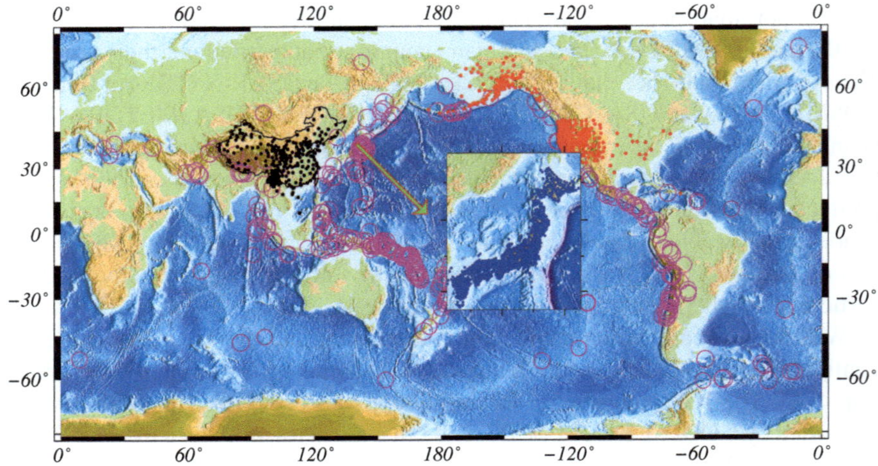

Fig. 2.3 Distribution of dense GNSS network and earthquakes. The black dots are CMONOC, blue dots are GEONET, and red dots are PBO. The purple circle is the epicenter location of earthquakes (Mw > 6.5) recorded by USGS (US Geological Survey) from 2010 to 2015

2.4 Solar and Geomagnetic Observations

Solar radiation and geomagnetic activity are the main driving factors of the temporal and spatial variations of the ionosphere. It is necessary to exclude the influence of the solar and geomagnetic activities before investigating seismic ionospheric disturbances. The influence of the solar and geomagnetic activities is usually global, and the characteristic of the disturbance is different from the seismic ionospheric disturbance. However, the solar and geomagnetic induced ionospheric anomalies and the ionosphere coupling mechanism have not yet been fully understood, and it is difficult to remove the influence from the theoretical model accurately. Currently, a more convenient method to eliminate the effect of solar radiation and geomagnetic activities is to select the relatively quiet periods based on solar geomagnetic activity indices. In the quiet period, the influence of the solar and geomagnetic activities on the ionosphere can be eliminated.

The intensity of solar activity is usually reflected through 10.7 cm solar radio flux, which is also called F10.7 index. The F10.7 radiation that is produced in the top of solar chromospheres and atmospheric coronal layer can reflect the general level of solar activity, and its change and sunspots and has a good correlation of ultraviolet radiation and visible light radiation changes. Unlike other solar indices, F10.7 radiation flux can be easily obtained by the ground observation stations and the observations are not affected by climate. Since 1947 the F10.7 radiation has accumulated about six solar activity cycles, so the F10.7 index has become the most common index of solar activity. The indices also provide valuable data for long-term climatology study.

Due to the important role of geomagnetic observation in disaster prevention, communication, and other aspects, many countries have their own geomagnetic observation network. However, due to the global change of geomagnetic field, local geomagnetic observation is often unable to monitor the change of geomagnetic field accurately. In order to solve this problem, the group called Intermagnet has been established under the advocate of Canada, Britain, and other countries with the aim to derive the global geomagnetic observation progress and the real-time global geomagnetic observations. So far, Intermagnet has more than 150 continuous geomagnetic observation stations distributed all over the world, and basically realizes the near-real-time continuous monitoring of global geomagnetic activity (http://www.intermagnet.org/). To describe various types of magnetic disturbances, a variety of geomagnetic indices are designed, such as Dst index, AE index, Kp index, Ap index. Dst index is composed of five uniform distribution of the geomagnetic observation in 20° from the north to the south, subtracting basic magnetic field from geomagnetic field and the mean of horizontal component H of geomagnetic daily quiet variations. Dst index shows the axial symmetry magnetic disturbance at dipole equatorial, which can reflect the geomagnetic field changes at middle latitude and near equator. Most of Dst disturbances are negative disturbances, meaning that the intensity of geomagnetic field decreases. The main reason is the equatorial current variation. When the value of Dst decreases among −50 to −30 nT, the weak magnetic storms is existed, while reaching −100 to −50 nT moderate storms are considered and less then −100 nT is considered as strong geomagnetic storms (Gonzalez et al. 1994). The Dst positive disturbance is mainly due to the effect of the solar wind pressure. The AE index is composed of the difference between the lower and upper envelope of the horizontal component of geomagnetic field H by 10–13 geomagnetic stations in Northern Hemisphere Aurora region, which reflects the changes in the magnetic field caused by the ionospheric current changes in the high latitude auroral zone. The Kp index, also known as the interplanetary magnetic index, is composed of geomagnetic activity in the geomagnetic field, which is characterized by the geomagnetic field horizontal component of the geomagnetic observation station. The Kp index with high value means a stronger magnetic disturbance. For the convenience of calculation, the Kp is usually converted to a linear relationship with the magnetic disturbance amplitude, which is the Ap index

2.5 Seismological Observations

The earthquake parameters also called source parameters, is the quantitative representation based on the observation of seismometers to analyze the space-time position, the energy, the fault rupture, and dynamic changes, including the basic parameters of the earthquake (earthquake occurrence time, epicenter location, focal depth, earthquake magnitude), focal mechanism and kinetic parameters, etc. The earthquake occurrence time is the moment when hypocenter has broken up. The epicenter can be calculated by using data from seismic stations to determine the origin of the waves. The earthquake magnitude is the parameter to reflect the strength of earthquake and the amount of energy. The time, epicenter location, and focal depth of earthquakes can be determined from independent five stations observations. In practice, the optimal estimation is usually obtained by statistical analysis methods with more stations observations. Compared with the time and space position of earthquake, the determination of the seismic energy is more complex. It is difficult to determine the magnitude of the releasing energy directly and accurately by the present technique. In seismology, the magnitude defined by the statistical law of the observed data is often be used to show the relative magnitude of seismic energy. Based on different seismological observation, the common magnitudes include near earthquake magnitude (ML), surface wave magnitude (Ms), body wave magnitude (MB, mB), shock duration magnitude (MD), moment magnitude (Mw), spectral magnitude (MT), etc. The seismic source mechanism parameters describe the mechanical process of the earthquake, which are usually characterized by the direction of the fault plane and the slip vector, so it is also called the fault parameter. At present, the most effective method to obtain the focal mechanism is using the P wave initial moving direction data. The kinematic parameters of earthquakes describe the fault dynamic rupture process, which mainly refer to the fault rupture direction, rupture velocity, fault duration, and fault distance. The inversion method of kinematic parameters mainly includes the source time function method and direct wave method. The earthquake parameters used in this book are the basic parameters and focal mechanism published by the United States Geological Survey (USGS) and Harvard University (http://www.globalcmt.org), and dynamic parameters are the fault model by eQuake-RC project (http://equake-rc.info).

The process of the earthquake is complex. Currently, the seismic wave that is mainly the elastic wave spreading around caused by the rupture of the earthquake can be measured. The seismic waveform recorded by the seismic instrument is the basic data to determine the seismic parameters. With the development of digital seismology, modern computer, and network communication technology, the real-time analysis of earthquake have been greatly developed. The data service center of American IRIS (Incorporated Research Institutions for Seismology) coordinates the seismic data of GSN (Global Seismographic Network), PASSCAL (Portable Array Seismic Studies of the Continental Lithosphere), FDSN (International Federation of Digital Seismograph Networks), OBSIP (The Ocean Bottom Seismograph Instrument Pool), EarthScope and other earthquake research projects. The center provides 3-D seismic wave observation data with high time resolution freely to global users (http://www.iris.edu). The basic parameters are important for the quick earthquake rupture determination and risk evaluations.

References

Calais, E., Minster, J. B., Hofton, M., et al. (1998). Ionospheric signature of surface mine blasts from Global Positioning System measurements. *Geophysical Journal International, 132*(1), 191–202.

Danilov, A. D. (2013). Ionospheric F-region response to geomagnetic disturbances. *Advances in Space Research, 52*(3), 343–366.

Ding, Z., Dai, L., Dong, M., et al. (2014). Progress in Ionospheric scatter radar: From traditional physique to EISCAT 3D. *Progress in Geophysics, 29*(5), 2379–2381.

Dvorkin, V., & Karutin, S. (2006). GLONASS: Current status and perspectives. In *The 3rd ALLAST Open Conference*, Hannover.

Ganguly, S., Brown, A., Dasgupta, A., et al. (2001). Ionospheric reconstruction using Faraday rotation data: A new technique. *Radio Science, 36*(4), 789–800.

Garriott, O. K., Rosa, A. V. D., & Ross, W. J. (1970). Electron content obtained from Faraday rotation and phase path length variations. *Journal of Atmospheric and Terrestrial Physics, 32*(4), 705–727.

Gonzalez, W. D., Joselyn, J. A., Kamide, Y., et al. (1994). What is a geomagnetic storm? *Journal of Geophysical Research Space Physics, 99*(A4), 5771–5792.

Hajj, G. A., & Romans, L. L. (1998). Ionospheric electron density profiles obtained with the global positioning system: Results from the GPS/MET experiment. *Radio Science, 33*(1), 175–190.

Heki, K., Otsuka, Y., Choosakul, N., Hemmakorn, N., Komolmis, T., & Maruyama, T. (2006). Detection of ruptures of Andaman fault segments in the 2004 great Sumatra earthquake with coseismic ionospheric disturbances. *Journal of Geophysical Research: Solid Earth, 111*(B9), B09313. https://doi.org/10.1029/2005jb004202.

Jin, R. (2016). Behaviors and characteristics of seismic ionospheric disturbance using GNSS observation. Ph.D. thesis, University of Chinese Academy of Sciences, China.

Jin, S. G., Wang, J., Zhang, H., & Zhu, W. (2004). Real-time monitoring and prediction of the total ionospheric electron content by means of GPS observations. *Chinese Astronomy and Astrophysics, 28*(3), 331–337.

Jin, S. G., Jin, R., & Li, J. H. (2014). Pattern and evolution of seismo-ionospheric disturbances following the 2011 Tohoku earthquakes from GPS observations. *Journal of Geophysical Research: Space Physics, 119*(9), 7914–7927.

Jin, S. G., Occhipinti, G., & Jin, R. (2015). GNSS ionospheric seismology: Recent observation evidences and characteristics. *Earth Science Reviews, 147,* 54–64. https://doi.org/10.1016/j.earscirev.2015.05.003.

Khan, A. (2016). Gravity waves and tropopause variations in Tibet from GNSS radio occultation observations. Ph.D. thesis, University of Chinese Academy of Sciences, China,.

Mendillo, M. (2006). Storms in the ionosphere: Patterns and processes for total electron content. *Reviews of Geophysics, 44*(4), RG4001, 1–47.

Mendizabal, J., Lagunilla, J. M., & Perez, R. B. (2008). *GPS and Galileo: Dual RF front-end receiver and design, fabrication, and test* (p. 208). New York: McGraw-Hill.

Pulinets, S., & Ouzounov, D. (2011). Lithosphere-Atmosphere-Ionosphere Coupling (LAIC) model—An unified concept for earthquake precursors validation. *Journal of Asian Earth Sciences, 41,* 371–382. https://doi.org/10.1016/j.jseaes.2010.03.005.

Rawat, V., Saraf, A. K., Das, J., Sharma, K., & Shujat, Y. (2011). Anomalous land surface temperature and outgoing long-wave radiation observations prior to earthquakes in India and Romania. *Natural Hazards, 59*(1), 33–46.

Rocken, C. (1997). Analysis and validation of GPS/MET data in the neutral atmosphere. *Journal Geophysical Research, 102,* 29849–29866.

Shah, M. (2017). Statistical analysis and modelling of seismic ionospheric anomalies from GNSS and other satellite observations. Ph.D. thesis, University of Chinese Academy of Sciences, China.

Tsurutani, B. T., Verkhoglyadova, O. P., Mannucci, A. J., et al. (2009). A brief review of "solar flare effects" on the ionosphere. *Radio Science, 44*(1), RS0A17.

Ware, R. (1992). GPS sounding of the earth atmosphere. *GPS World, 3,* 56–57.

Xiong, P., Shen, X. H., Bi, Y. X., Kang, C. L., Chen, L. Z., Jing, F., et al. (2010). Study of outgoing longwave radiation anomalies associated with Haiti earthquake. *Natural Hazards and Earth Systems Sciences, 10,* 2169–2178. https://doi.org/10.5194/nhess-10-2169-2010.

Chapter 3
GNSS Tropospheric Sounding

3.1 Atmospheric Refractivity

The Earth's atmosphere is a kind of inhomogeneous medium. With increase of height, atmospheric mass decreases. When the electromagnetic wave propagates through the atmosphere, the deviation from a straight line occurs due to the variation of the air mass or density, which is called the atmospheric refraction. The refractive index n of a substance can be expressed as (Jin et al. 2007),

$$n = \frac{c}{v} \tag{3.1}$$

where c and v are the speeds of the electromagnetic waves propagating in a vacuum and in some substance, respectively. The atmospheric refractivity N is defined as,

$$N = 10^6(n - 1) \tag{3.2}$$

The atmospheric refractivity can be further described as a function of meteorological parameters, including both hydrostatic and non-hydrostatic (dry and wet) components (Hopfield 1969) with the empirical formulas (Smith and Weintraub 1953; Hartmann and Leitinger 1984),

$$N = N_d + N_w \tag{3.3}$$

$$N_d = C_1 \frac{p}{T} \tag{3.4}$$

$$N_{\mathrm{w}} = C_2 \frac{e}{T} + C_3 \frac{e}{T^2} \tag{3.5}$$

where N_d and N_w are the dry and wet component refractivity, respectively, p is the pressure of the dry gas of the atmosphere in hPa (Hectopascal), e is partial pressure of the water vapor in hPa, T is absolute temperature in Kelvin (K), C_1, C_2 and

S. Jin et al., *GNSS Atmospheric Seismology*,
https://doi.org/10.1007/978-981-10-3178-6_3

C_3 are empirical constants as 77.624 K/hPa, −12.92 K/hPa, and 371,900 K^2/hPa, respectively.

3.2 GNSS Tropospheric Delay

The GNSS signals will be refracted when going through the neutral atmosphere, with the speed change and signal curve. The effects are called GNSS neutral atmospheric delay. Because the neutral atmospheric delay occurs mainly in the troposphere, the tropospheric delay is usually used instead of neutral atmospheric delay for the GNSS signal effects of the atmosphere below the ionosphere. This delay is one of the major error sources for GNSS positioning, which contributes a bias in height of several centimeters even when simultaneously recorded meteorological data are used in tropospheric models (Tregoning et al. 1998). Nowadays, GNSS has been widely used to determine the zenith tropospheric delay (ZTD) (Bevis et al. 1994; Emardson et al. 1998; Jin and Park 2005) through mapping functions (Niell 1996). The ZTD is the integrated refractivity along a vertical path through the neutral atmosphere:

$$ZTD = c\tau = 10^{-6} \int_0^{\infty} N(s)ds \tag{3.6}$$

where c is the speed of light in a vacuum, τ is the delay measured in units of time and N is the neutral atmospheric refractivity. The N is empirically related to standard meteorological variables as (Davis et al. 1985)

$$N = k_1\rho + k_2\frac{P_w}{Z_w T} + k_3\frac{P_w}{Z_w T^2} \tag{3.7}$$

where $k_i (i = 1, 2, 3)$ is constant, ρ is the total mass density of the atmosphere, P_w is the partial pressure of water vapor, Z_w is a compressibility factor near unity accounting for the small departures of moist air from an ideal gas, and T is the temperature in degrees Kelvin. The integral of the first term of Eq. 3.7 is the hydrostatic component (N_h) and the integral of the remaining two terms is the wet component (N_w). Thus, ZTD is the sum of the hydrostatic or "dry" delay (ZHD) and non-hydrostatic or "wet" delay (ZWD), due to the effects of dry gases and water vapor, respectively. The dry component ZHD is related to the atmospheric pressure at the surface, while the wet component ZWD can be transformed into the precipitable water vapor (PWV), which plays an important role in energy transfer and in the formation of clouds via latent heat, directly or indirectly influencing Numerical Weather Prediction (NWP) model variables (Bevis et al. 1994; Duan et al. 1996; Tregoning et al. 1998; Hernández-Pajares et al. 2001). Therefore, the Zenith Tropospheric Delay (ZTD) from GNSS is an important parameter of the atmosphere, which reflects the weather and climate processes, variations, and atmospheric vertical motions, etc. (Jin et al. 2009, 2011).

The ZHD can be precisely calculated by atmospheric models or surface meteorological parameters, and the zenith wet delay (ZWD) can be obtained by subtracting ZHD from ZTD, namely:

$$ZWD = ZTD - ZHD \tag{3.8}$$

There is a linear relationship between ZWD and precipitable water vapor (PWV) (Jin and Luo 2009)

$$PWV = \Pi \cdot ZWD \tag{3.9}$$

where Π is the conversion factor between ZWD and PWV. The physical sense of PWV represents the height at which water vapor in the atmosphere column of the unit bottom reaches saturation and condenses into liquid water. Π is a function of the mean temperature of the water vapor T_m:

$$\Pi = \frac{10^6}{\rho_w \cdot \frac{R}{m_w} \cdot \left(\frac{k_3}{T_m} + k_2 - \frac{m_w}{m_d} \cdot k_1\right)} \tag{3.10}$$

where ρ_{w} is the density of liquid water; R is the universal gas constant ($R = 8314$ Pa m^3 K^{-1} kmol^{-1}); m_w is the molar mass of water ($m_w = 18.02$ kg/kmol); m_{d} represents the molar mass of the dry atmosphere ($m_{\mathrm{d}} = 28.96$ kg/kmol); k_1, k_2, and k_3 are empirical physical constants ($k_1 = 77.604$ K/hPa, $k_2 = 70.4$ K/hPa, $k_3 = 3.776 \times 105$ K^2/hPa); and T_m is related to the temperature and vapor pressure at different altitudes in the atmosphere. In practice, an empirical formula for surface temperature and T_{m} is established by linear regression analysis using long-term radiosonde data.

3.3 Ground-GNSS Tropospheric Monitoring

3.3.1 2-D Tropospheric Monitoring

With double difference for two-frequency GNSS observations, some common error sources are eliminated, e.g., satellite and receiver clock errors, and tropospheric delays can be retrieved with high accuracy. For a pair of GPS observation stations and satellites, the ionosphere-free linear combination (LC) equation of double-differenced (DD) phase and code observations can be expressed as

$$\begin{aligned} LC_{kl}^{ij} &= \frac{1}{f_1^2 - f_2^2}(f_1^2 L1_{kl}^{ij} - f_2^2 L2_{kl}^{ij}) \\ &= \rho_{kl}^{ij} + ZTD_k(t)(m(z_k^i) - m(z_k^j)) - ZTD_l(t)(m(z_k^i) - m(z_k^j)) \\ &+ \frac{c}{f_1^2 - f_2^2}(f_1^2 N1_{kl}^{ij} - f_2^2 N2_{kl}^{ij}) + \varepsilon \end{aligned} \tag{3.11}$$

where ZTD is the zenith tropospheric delay,m is the mapping function, f_i is the frequency ($i = 1, 2$), Li and Ni are the double-differenced phase and ambiguity of the frequency f_i, respectively, ρ_{kl}^{ij} is the double-differenced pseudo-range, c is the velocity of light in vacuum, and ε_{kl}^{ij} is noise. Using the least square, the unknown parameters can be estimated, including ZTD, coordinate and ambiguity, etc. (Jin et al. 2010).

Nowadays, using the emerging Precise Point Positioning (PPP), high accuracy tropospheric delay can be estimated too, while the precise orbit and clock are provided by an external network such as IGS products. The quantity observed by the GPS receiver is the interferometric phase measurement of the distance from the GPS satellites to the receiver. The processing software solves or models the orbital parameters of the satellites, the transmitter and receiver positions, ionospheric delays, phase cycle ambiguities, and the clock drifts in addition to solving for the tropospheric delay parameters of interest. This requires the same type of GPS data processing software as that is used for high precision geodetic measurements, e.g., GAMIT software (King and Bock 2018), which solves for the ZTD and other parameters using a constrained batch least squares inversion procedure.

3.3.2 3-D Tropospheric Tomography

The tomography technique is used to reconstruct the detailed 3-D information of the studied objects through a certain mathematical constraint. In GNSS meteorology, the integral observations of tropospheric tomography are slant water vapor (SWV), with the movement of satellites in space and the rotation of Earth, and dense GNSS observations can retrieve the 3-D water vapor over the interested area with the tomography technique. To solve the integral problem of total water vapor, the discretized method is adopted in GNSS water vapor tomography. Based on the voxel-based method, the tropospheric zone over the interested area is discretized into the self-enclosed 3-D voxels in horizontal and vertical directions (Fig. 3.1). Due to the limited number of ground stations, the total number of tropospheric delay signals at single epoch is not enough and, therefore, it usually requires accumulating observation signals for a period time, which is called tomography window, so the inversion field is the average of the water vapor distribution over this period time. Assumed that the water vapor parameters of each voxel at epoch t are represented by $X(r, t)$, where r stands for the position of voxels, the slant water vapor (SWV) for signal p in a single tomography window can be described as

$$SWV^p = \sum_{r=1}^{n} x(r,t)dl_r^p \qquad (3.12)$$

where dl represents the intercept of the signal p at the r voxel.

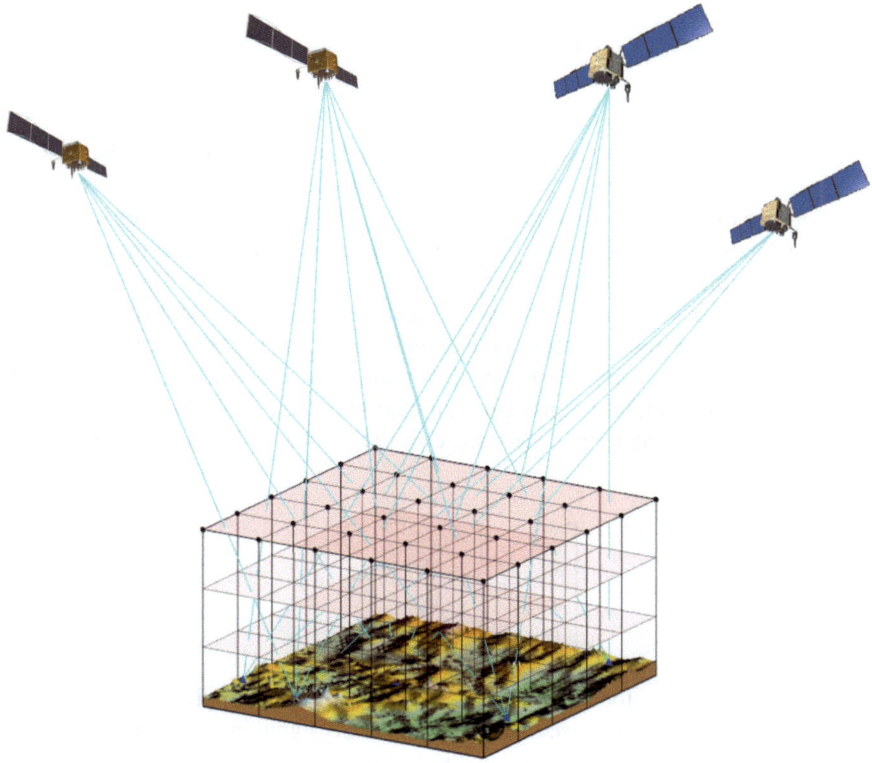

Fig. 3.1 The principle of GNSS tropospheric water vapor tomography

Although we assumed that the water vapor status remains stationary in a tomography window and the number of accumulated tomography signals will be much larger than the total number of voxels, the tropospheric tomographic observation equations are still ill-posed as some voxels always have no signal passing due to the limited spatial distribution of satellites and stations on the ground network. Thus, the observation equations cannot be solved directly by the least square (LS) estimation. In order to solve the inversion problem of the singular design matrix in the tomography equation, constraints are introduced into the observation model. It is supposed that there is a spatial correlation between the atmospheric water vapor in a specific voxel and surrounding voxels. The common methods of constraints include the horizontal constraint, vertical constraint, and boundary constraint (Jin et al. 2014; Rohm 2013; Dong and Jin 2018).

In the same layer of the grid, it is assumed that there is a Gaussian inverse distance weighted relationship between the water vapor density of a voxel and the other voxels. The Gaussian distance weighting function is expressed as

$$h_{i,j}^{i_1,j_1} = \frac{e^{\frac{d_{i_1,j_1}^2}{2\delta^2}}}{\sum_{i_1=1}^{nl}\sum_{j_1=1}^{nn} e^{\frac{d_{i_1,j_1}^2}{2\delta^2}}} \quad (3.13)$$

where (i, j) represents the voxel location calculated at a certain layer of the grid; (i_1, j_1) stands for a nearby voxel; d_{i_1,j_1} represents the distance from voxel (i, j) to (i_1, j_1); *nl* and *nn* are the number of the grid divided in the latitude and longitude directions; and δ is a Gaussian smoothing factor, which is determined according to the range of smoothing assumptions.

A priori water vapor distribution information is used to establish the vertical constraints to solve the alteration phenomenon of water vapor field inversion that upper water vapor density is smaller than water vapor density at the bottom. In addition, the top boundary constraint can also be added to force the water vapor density of voxels at the topper layer to zeros. Then, the tomography model can be described as

$$\begin{bmatrix} y \\ 0 \\ r \\ 0 \end{bmatrix} = \begin{bmatrix} A_{GNSS} \\ H \\ V \\ B \end{bmatrix} \cdot X + \varepsilon \quad (3.14)$$

where *y* is a column vector composed of all GNSS SWV in a single tomography window; A_{GNSS} is the design matrix composed of GNSS signal intercepts at each voxel; *H, V,* and *B* are the coefficient matrices for the horizontal, vertical, and top boundary constraints; *r* represents the prior water density provided by external observation techniques; and ε is the residual.

The design matrix of tomography equation established by GNSS technique and constraint equations is a large-scale sparse matrix and, in fact, the tomography solution is the inverse problem of the coefficient matrix. In order to ensure the physical sense of the solution, the results of the equations must be positive. The solving methods of tomography equations include the non-iterative algorithm, iterative algorithm, and the joint approach of both. Non-iterative methods, such as LS, singular value decomposition (SVD), etc., are not sensitive to initial values, and only approximate values can be obtained. The iterative reconstruction method requires a higher precision initial value to converge, and the accuracy of the corresponding solution is also higher. Bender et al. (2011) compared different iterative reconstruction algorithms, and the results showed that the multiplicative algebraic reconstruction technique (MART) has better iterative accuracy and iterative speed. The MART algorithm is given by the following expression (Jin and Park 2007; Jin et al. 2006):

$$x_j^{k+1} = x_j^k \left(\frac{\mathrm{SWV}_i}{A^i, x^k} \right)^{\frac{\lambda \cdot A_j^i}{\sqrt{A^i, A^i}}} \quad (3.15)$$

where SWV_i and A_i represent the ith observation of the GNSS tomography equations and the corresponding vector of equation coefficients, the inner product A^i, x^k is the back projection of ith GNSS tomography observation after the kth iteration, and λ is the relaxation parameter, which gives the weight correction rate. MART algorithm can be divided into two steps, first, the water vapor density of voxels (index j, j = 1, 2, ..., m) is corrected one by one for each observation (index i, i = 1, 2, ..., n), and then the next iteration is performed (index k, k = 1, 2, ..., p) until the solution converges. Here, the combined method is adopted, and SVD is implemented to obtain the approximate solution as the initial value of MART.

3.4 Space-Borne GNSS Radio Occultation

As one of the most advanced space exploration technologies in the twenty-first century, GNSS radio occultation can provide abundant information like the refractive index, the atmospheric pressure, and the atmospheric temperature. It is proved that the technique has the advantages of all-weather, high accuracy, high horizontal and vertical resolution, large data volume and low cost, etc., which give it a huge application value in the detection of ionospheric and neutral atmospheric parameters, numerical weather forecast, and climatology.

As the LEO satellite equipped with a GNSS receiver orbits around the Earth, an occultation event occurs (Fig. 3.2) when the received navigation signal from a setting (rising) GNSS satellite scan through successively deeper (higher) layers of the Earth's atmosphere until the GNSS signals descend below the Earth surface (rise above the atmosphere). The received GNSS signal is bent or delayed before arriving at the LEO satellite due to the Earth's atmosphere. The GNSS radio occultation technique takes advantage of the extremely precise phase and amplitude measurement of GNSS navigation signals that pass through the Earth's atmosphere to provide accurate retrieval of the vertical profiles of refractive index of the atmosphere. Consequently, the atmospheric properties such as air density, temperature, pressure, and humidity can be inferred (Kursinski et al. 1997; Rocken et al. 1997).

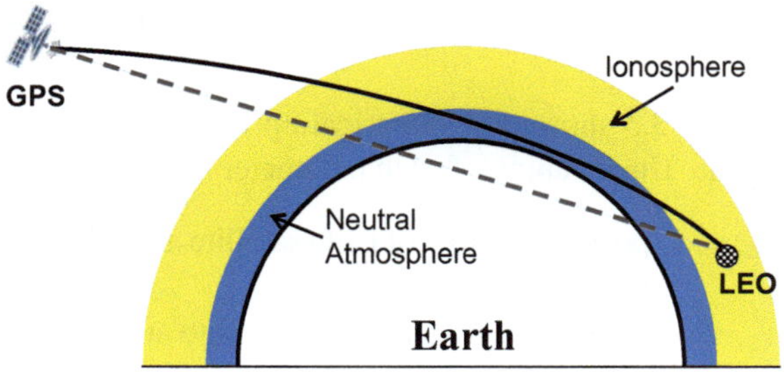

Fig. 3.2 Schematic diagram of GNSS radio occultation

3.4.1 Refraction Index

The refractive index n in a medium is defined as the ratio between the speed of light in a vacuum and the speed of light in the medium. In the neutral atmosphere, n is very close to one unity, such that it is usually expressed in terms of refractivity defined as $N = (n - 1) \times 10^6$. The refractivity at GNSS frequencies contains contributions from four major components, i.e., the dry neutral atmosphere (N^{dry}), water vapor (N^{vapor}), free electrons in the ionosphere (N^{iono}), and particulates (primarily liquid water and ice water content, N^{scatt}) through the following relationship (Kursinski et al. 1997; Hajj et al. 2002):

$$\begin{aligned} N &= N^{dry} + N^{vapor} + N^{iono} + N^{scatt} \\ &= 77.6\frac{P}{T} + 3.73 \times 10^5 \frac{P_w}{T^2} - \left(40.3 \times 10^7 \frac{n_e}{f^2} + O(\frac{1}{f^3})\right) + \left(1.4W_{liquid} + 0.6W_{ice}\right) \end{aligned} \quad (3.16)$$

where P and P_w are total pressure and water vapor partial pressure in mbar; T is temperature in Kelvin; n_e is electron number density per cubic meter; f is signal frequency in Hertz; and W_{liquid} and W_{ice} are referred to liquid water content and ice water content in gram per cubic meter, respectively.

Dry refractivity is proportional to molecular number density and is dominant below 60–90 km. The dry refractivity term is due to the polarizability of molecules in the atmosphere, i.e., the ability of an incident electric field to induce an electric dipole in the molecules. The moist refractivity term is due primarily to the large permanent dipole moment of water vapor and becomes significant in the lower troposphere, especially in the tropics and subtropics (Kursinski et al. 2000). The ionospheric term in Eq. 3.16 includes a first-order approximation ($1/f^2$) mainly due to free electrons in the ionosphere and becomes important above 60–90 km. The second-order term ($1/f^3$) is generally neglected (Kursinski et al. 1997). The scattering term given in Eq. 3.16 is due to liquid water droplets and ice crystals suspended in the atmosphere (Kursinski et al. 1997). For realistic suspensions of water or ice, the scattering term is small in comparison with the other terms and is therefore neglected in most RO applications (Kursinski et al. 1997).

3.4.2 Calculation of Total Refraction Angle

Figure 3.3 shows a geometrical optical relationship in an atmospheric occultation event. The meaning of each parameter in Fig. 3.3 is as follows:

$\vec{r_L}, \vec{r_G}$ position vectors of LEO satellite and GNSS satellite (from geocentric center to satellites);

$\vec{v_L}, \vec{v_G}$ velocity vectors of LEO satellite and GNSS satellite;

$\vec{e_L}, \vec{e_G}$ direction vectors of GNSS signals at LEO satellite and GPS satellite;

ϕ_L, ϕ_G included angles between $-\vec{r_L}$ and $\vec{e_L}$, $-\vec{r_G}$ and $\vec{e_G}$, respectively;

η_L, η_G included angles between $\vec{v_L}$ and $\vec{r_L}$, $\vec{v_G}$ and $\vec{r_G}$, respectively;
φ_L, φ_G included angles between $\vec{e_L}$ and $\vec{v_L}$, $\vec{e_G}$ and $\vec{v_G}$, respectively;
α total refraction angle;
r_0 the shortest distance from signal path to geocentric center; and
a impact radius.

Due to the effect of the Earth's atmospheric refraction, the GNSS signals on both frequencies will produce an additional phase delay:

$$\Delta L_i = L_i - r_{LG} = \int n_i ds - r_{LG} \tag{3.17}$$

where i takes the values of 1 and 2, and r_{LG} represents the geometric distance between GNSS and LEO satellites. Then the following equation can be obtained after derivation on time:

$$\frac{d\Delta L_i}{dt} + \frac{dr_{LG}}{dt} - \frac{dL_i}{dt} = 0 \tag{3.18}$$

The first term at the left side is the derivative of the additional phase delay to time, commonly referred to as additional Doppler. The Doppler shift caused by the Earth's atmosphere is

$$\frac{\Delta f_i}{f_i} = \frac{1}{c}\frac{dL_i}{dt} \tag{3.19}$$

And there is a relationship between Doppler shift and satellite velocity (Jin et al. 2014):

$$\frac{dL_i}{dt} = |\vec{e}_L| \cos(\varphi_L(a)) - |\vec{e}_L| \cos(\varphi_G(a)) \tag{3.20}$$

According to the geometric relationship in Fig. 3.3, we can get the expression of φ_L and φ_G:

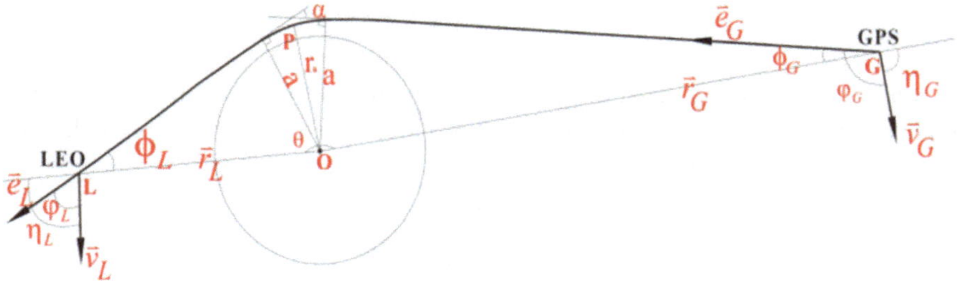

Fig. 3.3 Geometrical schematic diagram of atmospheric GNSS RO inversion

$$\varphi_L(a) = \eta_L - \phi_L = \eta_L - \arcsin\left(\frac{a}{|\vec{r}_L|}\right) \tag{3.21}$$

$$\varphi_G(a) = \pi - \eta_G + \phi_G = \pi - \eta_G + \arcsin\left(\frac{a}{|\vec{r}_G|}\right) \tag{3.22}$$

Then the atmospheric Doppler observation equation is obtained:

$$\frac{\Delta f_i}{f_i} = \frac{f_i}{c}|\vec{e}_L|\cos\left(\eta_L - \arcsin\left(\frac{a}{|\vec{r}_L|}\right)\right) - |\vec{e}_L|\cos\left(\pi - \eta_G + \arcsin\left(\frac{a}{|\vec{r}_G|}\right)\right) \tag{3.23}$$

Under the assumption of spherical symmetry, according to Snell's law (assuming that the refractive index is 1 at GNSS and LEO satellites):

$$|\vec{r}_L|\sin\phi_L = |\vec{r}_G|\sin\phi_G = a \tag{3.24}$$

Using the occultation observation data, as well as the position and velocity information of the satellites, the collision radius and total refraction angle can be solved according to the above formula. Above calculations are based on certain approximate assumptions, to obtain more accurate total refraction angle, a series of corrections are needed. The main corrections include the Earth's oblateness correction (Syndergaard 1998), elimination of ionospheric effects (Vorob'ev and KrasIl'NIKova 1994) and optimization of observation data (Hocke 1997).

3.4.3 *Abel Inversion*

Under the assumption of spherically symmetric atmosphere, the relationship between the instantaneous differential refraction angle $d\alpha$ and the differential refractive index dn of the spherical atmospheric layered signal is as follows (Jin et al. 2014):

$$d\alpha = -\frac{dn}{n}\frac{a}{\sqrt{n^2r^2 - a^2}} \tag{3.25}$$

Since the refractivity generally decreases at higher altitudes, to allow the bending angle to be positive values, a negative sign is added. Then the expression of the total refraction angle is obtained by integrating the signal path:

$$\alpha(a) = -2a\int_a^\infty \frac{1}{n}\frac{dn}{dx}\frac{ax}{\sqrt{x^2 - a^2}} \tag{3.26}$$

where $x = nr$, r is the distance from the center of curvature of a ray path, and the integral is over the portion of the atmosphere above r_0 (the point on the ray path that is closest to the Earth's center).

Equation 3.26 provides the forward calculation of bending angle α given the refractive index profile $n(x)$. By inverting the equation through the Abel transformation, the $n(x)$ can be expressed as a function of α (Fjeldbo et al. 1971):

$$n(x) = exp\left[\frac{1}{\pi}\int_0^{\infty}\frac{\alpha(a)}{\sqrt{a^2 - x^2}}da\right] \tag{3.27}$$

Note that Eq. 3.20 embeds the assumption of local spherically symmetric atmosphere assumption, i.e., the refractive index only varies along radius direction. However, the ellipsoidal shape of the Earth (with an equatorial radius roughly 20 km larger than its polar radius) and horizontal gradients in atmospheric structure produce nonspherical symmetry in the refractive index field. Moreover, the ray paths for a given occultation event do not scan the atmosphere vertically nor are they coplanar. Therefore, measurement of $\alpha(a)$ will be affected by the tangential refractivity gradients and occultation geometry, and Eq. 3.27 could introduce systematic errors into the retrieved refractive index profile $n(x)$.

3.4.4 Inversion of Atmospheric Parameters

Firstly, we define the perigee height:

$$z = r_t - r_e \tag{3.28}$$

where r_t is the radius vector at the perigee, and r_e is the radius of the Earth in the occultation plane. Equation 3.16 shows the dependency of atmospheric refractivity on various atmospheric properties in GNSS RO frequencies. In the neutral atmosphere, which includes the stratosphere and the troposphere (below ~70 km), the refractivity primarily depends on the dry atmosphere and water vapor as first described by Smith and Weintraub (1953). For realistic suspensions of water or ice, the contribution due to the liquid water and ice water content is generally small in comparison with other terms and will be neglected here. Therefore, the refractivity variation with tangent radius r becomes a simple a function of atmospheric total pressure P (in hPa), water vapor partial pressure Pw (in hPa) and temperature T (in Kelvin), such as (Jin et al. 2014):

$$N(z) = (n-1)\times 10^6 = c_1\frac{P(z)}{T(z)} + c_2\frac{P_w(z)}{T(z)^2} \tag{3.29}$$

The constants are $c_1 = 77.6$ (K hPa^{-1}) and $c_2 = 3.73 \times 10^5$ (K^2/hPa^{-1}). The second item on the right is the water vapor item, which is usually negligible. Then Eq. 3.22 can be simplified as

$$N(z) = (n-1) \times 10^6 = k_1 \frac{P(z)}{T(z)} \tag{3.30}$$

The expression of the atmospheric density profile is obtained by combining the ideal gas state equation:

$$\rho(z) = \frac{M}{R}\frac{P(z)}{T(z)} = \frac{M}{c_1 R} N(z) \tag{3.31}$$

where M is the average molecular weight of dry air, in homogeneous layer (usually below 100 km), $M = 28.964$ kg/kmol, and R is the ideal gas universal constant with a value of 8314 J/(K kmol). The atmospheric static equilibrium equation is

$$dp = -\rho g dz \tag{3.32}$$

Integrating Eq. 3.32, we can get the atmospheric pressure:

$$P(z) = \int_{z_0}^{\infty} \rho(z) g(z) dz = \int_{z_0}^{z_u} \rho(z) g(z) dz + P_u \tag{3.33}$$

where $g(z)$ is the local gravitational acceleration, z_u is the upper boundary, which is usually set at 35 km, and P_u is the pressure at z_u, usually obtained from other databases.

If the neutral atmosphere is approximated as dry air, the temperature profile can be derived from the following equation:

$$T_d(z) = \frac{c_1 P(z)}{N(z)} \tag{3.34}$$

However, the water vapor content in the low troposphere is relatively rich, so the atmospheric density is usually rewritten as

$$\rho(z) = \rho_d(z) + \rho_w(z) = \frac{M_d P(z)}{T(z) R} + \frac{(M_w - M_d) P_w(z)}{T(z) R} \tag{3.35}$$

If the atmospheric temperature profile is known, then the profile of wet pressure is obtained as

$$P_w(z) = T(z)^2 \frac{N(z) - k_1 P(z)/T(z)}{c_2} \tag{3.36}$$

3.5 Other GNSS RO

The geometric optics technique has been adapted from the space-borne RO to the airborne measurements (Healy et al. 2002; Mousa and Tsuda 2004; Xie et al. 2008). A number of field campaign experiments have also been conducted, and preliminary comparisons and analysis of measurements with radiosonde observations and Numerical Weather Prediction (NWP) model analyses have been reported (Aoyama et al. 2004; Muradyan et al. 2011). The geometric optics retrieval technique still has challenges due to the multipath effect in the lower troposphere. Therefore, the more advanced radio-holographic retrieval method is thus needed to derive more accurate and higher vertical resolution RO soundings (e.g., Xie et al. 2012). Contrary to the space-borne RO observation, with a GNSS receiver inside the atmosphere, the receiver is either static (e.g., on mountain-top) or moving (e.g., airborne, in the order of ~0.25 km/s) much slower than the GNSS occultation satellite (~3.87 km/s) and the LEO satellites (~7.8 km/s). A typical mountain-top or airborne radio occultation generally takes ~20–30 min to sample the atmosphere from the altitude (~5–10 km) of the receiver down to the surface. The sampling rate requirement for the airborne RO measurements is thus lower than that for the space-based occultation technique, such that it can be satisfied by an off-the-shelf GPS receiver (Jin et al. 2014). Xie et al. (2008) demonstrated significant representative errors that can be induced by the combination of limb sounding geometry and the horizontal tangent point drifting. Therefore, it is important to note the airborne RO bending or refractivity profile derived from an occultation measurement should not be treated as local or in situ measurement. Various data assimilation techniques used in the space-borne RO community could also be adapted to maximize the benefit of airborne RO measurements for regional numerical weather forecasting.

References

Aoyama, Y., Shoji, Y., Mousa, A., Tsuda, T., & Nakamura, H. (2004). Temperature and water vapor profiles derived from downward-looking GPS occultation data. *Journal of the Meteorological Society of Japan, 82,* 433–440.

Bender, M., Dick, G., Ge, M., Deng, Z., Wickert, J., Kahle, H. G., et al. (2011). Development of a GNSS water vapour tomography system using algebraic reconstruction techniques. *Advances in Space Research, 47,* 1704–1720.

Bevis, M., Businger, S., Chiswell, S., et al. (1994). GPS meteorology: Mapping Zenith wet delays onto precipitable water. *Journal of Applied Meteorology, 33,* 379–386.

Davis, J. L., Herring, T. A., Shapiro, I., Rogers, A., & Elgered, G. (1985). Geodesy by radio interferometry: Effects of atmospheric modeling errors on esimatetes of baseline length. *Radio Science, 20*(6), 1593–1607.

Dong, Z. N., & Jin, S. G. (2018). 3-D water vapor tomography in Wuhan from GPS, BDS and GLONASS observations. *Remote Sensing, 10*(1), 62. https://doi.org/10.3390/rs10010062.

Duan, J., Bevis, M., Fang, P., et al. (1996). GPS meteorology: Direct estimation of the absolute value of precipitable water. *Journal of Applied Meteorology, 35,* 830–838.

Emardson, T. R., Elgered, G., & Johansson, J. M. (1998). Three months of continuous monitoring of atmospheric water vapor with a network of GPS receivers. *Journal of Geophysical Research, 103,* 1807–1820.

Fjeldbo, G. F., Eshleman, V. R., & Kliore, A. J. (1971). The neutral atmosphere of Venus as studied with the Mariner V radio occultation experiments. *Astronomical Journal, 76,* 123–140.

Hajj, G. A., Kursinski, E. R., Romans, L. J., et al. (2002). A technical description of atmospheric sounding by GPS occultation. *Journal of Atmospheric and Solar-Terrestrial Physics, 64*(4), 451–469.

Hartmann, G., & Leitinger, R. (1984). Range errors due to ionospheric and tropospheric effects for signal frequencies above 100 MHz. *Bulletin Géodésique, 58,* 109–136.

Healy, S. B., Haase, J., & Lesne, O. (2002). Abel transform inversion of radio occultation measurements made with a receiver inside the Earth's atmosphere. *Annales Geophysicae, 20*(8), 1253–1256.

Hernández-Pajares, M., Juan, J. M., Sanz, J., et al. (2001). A new strategy for real-time integrated water vapor determination in WADGOPS networks. *Geophysical Research Letters, 28*(17), 3267–3270.

Hocke, K. (1997) Inversion of GPS meteorology data. *Annales Geophysicae, 15*(4), 443–450 (Springer, Berlin/Heidelberg).

Hopfield, H. (1969). Two-quartic tropospheric refractivity profile for correcting satellite data. *Journal of Geophysical Research, 74*(18), 4487–4499.

Jin, S., & Luo, O. F. (2009). Variability and climatology of PWV from global 13-year GPS observations. *IEEE Transactions on Geoscience and Remote Sensing, 47,* 1918–1924.

Jin, S. G., & Park, P. H. (2005). A new precision improvement of zenith tropospheric delay estimates by GPS. *Current Science, 89*(6), 997–1000.

Jin, S., & Park, J. U. (2007). GPS ionospheric tomography: A comparison with the IRI-2001 model over South Korea. *Earth, Planets and Space, 59,* 287–292.

Jin, S., Park, J. U., Wang, J. L., Choi, B. K., & Park, P. H. (2006). Electron density profiles derived from ground-based GPS observations. *The Journal of Navigation, 59,* 395–401.

Jin, S. G., Park, J., Cho, J., & Park, P. (2007). Seasonal variability of GPS-derived Zenith Tropospheric Delay (1994–2006) and climate implications. *Journal Geophysical Research, 112,* D09110. https://doi.org/10.1029/2006JD007772.

Jin, S. G., Luo, O. F., & Gleason, S. (2009). Characterization of diurnal cycles in ZTD from a decade of global GPS observations. *Journal of Geodesy, 83*(6), 537–545. https://doi.org/10.1007/s00190-008-0264-3.

Jin, S. G., Luo, O. F., & Ren, C. (2010). Effects of physical correlations on long-distance GPS positioning and zenith tropospheric delay estimates. *Advances in Space Research, 46*(2), 190–195. https://doi.org/10.1016/j.asr.2010.01.017.

Jin, S. G., Feng, G. P., & Gleason, S. (2011). Remote sensing using GNSS signals: Current status and future directions. *Advances in Space Research, 47*(10), 1645–1653. https://doi.org/10.1016/j.asr.2011.01.036.

Jin, S.G., Cardellach, E., Xie, F. (2014). *GNSS remote sensing: Theory, methods and applications* (276 pp). Springer, Netherlands.

King, R. W., & Bock, Y. (2018). *Documentation for the GAMIT GPS analysis software*. Cambridge Mass: Massachusetts Institute of Technology.

Kursinski, E. R., Hajj, G. A., Schofield, J. T., Linfield, R. P., & Hardy, K. R. (1997). Observing the earth's atmosphere with radio occultation measurements using the Global Positioning System. *Journal Geophysical Research, 102*(D19), 23429–23465.

Kursinski, E. R., Healy, S. B., & Romans, L. J. (2000). Initial results of combining GPS occultations with ECMWF global analyses within a 1DVar framework. *Earth Planets Space, 52,* 885–892.

Mousa, A., & Tsuda, T. (2004). Inversion algorithms for GPS downward looking occultation data: Simulation analysis. *Journal of the Meteorological Society of Japan, 82,* 427–432.

Muradyan, P., Haase, J. S., Xie, F., Garrison, J. L., & Voo, J. (2011). GPS/INS navigation precision and its effect on airborne radio occultation retrieval accuracy. *GPS Solutions, 15*(3), 207–218. https://doi.org/10.1007/s10291-010-0183-7.

Niell, A. E. (1996). Global mapping functions for the atmospheric delay at radio wavelengths. *Journal of Geophysical Research, 101*(B2), 3227–3246.

Rocken, C., Anthes, R., Exner, M., Ware, R., Feng, D., Gorbunov, M., et al. (1997). Verification of GPS/MET data in the neutral atmosphere. *Journal of Geophysical Research, 102,* 29849–29866.

Rohm, W. (2013). The ground GNSS tomography–unconstrained approach. *Advances in Space Research, 51,* 501–513.

Smith, E. K., & Weintraub, S. (1953). The constants in the equation for atmospheric refractive index at radio frequencies. *Proceedings of the IRE, 41,* 1035–1037.

Syndergaard, S. (1998). Modeling the impact of the Earth's oblateness on the retrieval of temperature and pressure profiles from limb sounding. *Journal of Atmospheric and Solar-Terrestrial Physics, 60*(2), 171–180.

Tregoning, P., Boers, R., & O'Brien, D. (1998). Accuracy of absolute precipitable water vapor estimates from GPS observations. *Journal of Geophysical Research, 103*(28), 701–710.

Vorob'ev, V. V., & KrasII'NIKova, T. G. (1994). Estimation of the accuracy of the atmospheric refractive index recovery from Doppler shift measurements at frequencies used in the NAVSTAR system. *Izvestiya Atmospheric and Oceanic Physics, 29*(5), 602–609.

Xie, F., Haase, J. S., & Syndergaard, S. (2008). Profiling the atmosphere using the airborne GPS radio occultation technique: A sensitivity study. *IEEE Transactions on Geoscience and Remote Sensing, 46*(11), 3424–3435.

Xie, F., Wu, D. L., Ao, C. O., Mannucci, A. J., & Kursinski, E. R. (2012). Advances and limitations of atmospheric boundary layer observations with GPS occultation over southeast Pacific Ocean. *Atmospheric Chemistry and Physics, 12,* 903–918.

Chapter 4
GNSS Ionospheric Sounding

4.1 Ionosphere

The Earth's ionosphere ranges from about 60 to 1000 km in the atmosphere (Li and Huang 2005). Different from other Earth's spheres, the ionosphere contains a large number of ions and free electrons under the Sun radiation and cosmic ray excitation effect. The ionosphere and the ionospheric formation theory informed by Hulburt and Chapmann marked the start of modern ionospheric theory (Xiong et al. 1997). The ionospheric steady state is formed on the combined action of the light chemistry, thermodynamics, kinetics, electrodynamics, and other process. It not only has the complex spatial distribution structure but also has the time-varying characteristics with different time scales and controlled by multiple mechanisms.

From the ground to about 1000 km height, the Earth's atmosphere, namely, the ionosphere, has vertical layered structure due to the gravity. The neutral atmosphere can be approximately divided into the troposphere, the stratosphere, the middle layer, and the thermal layer in the vertical structure according to the temperature. In the ionospheric study, the ionosphere is usually divided into D, E1, E2, Es, F1, F2 layer and the topside ionosphere according to the electron density. Each layer presents different characteristics due to their different maintenance mechanism. The D layer at the bottom of the ionosphere is mainly produced by the Lyman-α radiation effect at the wavelength of 121.5 nm. There are still a large number of neutral particles in the D layer, and the frequent collisions with free electron lead to the strong absorption of electromagnetic waves in the area. At night, without the effect of solar radiation, the electron density of D layer will decrease sharply to approximate 0. The E layer is mainly formed by the ionization of far X rays and far ultraviolet rays on the oxygen molecule. During the daytime, the electron density can be up to 10^{11} el/m^3, while the electronic density decreases to the level of 10^9 el/m^3 at night. And the Es layer is an occasional layer. It lasts for a few minutes to a few hours. Its electron density is two times higher than the upper and lower regions, while the causes are still one of the hot spots in the ionospheric study. The F layer is mainly produced from ionization of the middle-strong ultraviolet solar radiation on single oxygen atoms. During the

S. Jin et al., *GNSS Atmospheric Seismology*,
https://doi.org/10.1007/978-981-10-3178-6_4

day, the F layer is usually further divided into F1 and F2 layers. Because the F2 layer has the highest free electron density, we usually describe the ionospheric change characteristics by the peak electron density $N_m F2$ and its corresponding height $h_m F2$ of the F2 layer.

The ionosphere not only has an obvious vertical structure but also has a complicated horizontal distributed structure. Ionospheric electron content in low latitude area is significantly higher than that in high latitude area, and in the dayside is higher than that of in the nightside. This distribution is influenced by the geomagnetic latitude and the solar local time. Essentially, it is derived from the difference of the regional geomagnetic field and the difference of the solar radiation. In addition, there are other local distributed structures for local factors, such as the double-crest structure caused by the eastward electric field over the equator, the regional disturbances caused by the earthquake, the lightning, and other events.

4.2 Ionospheric Delay

The theoretical basis of the ionospheric sounding is the propagation theory of the radio waves in the ionized medium under the external magnetic field, namely, the magnetoionic theory. The phase refraction index of the collisionless cold magnetic plasma can be expressed by the phase refraction formula of Appleton–Hartree:

$$\begin{aligned} n_p^2 &= 1 - \frac{X}{1 - iZ - \frac{Y_T^2}{2(1-X-iZ)} \pm \sqrt{\frac{Y_T^4}{4(1-X-iZ)^2} + Y_L^2}} \\ X &= \frac{\omega_p^2}{\omega^2} = \frac{N_e e^2}{4\pi^2 \varepsilon_0 m f^2} \\ Y_L &= \frac{\omega_H}{\omega}\cos\theta = \frac{\mu_0 H_0 |e|}{2\pi m}\frac{1}{f}\cos\theta \\ Y_T &= \frac{\omega_H}{\omega}\sin\theta = \frac{\mu_0 H_0 |e|}{2\pi m}\frac{1}{f}\sin\theta \\ Z &= \frac{\nu_e}{\omega} \end{aligned} \tag{4.1}$$

where the parameters are as follows:

- ω_p the plasma angular frequency, namely, $[(N_e e^2)/(m\varepsilon_0)]^{0.5}$;
- ω the radio signal angular frequency;
- N_e electron density;
- e the electric charge of a single electron, namely, 1.6022×10^{-19} C;
- ε_0 dielectric constant in vacuum, namely, 8.8542×10^{-12} C²/N m;
- m electronic mass, namely, 9.1096×10^{-31} kg;
- f the frequency of radio signal, namely, $\omega/2\pi$;
- θ the angle between the geomagnetic field and the vector of the signal wave;

ω_H electronic magnetic rotation angular frequency, namely, $(\mu_0 H_0|e|)/m$;
H_0 geomagnetic intensity;
μ_0 magnetic conductivity in vacuum, namely, 12.57×10^{-7} H/m; and
ν_e electron collision frequency

Here the positive and negative signs before the root sign indicate different polarization states, that is to say, there are two kinds of radio signal propagation modes in plasma. The positive sign corresponds to the ordinary wave, and the opposite state is the extraordinary wave. The electron collision frequency is mainly related to the electron temperature, the neutral particle, and the electron density. For example, the electronic collision frequency, calculated by using the empirical formula of Kelley, is less than the magnitude of 10^4. The current frequencies of several major GNSS signals are between 1 and 2 GHz. For radio signals within the GNSS carrier frequency range, the imaginary part Z of Appleton–Hartree phase refraction formula is much less than 1. In the discussion of the ionospheric effect on the GNSS signals, the influence of the imaginary part of the refractive index is usually negligible.

We expand Eq. 4.1 in power series (Fleury et al. 2009) and thus yields

$$n_p = 1 - \frac{1}{2}X \pm \frac{1}{2}X|Y_L| - \frac{1}{8}X^2 - \frac{1}{4}X(Y_T^2 + 2Y_L^2) + O\left(\frac{1}{f^5}\right) \tag{4.2}$$

As the magnitude of the plasma frequency is at 10^7 Hz and the magnitude of the electron magnetic rotation frequency is at 10^6 Hz, X, Y_L, Y_T is much less than 1. In general, after ignoring the high order small quantity in the refraction formula, the refractive index of the ionosphere can be expressed as

$$n_p \approx 1 - \frac{X}{2} = 1 - \frac{N_e e^2}{8\pi^2 \varepsilon_0 m f^2} \approx 1 - \frac{40.3 N_e}{f^2} \tag{4.3}$$

The radio signal propagation speed depends on the refractive index of the medium, so, for the GNSS radio signals, the ionospheric effects mainly depend on the electron density and the signal frequency, which is the theoretical basis of GNSS ionospheric detection.

For the wave packet composed of the signal with similar frequency, the relationship between the group index and the phase index can be expressed as (Hofmann-Wellenhof et al. 1992)

$$n_g = n_p + f\frac{dn_p}{df} \tag{4.4}$$

Combined with Eq. 4.3, the relationship between group velocity and group refraction index can be expressed as

$$v_g = \frac{c}{n_g} \approx \frac{c}{1 + \frac{40.3N_e}{f^2}} \approx c\left(1 - \frac{40.3N_e}{f^2}\right) \tag{4.5}$$

where c is the speed of radio signal propagation in vacuum. Supposed that the propagation time of the ranging codes is δt after eliminating all the errors except the ionospheric delay, and the distance between the satellite and the ground is

$$D = \int\limits_{\delta t} v_g dt = \int\limits_{\delta t} c\left(1 - \frac{40.3N_e}{f^2}\right)dt = c\delta t - \frac{40.3}{f^2}\int\limits_{s} N_e ds \tag{4.6}$$

where $c\delta t$ is the result of distance measurement using the GNSS ranging code, namely, the pseudo-range; s is the signal propagation path. The second item at the right side of Eq. 4.6 indicates the ionospheric delay from GNSS pseudo-range measurement. We usually refer to the integration of the electron density along the signal path, that is, the number of electrons in the cylinder with the unit area along the signal path, as the total electron content. As a result, the ionospheric delay of GNSS pseudo-range measurement can be simplified as

$$\delta Pion = -\frac{40.3TEC}{f^2} \tag{4.7}$$

The phase velocity of the electromagnetic wave with single frequency, that is, the relationship between the phase velocity and the ionospheric refractive index is described as

$$v_p = \frac{c}{n_p} = \frac{c}{1 - \frac{40.3}{f^2}N_e} \approx c\left(1 + \frac{40.3}{f^2}N_e\right) \tag{4.8}$$

Similar to the derivation of the ionospheric delay of GNSS pseudo-range measurements, the ionospheric delay of the ionospheric carrier phase measurements can be represented as

$$\delta Lion = \frac{40.3TEC}{f^2} \tag{4.9}$$

4.3 Ground-GNSS TEC Estimation

GNSS observation equation, namely, the expression of GNSS observations, is the foundation of GNSS navigation and other GNSS applications including ionospheric sounding. The ionospheric refraction formula quantifies the influence of the ionosphere on the radio signals. The integral of refracted index along the signal path is derived to obtain the total ionospheric delay, and therefore, we can inform the relationships between GNSS observations and ionospheric parameters such as electron content, electron density.

4.3.1 GNSS Observation Equations

The Global Navigation Satellite System (GNSS) contains US's GPS, China's BDS, Russia's GLONASS, and EU's Galileo. Here we take GPS as an example. GPS consists of a constellation of up to 30 operating satellites in six circular orbits 20,200 km above the Earth at an inclination angle of 55° with a 12-h (sidereal) period. Each GPS satellite broadcasts two spread spectrum L-band radio signals with the frequencies $f_1 = 1.57542$ GHz and $f_2 = 1.2276$ GHz. As the ionosphere is a dispersive medium, dual-frequency GPS receivers are able to evaluate the ionospheric effect by measuring the modulations on the codes and the carrier phases. The equations of dual-frequency GPS carrier phase (L) and code observations pseudo-range (P) are expressed as

$$L_{k,j}^{i} = \lambda_k \varphi_{k,j}^{i} = \rho_{0,j}^{i} - d_{ion,k,j}^{i} + d_{trop,j}^{i} + c(\tau^i - \tau_j) - \lambda_k (b_{k,j}^{i} + N_{k,j}^{i}) \quad (4.10)$$

$$P_{k,j}^{i} = \rho_{0,j}^{i} + d_{ion,k,j}^{i} + d_{trop,j}^{i} + c(\tau^i - \tau_j) + d_{q,k}^{i} + d_{q,k,j} + \varepsilon_j^i \quad (4.11)$$

where subscript k stands for the frequency ($k = 1, 2$), superscript i and subscript j represent the satellite and ground-based GPS receiver, respectively, and other parameters are as follows:

ρ_0	the true distance between the GPS receiver and satellite;
d_{ion} and d_{trop}	the ionospheric and tropospheric delays;
c	the speed of light in vacuum space;
τ	the satellite or receiver clock offset;
b	the phase delay of satellite and receiver instrument bias;
d_q	the code delay of satellite and receiver instrument bias;
λ	the carrier wavelength;
φ	the total carrier phase between the satellite and receiver;
N	the ambiguity of the carrier phase; and
ε	other residuals.

4.3.2 GNSS Ionospheric Observations

We implement the frequency difference on GNSS observations from Eqs. 4.10 and 4.11, and thus yields

$$P_{i,12}^{j} = P_{i,1}^{j} - P_{i,2}^{j} = -40.3TEC\left(\frac{1}{f_1^2} - \frac{1}{f_2^2}\right) + DCB_i^j + \varepsilon_P$$

$$L_{i,12}^{j} = L_{i,1}^{j} - L_{i,2}^{j} = -(\lambda_k N_{i,1}^{j} - \lambda_k N_{i,2}^{j}) + 40.3TEC\left(\frac{1}{f_1^2} - \frac{1}{f_2^2}\right) + DCB_i^j + \varepsilon_L \quad (4.12)$$

Herein $DCB_i^j = (d_{i,1} + d_1^j) - (d_{i,2} + d_2^j)$, i.e., the differential code bias DCB, $L_{i,k}^j = \lambda_k \varphi_{i,k}^j$, ε_P and ε_L are the sum of different multipath error and other frequency-related items of pseudo-range observation and carrier phase observation, respectively, which are called the residual error. Usually, we call $P_{i,12}^j$, $L_{i,12}^j$ the GNSS ionospheric observation. For a continuous observation arc, that is, the relative ambiguity and differential code bias are constant, regardless of ε_P and ε_L, changes of GNSS ionospheric measurements directly reflect changes of the total electron content along the ray path. Compared with the GNSS carrier phase observation, the GNSS pseudo-range observation is not affected by the integer ambiguity constant, while its residual error is larger. In the acquisition of high precision TEC observation, it is usual to combine both advantages into a new ionospheric observation with no integer ambiguity and smaller noise. One method is based on the pseudo-range observation, which uses the smoothness of the variation rate of the high precision carrier phase observation, i.e., the carrier phase smoothing pseudo-range technique. It can be achieved as follows:

$$\begin{aligned} P_{i,12,sm}^j(t) &= \omega_t P_{i,12}^j(t) + (1 - \omega_t) P_{i,12,prd}^j(t), \ (t > 1) \\ P_{i,12,prd}^j(t) &= P_{i,12,sm}^j(t-1) + [L_{i,12}^j(t-1) - L_{i,12}^j(t)], \ (t > 1) \end{aligned} \tag{4.13}$$

Firstly, we estimate the pseudo-range observation of the present epoch $P_{i,12,prd}^j(t)$ through the smoothing value of the previous epoch, i.e., change ratio of carrier phase observation, and then obtain the weighted average with the pseudo-range observation at present epoch. The method can effectively suppress the influence of Gaussian noise, and thus improves the accuracy of the ionospheric observation. The other method is based on the carrier phase observation, which uses the constant characteristics of relative integer ambiguity in the arc, and eliminates the integer ambiguity to achieve high precision absolute TEC monitoring. It can be achieved as follows:

$$L_{i,12}^j = 40.3TEC\left(\frac{1}{f_1^2} - \frac{1}{f_2^2}\right) + \frac{\sum P_{i,12}^j + L_{i,12}^j}{Nep} - DCB_i^j \tag{4.14}$$

Here Nep is the number of epochs in an arc. Figure 4.1 shows TEC time series of the pseudo-range observation, carrier phase observation, and carrier smoothed pseudo-range observation of the satellite G01 at the SHAO station of IGS on the first day of 2016. The results show that both methods can suppress noise and obtain the TEC with high accuracy, and acquire ionospheric sounding without the effect of the integer ambiguity. It should be noted that, unlike the pseudo-range measurements, the GNSS carrier phase observations are affected by the cycle slip. It is necessary to detect and repair the cycle slip before smoothing the relative ambiguity or the pseudo-range observation. After obtaining the "clean" GNSS ionospheric observation, the total electron content along the signal path can be obtained directly from Eqs. 4.12–4.14.

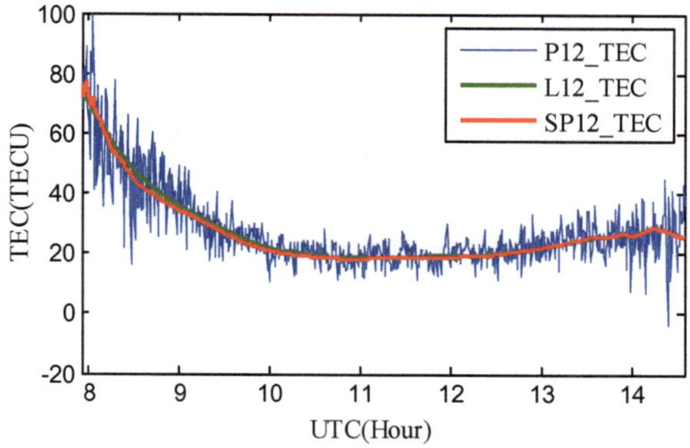

Fig. 4.1 TEC time series based on the pseudo-range observation, carrier phase observation, and carrier smoothed pseudo-range observation of the satellite G01 at the SHAO station of IGS on the first day of 2016 (DCB is corrected here)

Normally, Eq. 4.15 can be used to monitor the relative changes of GNSS ionosphere. For absolute TEC variation, the differential code bias DCB_i^j is required.

$$
\begin{aligned}
TEC &= \frac{f_1^2 f_2^2}{40.3(f_1^2 - f_2^2)}(P_{i,12,sm}^j - DCB_i^j) \\
&= \frac{f_1^2 f_2^2}{40.3(f_2^2 - f_1^2)}\left(L_{i,12}^j - \frac{\sum P_{i,12}^j + L_{i,12}^j}{Nep} + DCB_i^j\right)
\end{aligned} \tag{4.15}
$$

4.3.3 Cycle Slip Detection

Based on the methods of GNSS ionospheric detection introduced in the second chapter, the TEC time series are obtained by dual-frequency observations provided by GNSS continuous tracking stations. Since the carrier phase observations have higher precision, we use the carrier phase observation to obtain the TEC. Though the carrier phase observations are influenced by integer ambiguity and hardware delay, both of them are constant for a continuous observation arc, which do not affect the extraction of ionospheric disturbance signals. In practical observations, the signals lose lock when the GNSS signals are blocked or disturbed, and the receiver counter cannot work continuously. When the signals get the lock again, the whole cycle count will be wrong, resulting in the appearance of cycle slip in the sequence of carrier phase observations. The continuity of the GNSS observation arc is destroyed by the generation of the cycle slip, making the integer ambiguity of the carrier phase observation value no longer a constant. So the influence of the cycle slip must be removed from the TEC sequence estimation from the dual-frequency GNSS carrier phase observations before discussing the ionospheric disturbances. One method is that we only detect cycle slip in the GNSS carrier phase observations without repair, and take it as a new observation arc when cycle slip happens. In another method, we combine different methods of cycle slip detection, and then detect and repair cycle slip. The former one is easy to operate, but it can split the arc segment. For low elevation observations, the arc segment is divided into several short observation arc segments, which is difficult to analyze the disturbance in long time period. In order to ensure the integrity of the disturbance observations, we combined the TurboEdit algorithm and the second-order ionospheric residual method to detect and repair the cycle slip in the GNSS observation arc (Blewitt 1990; Cai et al. 2013). TurboEdit cycle slip detection is based on the M-W combination observations as follows:

$$L_{MW} = \frac{f_1 L_1 - f_2 L_2}{f_1 - f_2} - \frac{f_1 P_1 + f_2 P_2}{f_1 + f_2} = \lambda_{WL} N_{WL} \tag{4.16}$$

where $\lambda_{WL} = c/(f_1 - f_2)$ is the wavelength of wide lane observations, $N_{WL} = N_1 - N_2$ is the wide lane ambiguity. It is easy to derive from the last formula

$$N_{WL} = \phi_1 - \phi_2 - \frac{f_1 P_1 + f_2 P_2}{\lambda_{WL}(f_1 + f_2)} \tag{4.17}$$

The average error and the standard error in the GNSS arc segment at the kth epoch can be expressed as

$$\begin{aligned} \bar{N}_{WL}(k) &= \bar{N}_{WL}(k-1) + \frac{1}{k}[N_{WL}(k) - \bar{N}_{WL}(k-1)] \\ \sigma(k) &= \sqrt{\sigma^2(k-1) + \frac{1}{k}[(\bar{N}_{WL}(k) - \bar{N}_{WL}(k-1))^2 - \sigma^2(k-1)]} \end{aligned} \tag{4.18}$$

When there is no cycle slip, $\bar{N}_{WL}$ is a constant in theory, while its time variation is similar to the variation of the white noise in the actual observation. When

$$\begin{aligned} &\left|N_{WL}(k) - \bar{N}_{WL}(k-1)\right| \geq 4\sigma(k) \\ &|N_{WL}(k+1) - N_{WL}(k)| \leq 1 \end{aligned} \tag{4.19}$$

it is considered that there is no cycle slip at kth epoch. The difference of the cycle slips between the two arc segments is

$$dN_{WL} = dN_1 - dN_2 = N_{WL}(k) - \bar{N}_{WL}(k-1) \tag{4.20}$$

Here the threshold 1 of the difference between the kth epoch and the $(k+1)$th epoch is an experienced value in Eq. 4.19. If the difference is large, the observation at the kth epoch needs to be removed as gross error. Although the method can be more accurate in the detection of cycle slip, it cannot obtain the solution of cycle slips of two arc segments independently. The second-order ionospheric residual method for cycle slip detection is based on geometric independent linear combination observations

$$\phi_{GF} = L_1 - L_2 = \lambda_1 N_1 - \lambda_2 N_2 + \left(\frac{f_1^2}{f_2^2} - 1\right) I_1 \tag{4.21}$$

where I_1 indicates the ionospheric delay of the L_1 band. We define the ionospheric residual phase as

$$\phi_{PIR} = N_1 - \frac{\lambda_2 N_2}{\lambda_1} + I_R \tag{4.22}$$

where $I_R = (\gamma - 1)I_1/\lambda_1$, and thus yields

$$\begin{aligned} dN_1 - \frac{\lambda_2}{\lambda_1} dN_2 = & [\phi_{PIR}(k) - 2\phi_{PIR}(k-1) + \phi_{PIR}(k-2)] \\ & - [I_R(k) - 2I_R(k-1) + I_R(k-2)] \end{aligned} \tag{4.23}$$

Here the second term in the above formula is the second-order temporal variation of the ionosphere. For observation sequences within 30-s sampling interval, three adjacent epochs span 1 min and the second-order temporal variation is approximated to 0. So the combination of Eqs. 4.18 and 4.23 can solve the whole cycle of the two bands to repair the cycle slip. After eliminating the influence of cycle slip and get the "clean" TEC sequence based on the carrier phase observation, we can use the above the mentioned filter to extract the co-seismic ionospheric disturbance signals.

In the previous section, we introduce the basic principle of GNSS detection for the ionospheric electron content. High precision TEC sequences can be obtained directly from the previous section. However, the influence of the constant of differential code bias is often included. In addition, the spatial positions of the signal path will change over the motion of the GNSS satellites, and the calculation for absolute total electron content in a particular position is usually realized by ionospheric modeling. Although the two-dimensional ionospheric modeling for GNSS TEC is not sensitive to the vertical structure of the ionosphere, it is appropriate to reflect the characteristics of the seismic ionosphere. Three-dimensional tomography and ionospheric occultation technique can inverse the ionospheric vertical profile. However, we cannot conduct the systematic research on the seismic ionospheric disturbance due to the lack of the resolution at present, so we only discuss the two-dimensional TEC modeling here.

4.3.4 2-D Ionospheric Estimation

GNSS satellite orbit is usually located at about 20,000 km height from the ground, while the height of GEO satellite orbit of BeiDou Satellite Navigation (BDS) constellation is about 36,000 km. The delay caused by the electronic content along the GNSS satellite signal path is actually the total effect of the signal path from the ground receiver to the GNSS satellite. However, as shown in Fig. 4.2, due to the vertical structure of the Earth's atmosphere/ionosphere, the free electrons are mainly distributed in the height of 200–600 km. In order to simplify the GNSS ionospheric model, the free ionospheric electrons are usually assumed to concentrate on a thin layer at the peak electron density height, that is, a single-layer model is used to

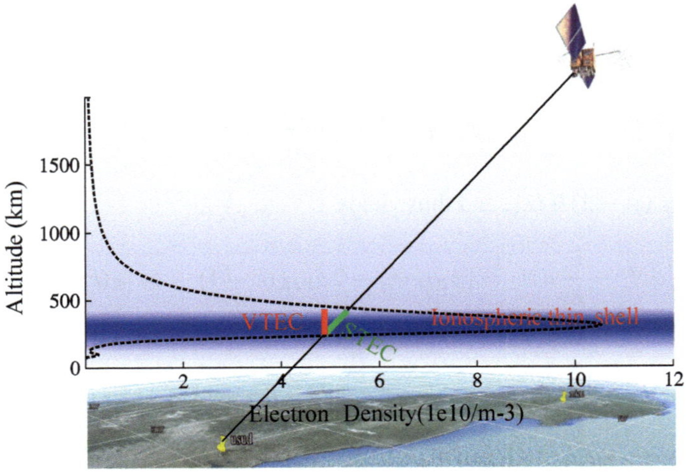

Fig. 4.2 Hypothesis of single-layer ionosphere and the exhibition of vertical TEC and slant TEC. Here the vertical electron density profile over USUD of the IGS station is calculated by the IRI model (the corresponding time is UT20:00 on January 1, 2015)

characterize the electron content of the whole ionosphere. Based on single-layer ionospheric model, we can conveniently use the intersection point of the signal path and the thin layer, i.e., the puncture point to characterize the corresponding observation area. However, for the same epoch and position, the ionospheric total electron content of the signal path changes over different directions. The direction of the signal path for an observation station is mainly characterized by the satellite elevation angle and the azimuth angle of the satellite. Since the ionospheric free electrons are mainly concentrated in the 200–600 km, the ionospheric delay of the GNSS signal path is mainly composed of free electrons in the horizontal scale in the region within several hundred kilometers. This scale is less than horizontal correlation distance of the ionosphere in normal (Yue et al. 2007). The horizontal gradient can be ignored when considering the electron content in the small area. Thus, in the study of GNSS ionospheric sounding, we generally only use the satellite elevation angle to characterize the signal path direction for the GNSS ionosphere detection. In Fig. 4.2, Vertical Total Electron Content (VTEC) in zenith direction is introduced to be distinguished from the Slant Total Electron Content (STEC) along the slant path. Unlike STEC, VTEC is independent of the satellite elevation angle and is corresponding to the puncture point position and the epoch, which can reflect the general characteristics of ionospheric electron content changes over the station.

With the aid of the ionospheric electron content mapping function, the STEC along the GNSS slant path can be converted into path-independent VTEC, and then the inversion of the ionospheric model is carried out. The definition based on the single-layer ionosphere mapping function is given as follows:

$$mf(zen) = \frac{sTEC(\theta, \lambda, t, zen)}{vTEC(\theta, \lambda, t)} \tag{4.24}$$

Here zen is the zenith distance of the puncture point of the signal path, θ, λ indicate the latitude and longitude of the puncture point, respectively, and t is the observation epoch. Scholars have made many attempts for the realization of ionosphere mapping function, for instance, mapping function in the Klobuchar model is applied to the GPS broadcast ephemeris (Klobuchar 1987)

$$mf(zen) = 1 + 2\left(\frac{zen + 6}{96}\right)^3 \tag{4.25}$$

Clynch inform the Q-factor mapping function

$$mf(zen) = \sum_{i=0}^{3} a_i \left(\frac{z}{90}\right)^{2i} \tag{4.26}$$

The piecewise mapping function is proposed by Ou (1996)

$$mf(ele) = \frac{P}{\cos(zen)}, \begin{cases} \sin(5^\circ + 55^\circ), ele < 5^\circ \\ \sin(ele + 55^\circ), 5^\circ \leq ele < 40^\circ \\ 1, ele \geq 40^\circ \end{cases} \tag{4.27}$$

where ele is the satellite elevation angle of the signal path at the station. The most convenient one is the cosine mapping function (Manucci et al. 1993)

$$mf(zen) = 1/\cos(zen) \tag{4.28}$$

Essentially, these mapping functions are based on the hypothesis that ionospheric free electrons in the thin layer are approximately uniformly distributed. In other words, the observed STEC increases with the increase of the intersection signal path length. The effect of all the projection functions almost has no difference when the thin layer height of the single-layer ionospheric model is 350–450 km and the satellite elevation angle is larger than 15–20° (Schaer 1999). Taking the accuracy and simplicity of the calculation into account, Eq. 4.28 is the most widely used ionospheric mapping function. Figure 4.3 shows the geometric relationship between the zenith distance of puncture points of the single-layer ionosphere model and elevation angle of the receiver stations. The station coordinates can be achieved from GNSS observation files and GNSS satellite positions can be determined by orbit parameters from broadcast ephemeris in the navigation message. For the post GNSS ionospheric research, satellite orbit positions can be obtained by the interpolation of the precise ephemeris (igscb.jpl.nasa.gov). The station locations from the observation files and the satellite locations derived from the ephemeris are usually the 3-D coordinates in the Earth-fixed system of WGS84. By coordinate conversion, we can infer the

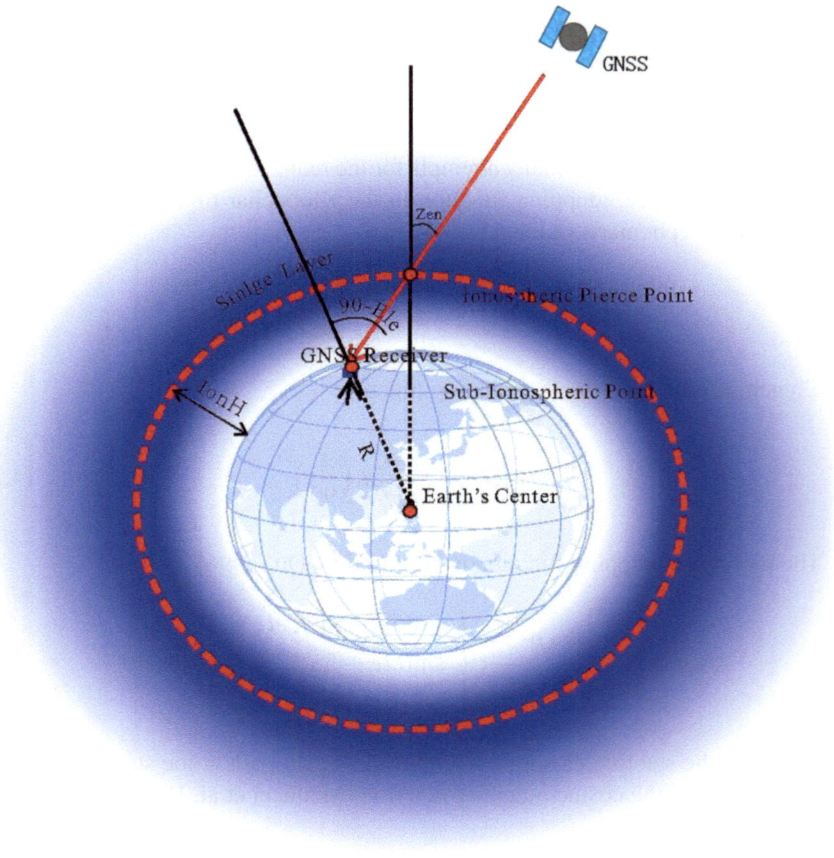

Fig. 4.3 Geometrical relationship between the satellite elevation angle and zenith distance of the puncture point of the signal path under the hypothesis of thin ionospheric layer. The red arrow indicates the GNSS signal path, and the red dashed line indicates the hypothetical thin layer

topocentric coordinates of the navigation satellites, which take the location of the station as the center.

$$\begin{bmatrix} N \\ E \\ U \end{bmatrix} = \begin{bmatrix} -\sin\theta\cos\lambda & -\sin\theta\sin & \cos\theta \\ -\sin & \cos\lambda & 0 \\ \cos\theta\cos\lambda & \cos\theta\sin\lambda & \sin\theta \end{bmatrix} \begin{bmatrix} X_{sat} - X_{rcv} \\ Y_{sat} - Y_{rcv} \\ Z_{sat} - Z_{rcv} \end{bmatrix} \tag{4.29}$$

Here θ, λ are the latitude and longitude in the coordinate system of WGS84; $\begin{bmatrix} X_{sat} & Y_{sat} & Z_{sat} \end{bmatrix}$ are the receiver coordinates and satellite coordinates in the coordinate system of WGS84; $[\, N \;\; E \;\; U \,]$ are the north, the east, and the vertical components of navigation satellite in the topocentric coordinate system. The satellite elevation angle can be obtained from the topocentric coordinates of the satellite

$$ele = \operatorname{atan}\left(\frac{U}{\sqrt{E^2 + U^2}}\right) \tag{4.30}$$

Using sine theorem, zenith distance at the puncture point can be obtained from satellite elevation angle

$$zen = a\sin\left(\frac{R\cos(ele)}{R + IonH}\right) \tag{4.31}$$

After determining the zenith distance at the puncture point of the signal path, the elevation angle-independent VTEC can be converted from the observed STEC through the mapping function above. The temporal and spatial variation function based on GNSS ionospheric observation can be described as

$$\begin{aligned} VTEC(\theta, \lambda, t) &= \left[\frac{f_1^2 f_2^2}{40.3(f_1^2 - f_2^2)}(P_{i,12,sm,t}^j - DCB_{i,t}^j)\right] mf(zen_{\theta,\lambda,t}) \\ &= \left[\frac{f_1^2 f_2^2}{40.3(f_2^2 - f_1^2)}\left(L_{i,12,t}^j - \frac{\sum P_{i,12}^j + L_{i,12}^j}{Nep} + DCB_{i.t}^j\right)\right] mf(zen_{\theta,\lambda,t}) \end{aligned} \tag{4.32}$$

Although receiver type and the precision of the carrier phase measurement are slightly different, the current receivers can reach the level of submillimeter (0.2–0.3 mm) (Li and Huang 2005). The integer ambiguity and hardware delay are not considered when computing the relative variation, as they are constant for the same arc segment from the carrier phase and the pseudo-range observation. From the error propagation law, we can see that the relative accuracy of the GNSS STEC sequence based on the carrier phase observation is less than 0.004TECU. In the analysis of the actual seismic ionospheric disturbance, it is often needed to determine the location of the disturbance, so the relative TEC variation sequence calculated by Eq. 4.32 is more convenient to analyze the spatial and temporal characteristics of the disturbance. However, the accuracy can be directly obtained from the observation error and the error propagation law by Eq. 4.15, while the error caused by the mapping function in Eq. 4.32 cannot be directly determined. The size of the error is mainly influenced by the horizontal gradient and the vertical structure of the local ionospheric TEC. The error size is different over the region, however, as the GPS orbital period is about 12 h, the time and space location in a particular area is basically coincident for a few days. We usually obtain the filter residuals of relative TEC sequence through continuous observations on the same station and satellite for several days, and then estimate whether the observation error and the mapping function error affect the seismic ionospheric disturbance detection.

We can establish the global or regional ionospheric model based on the observations of total electron content of many GNSS stations. The temporal and spatial distribution mathematical models of current mainstream are spherical harmonic function model, polynomial model, grid interpolation model, generalized trigonometric series model, spherical model, etc.

4.3.4.1 Spherical Harmonic Function Model

Spherical harmonic functions compose of orthogonal system, and any spherical function can be expanded as a linear combination of spherical harmonic functions. The spherical harmonic function of the ionospheric total electron content is expressed as follows (Schaer 1999):

$$VTEC(\theta, s) = \sum_{n=0}^{n_{\max}} \sum_{m=0}^{n} \tilde{P}_{nm}(\sin(\theta))[(\tilde{C}_{nm}\cos(ms) + \tilde{S}_{nm}\sin(ms)] \tag{4.33}$$

where the parameters indicate as follows:

θ the geodetic latitude or geomagnetic latitude of the puncture point;

$s = \lambda - \lambda_{Sun}$ the Sun time angle of the puncture point, λ, λ_{Sun} are the geodetic longitude or geomagnetic longitude of the Sun at the corresponding epoch and puncture point longitude;

$n_{\max}$ the highest order of spherical harmonic expansion;

$\tilde{P}_{nm} = N_{nm}P_{nm}$ the normalized associated Legendre function with n order and m time, where N_{nm} is the normalization factor, P_{nm} is the classical Legendre function. The realization of the normalization function is given as follows:

$$N_{nm} = \sqrt{\frac{(n-m)!(2n+1)(2-\delta_{0m})}{(n+m)!}} \tag{4.34}$$

where δ is the Kronecker function that is 1 when the first and second subscripts are the same and cannot be 0 at the same time; and

$\tilde{C}_{nm}, \tilde{S}_{nm}$ the unknown spherical harmonic coefficients, i.e., the regional or global ionospheric electron content model parameters.

The mathematical model is mainly applied to the global $vTEC$ establishment, such as the GIM (Global Ionospheric Map) released by the European orbit center CODE, which is based on the model. The essence of this mathematical model is that the temporal and spatial distribution of the ionosphere is a superposition of different temporal and spatial scales. Different orders represent different time-varying characteristics and spatial distribution of electron content in different scales. For instance, C_{00} is the average of the global ionospheric total electron content at the corresponding epoch. For the $n_{\max}$ order, the spherical harmonic function can reflect the variation in the latitude $2\pi/n_{\max}$ and longitude $2\pi/m_{\max}$. The total ionospheric electron content is decomposed into nCS (i.e., the number of parameters of the ionosphere model to be solved) ionospheric electron content variation terms at different scales.

$$nCS = (n_{\max}^2 + 1)^2 - (n_{\max} - m_{\max})(n_{\max} - m_{\max} + 1) \tag{4.35}$$

Spherical harmonic function model can also be applied to the establishment of regional ionospheric model. Zhang et al. establish the regional ionospheric model in China by utilizing the four-order harmonic function, and the fitting residual error is less than 5 TECU (Zhang 2006). However, the ionospheric spherical harmonic coefficients based on the regional GNSS observations are not applicable to solve the global ionospheric electron content. For the region far away from the measurement area, there is no observed constraint in the solution of the spherical harmonic function, and the model result is obviously far away from the actual ionospheric electron content. On the other hand, the regional ionospheric modeling is usually aimed at detecting the small-scale spatial and temporal variations of the ionosphere, and the spherical harmonic function with high order will significantly increase the computation and may cause the problem of the ill-conditioned equation.

4.3.4.2 Triangular Grid Model

The triangular grid model is the model for the global ionospheric map released by JPL. Its principle is that a series of grid points are considered as the unknowns, according to the spatial correlation of the unknown point grid and the puncture point responding to the GNSS observation, the measurement equation is established by using the linear interpolation method, and thus yields (Sardon et al. 1994; Mannucci et al. 1998):

$$VTEC_{IPP} = \sum_{i=V_A,V_B,V_C} W(\theta, \lambda, i) vTEC_i + DCB_r^s \tag{4.36}$$

Herein $W(\theta, \lambda, i)$ is the weight of $vTEC$ at the IPP location through the interpolation of the grid nodes V_A, V_B, V_C. However, the method of random process is used to describe the $vTEC$ change of grid nodes, and the renewal and the transfer of the variance in the random process are realized by the Kalman filter algorithm. The model is not only applied to the global ionospheric electron content distribution fitting but also is often used for the establishment of a high spatial and temporal ionospheric model based on the regional dense GNSS observation network (Aa et al. 2015).

4.3.4.3 Polynomial Model

The polynomial model is a mathematical model based on the two-dimensional Taylor expansion at the central point of the survey area, which is commonly used to model the regional ionosphere. The model is also called surface fitting model (Li and Huang 2005), and its expression is as follows (Komjathy 1997):

$$VTEC(\theta, \lambda) = \sum_{n=0}^{n_{\max}} \sum_{n=0}^{m_{\max}} E_{nm}(\theta - \theta_0)^n (s - s_0)^m \tag{4.37}$$

Herein E_{nm} is the Taylor expansion coefficient of the ionosphere to be solved; $\theta - \theta_0$ indicates the difference of latitude between the puncture point and the center of the measurement area; $s - s_0$ is the difference of solar hour angle between the puncture point and the center of the measurement area. Compared with the spherical harmonic function model, the amount of calculation of the polynomial model is smaller, while it is not suitable for the ionosphere modeling in large regions. The results show that the accuracy of the polynomial model at the regional boundary is significantly poorer when fitting the ionospheric electron content distribution in the large area (Liu et al. 2008).

The model parameters of the ionospheric electron content can be inverted by the combination of mathematical model of the temporal and spatial distribution and $vTEC$from GNSS observations (Eq. 4.32). The forms of expression are different for different ionospheric parametric methods, but the inversion methods are similar, which estimates the corresponding parameters by the least square method.

4.3.4.4 Generalized Trigonometric Series Model

The main factors affecting the spatial and temporal characteristics of VTEC are geomagnetic latitude and local time, namely, the solar radiation intensity on the geomagnetic field and the ionosphere. The generalized trigonometric function model is the ionospheric model with spatial and temporal characteristics, where the VTEC is expressed as the sum of the effect of several local time and geomagnetic latitudes according to the variation characteristics. It can be expressed as (Yuan and Ou 2004)

$$\begin{aligned} VTEC = a_1 + \sum_{i=1}^{N_2} [a_{i+1}\phi_m^i] + \sum_{i=1}^{N_3} [a_{i+N_2+1}h_m^i] + \sum_{i=1}^{N_i N_j} [a_{i+N_2+N_3+1}\phi_m^i h^j] \\ + \sum_{i=1}^{N_4} [a_{2i} + N_2 + N_3 + N_i + N_{j-1}\cos(ih) + a_{2i} + N_2 + N_3 + N_i + N_{j-1}\sin(ih)] \end{aligned} \tag{4.38}$$

where the first term to the fifth term in the right side of the equation are variations related with the latitude, variations related with local time, variations related with the latitude and the local time, and the sum of the periodic variations related with local time, respectively. Here $\phi_m = \theta + 0.064\cos(\lambda - 1.617)$, where θ indicates the geographical latitude, λ indicates the geographical longitude; $h = 2\pi(t - 14)/24$, where t is the local time corresponding to the puncture point. Generalized triangular coefficient model is appropriate for ionosphere modeling in a small region. However, it is not suitable for ionospheric modeling in a large area, as the basis functions are not orthogonal, the calculation results are not stable, and the components are not strictly corresponding to the periodic variations in the ionosphere (Yuan and Ou 2005; Liu et al. 2008).

Besides the parametric methods informed above using the spatial and temporal characteristics of ionospheric parameters, Felten et al. proposed the three-dimensional Chapman method to describe the ionospheric electron density and carried out the integration of the profiles to obtain electron content (Feltens 1998). Liu et al. proposed the parameterized spherical cap harmonic model appropriate for temporal and spatial characteristics in the polar ionosphere (Liu et al. 2011). For a specific local or global area, these mathematical models can well reflect the spatial and temporal characteristics of the ionosphere when selecting appropriate parameters. The essential reason for the effects on the accuracy of the ionospheric model is the density and the spatial distribution of the GNSS ionospheric observation.

4.3.5 3-D GNSS Ionospheric Tomography

Both the relative GNSS TEC sequences based on the arc segments and the local or global TEC modeling cannot well reflect the vertical structure and temporal variations of the GNSS ionospheric free electron. On the one hand, the three-dimensional ionospheric tomography becomes possible with the establishment of the dense regional intensive continuous GNSS observation stations. On the other hand, with the development of satellite technology, occultation observation from the onboard GNSS receivers GNSS has gradually become one of the important means to detect ionospheric vertical structure.

Ionospheric tomography technique was firstly introduced in 1986. Austen et al. firstly proposed to use the tomography technique to invert the ionospheric electron density (Austen et al. 1988). However, the rays used in the tomography are less due to the sparse distribution of the stations and satellite constellations, making the tomography a rank-deficient inversion. Regional ionospheric observation has been realized to solve the problem informed above in part because of the development of the GNSS system and the establishment of the dense regional GNSS station network. Figure 4.4 shows the geometry sketch map of the ionospheric tomography. We carry out discretization on the ionospheric space and select the appropriate basis functions to describe the ionospheric electron density, and then, the STEC along the signal path can be obtained by the integration of electron density along the path

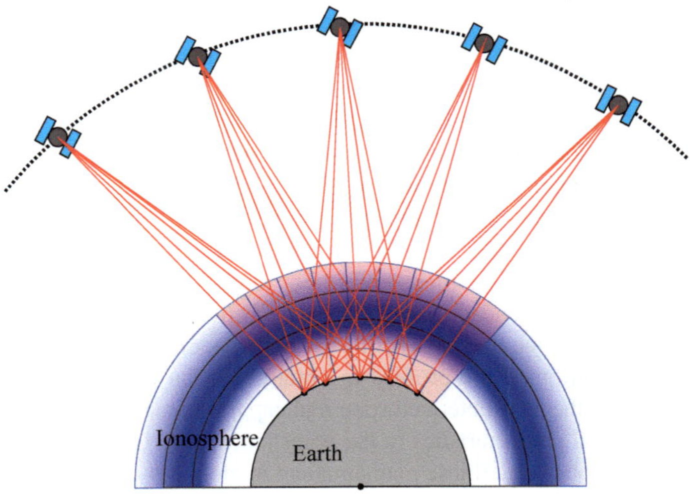

Fig. 4.4 Principle of GNSS tomography

$$
\begin{aligned}
STEC_i &\approx \int_{rcv}^{sat} \sum_{j=1}^{N_b} w_{ij} b_{ij}(\theta, \lambda, h) ds \\
&= \sum_{j=1}^{N_b} w_{ij} \int_{rcv}^{sat} b_{ij}(\theta, \lambda, h) ds \\
&= \sum_{j=1}^{N_b} x_j B_{ij}
\end{aligned}
\tag{4.39}
$$

where N_b is the number of the base functions; θ, λ, h are the latitude, the longitude, and the altitude, respectively; b_{ij} is the electron density function at the responding location; w_{ij} is the coefficient of the weight; the subscript i is the observation serial of the STEC. Considering the measurement noise and the discretization error yields

$$
STEC_i = \sum_{j=0}^{N_b} B_{ij} x_{ij} + \varepsilon_i \tag{4.40}
$$

This formula is observation equation for the ionospheric tomography. The key is the selection of basis functions, that is, how to model the electron density of the corresponding position. The simplest method is to discretize the ionosphere into grids (see Fig. 4.4) and suppose the electron density in a grid is a constant, which is called, namely, pixel-based tomography

$$
STEC_i \approx \sum w_{ij} l_{ij} b_{ij} \tag{4.41}
$$

Here l_{ij} is the length of the ith signal in the jth grid, b_{ij} is a constant to be solved, namely, the electron density of the grid. When referring to the function-based ionospheric tomography, b_{ij} is a set of functions.

The tomographic inversion mathematical algorithm is mainly divided into series expansion such as iterative reconstruction, regularization method, singular value decomposition, and least squares fitting and transform algorithms such as Fourier transform, wavelet transform (see Chen 2012). The above methods can be used to invert the three-dimensional electron density distribution in the ionosphere with the premise of dense GNSS observations with different angles. For the limited current dense regional GNSS network, the tomography is difficult to be applied in most areas.

4.4 Space-Borne GNSS Ionospheric Monitoring

In this section, we aim at introducing the principle and the method of GNSS-RO technique. As shown in Fig. 4.5, the GPS signals will be bent during traveling from GPS transmitters to LEO satellites, when the signal goes through Earth's atmosphere and ionosphere. As the movements of GPS and LEO satellites, the signal paths finish scanning the atmosphere and ionosphere, and we call it one occultation event. If the scanning is up to down, we call it setting occultation, and on the contrary, we call it rising occultation. According to the occultation altitude (the distance of OP as shown in Fig. 4.5 (Jin et al. 2014), here P is the closest point on the signal path to Earth's geometric center), occultation events can be divided into two categories: atmosphere occultation events (lower than 90 km); ionosphere occultation events (90 km to LEO altitude). Atmosphere occultation events' time duration is 1–3 min, and ionosphere occultation events' time duration is 10–20 min (Wu et al. 2008). Benefit from the improvement of GPS-RO technique and the successful of GPS-RO missions, a huge number of occultation observations are available. It is a useful database not only for military and civil users but also for Earth's space environment researchers.

Here we focus on introducing ionospheric GPS-RO sounding. Ionospheric GPS-RO inversion method, error analysis, and scientific density will be introduced as follows.

Figure 4.6 is the geometric diagrammatic sketch of one moment for GPS-RO. The GPS satellites, LEO satellites, and the ray path locate in the same plane, which is called occultation plane. Points O, L, G, and P are refraction center, the LEO satellite's location, the GPS satellite's location, and the closest point on the ray path. Point O is coincident with Earth's center assuming that the Earth is a sphere. The other parameters shown in Fig. 4.6 are explained as follows (Jin et al. 2014):

$\vec{e}_L, \vec{e}_G$ ray path direction vectors in the point of LEO and GPS satellites;

$\vec{v}_L, \vec{v}_G$ projection of velocity vectors of LEO and GPS satellites on the occultation plane;

$\vec{r}_L, \vec{r}_G$ location vectors of LEO and GPS satellites referring to the refraction center;

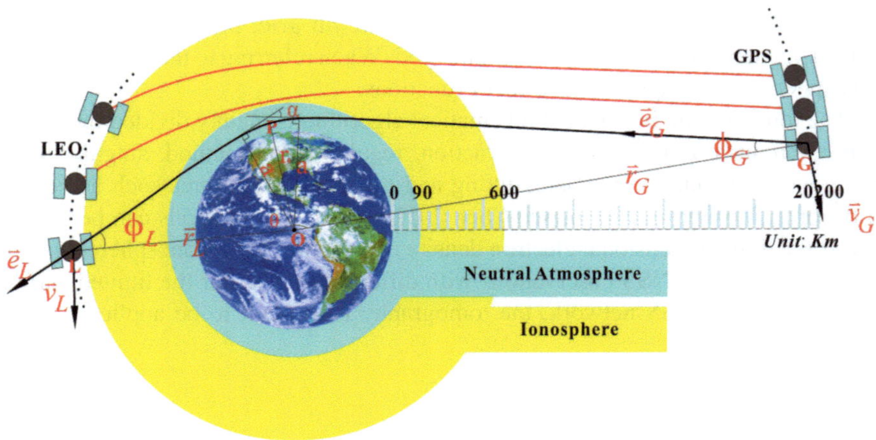

Fig. 4.5 Scheme of GPS-LEO

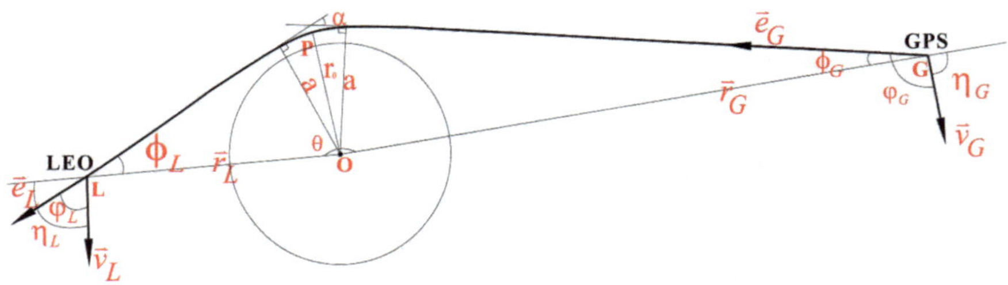

Fig. 4.6 Instantaneous geometric relationship of one occultation event

ϕ_L	angle between $\vec{r}_L$ and $\vec{e}_L$ (exit direction of signal);
ϕ_G	angle between $\vec{r}_G$ and $\vec{e}_G$ (incident direction of signal);
φ_L	angle between $\vec{v}_L$ and $\vec{e}_L$ (exit direction of signal);
φ_G	angle between $\vec{v}_G$ and $\vec{e}_G$ (incident direction of signal);
η_L	angle between $\vec{r}_L$ and $\vec{v}_L$;
η_G	angle between $\vec{r}_G$ and $\vec{v}_G$;
r_0	distance between perigee and refraction center;
a	impact parameter, which is the perpendicular distance between refraction center and asymptote of incident or exit direction of signal;
α	bending angle; and
θ	angle between $\vec{v}_L$ and $\vec{r}_G$.

4.4.1 Ionosphere Inversion Based on Doppler

According to Fermat's principle, the ray path range $\widehat{LG}$ can be defined as

$$L = \int nds \tag{4.42}$$

Due to the refraction of Earth's atmosphere and ionosphere, excess phases exist in both carrier phases. Here we use ΔL_i to describe the excess phase.

$$\Delta L_i = L_i - r_{LG} = \int nds - r_{LG} \tag{4.43}$$

r_{LG} is the geometric distance between the GPS satellite and LEO satellite. Subscript i is the frequency ($i = 1, 2$).

Differentiating Eq. 4.43, one can get

$$\frac{d\Delta L_i}{dt} = \frac{dL_i}{dt} - \frac{dr_{LG}}{dt} \tag{4.44}$$

As we know (Hocke 1997),

$$\frac{dL_i}{dt} = c\frac{\Delta f_i}{f_i} \tag{4.45}$$

Here, Δf_i is the Doppler shift of f_i. The relationship between satellites' velocities and $\frac{dL_i}{dt}$ is expressed as follow (Syndergaard 1998),

$$\frac{dL_i}{dt} = |\vec{v}_L| \cos \varphi_L(a) - |\vec{v}_G| \cos \varphi_G(a) \tag{4.46}$$

Using Eqs. 4.44 and 4.46, we can get

$$\frac{d\Delta L_i}{dt} + \frac{dr_{LG}}{dt} - (|\vec{v}_L| \cos \varphi_L(a) - |\vec{v}_G| \cos \varphi_G(a)) = 0 \tag{4.47}$$

According to the geometrical relationship shown in Fig. 4.6,

$$\varphi_L(a) = \eta_L - \arcsin\left(\frac{a}{|\vec{r}_L|}\right) \tag{4.48}$$

$$\varphi_G(a) = \pi - \eta_G + \arcsin\left(\frac{a}{|\vec{r}_G|}\right) \tag{4.49}$$

$$\alpha = \theta - \arccos\left(\frac{a}{|\vec{r}_L|}\right) - \arccos\left(\frac{a}{|\vec{r}_G|}\right) \tag{4.50}$$

$\frac{r_{LG}}{dt}, |\frac{d\vec{r}_L}{dt}|, |\frac{d\vec{r}_G}{dt}|, \theta, |\vec{r}_L|, |\vec{r}_G|$ can be derived from satellites' locations and velocities. $\frac{d\Delta L_i}{dt}$ is the derivative with the time of excess phase. Using Eqs. 4.47–4.49, the impact parameter can be derived by iteration. After getting the parameter a, bending angle α can be computed by Eq. 4.50.

With the approximation of spherical symmetry of Earth's atmosphere and ionosphere, using the method of Abel inversionFjeldbo and Kliore 1971; Kursinski et al. 1997),

$$n(a_0) = \exp\left[\frac{1}{\pi}\int_{a_0}^{\infty} \frac{\alpha(a)}{\sqrt{a^2 - a_0^2}} da\right] \tag{4.51}$$

Here a_0 the corresponding impact parameter of current radio occultation observation. The refractive index of Earth ionosphere can be expressed as

$$n = 1 - \frac{40.3}{f_i^2} Ne \tag{4.52}$$

(Bassiri and Hajj 1992; Hofmann-Wellenhof et al. 1992; Li and Huang 2005; Liu et al. 2006). Ne is the electron density. Then, Ne can be derived with

$$Ne = \frac{f_i^2}{40.3}(1 - n) \tag{4.53}$$

This method needs high precision orbit determination data and its processing procedure is complex. Usually, it will be used for atmosphere inversion and planetary radio occultation.

4.4.2 Ionosphere Inversion Based on TEC

Ionosphere inversion using TEC data is based on the fact that the bending angle due to the ionosphere is very small (Hajj and Romans 1998), especially for upper ionosphere. Considering another assumption used widely in GPS-RO inversion, which is the assumption of spherical symmetry, the signal path traveling from GPS transmitter to the LEO satellite can be modeled as a straight line as shown in Fig. 4.7.

The first step of ionosphere inversion based on TEC is to compute the slant total electron content along the signal path with GPS dual-frequency or single-frequency data. This technique has been developed in the field of ground-GNSS ionosphere sounding. TEC can be derived by Eq. 4.54 with GPS dual-frequency data.

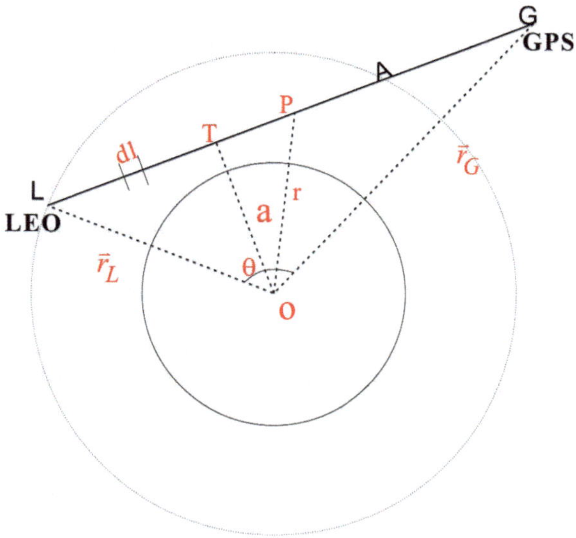

Fig. 4.7 Scheme of occultation geometry with straight line

$$
\begin{aligned}
TEC &= \frac{f_1^2 f_2^2}{40.28(f_1^2 - f_2^2)}(L_1 - L_2 - (N_1\lambda_1 - N_2\lambda_2 + b_1 - b_2)) + \varepsilon_L \\
&= \frac{f_1^2 f_2^2}{40.28(f_2^2 - f_1^2)}(P_1 - P_2 - (B_1 - B_2)) + \varepsilon_P
\end{aligned} \tag{4.54}
$$

L_1, L_2, P_1, and P_2 are GPS carrier phase and pseudo-range observables. N_1, N_2, λ_1, and λ_2 are ambiguities, and wavelengths for L_1, L_2. b_1, b_2, B_1, and B_2 are instrument biases. ε_L, ε_P are noise. Here, the effect of propagation path difference and high order items of refraction has been ignored. TEC error will increase in solar maximum year (Syndergaard 2000; Datta-Barua et al. 2008).

Another method to compute TEC is based on L_1 and P_1.

$$
TEC = \frac{f_1^2}{80.56}(P_1 - L_1) + N + \varepsilon \tag{4.55}
$$

N is the ambiguity. ε is the measuring error. The precision of TEC which is derived with single-frequency GPS data is lower because of greater noise of P_1, although it has no errors caused by propagation path difference.

According to the definition of TEC, and with the assumption mentioned above, TEC can be expressed as the Abel transform of the electron density:

$$TEC(a) = \left[\int_a^{r_{LEO}} + \int_a^{r_{GPS}} Ne(r)dl\right] = \left[\int_a^{r_{LEO}} + \int_a^{r_{GPS}} \frac{Ne(r)r}{\sqrt{r^2 - a^2}}dr\right] \approx 2\int_a^{r_{LEO}} \frac{Ne(r)r}{\sqrt{r^2 - a^2}}dr$$
$$(l = \overline{LP} = \sqrt{r^2 - a^2} \Rightarrow dl = \frac{r}{\sqrt{r^2 - a^2}}dr) \quad (4.56)$$

r distance between the Earth geometrical center and one point on the ray path;
r_{LEO} distance between the Earth geometrical center and the LEO satellite; and
r_{GPS} distance between the Earth geometrical center and the GPS satellite.

The electron content above the LEO satellite is neglected and considering the geometric symmetry in Eq. 4.56. Using Abel inversion, the following yield can be derived.

$$Ne(r) = -\frac{1}{\pi}\int_r^{r_{LEO}} \frac{\frac{dTEC(a)}{da}}{\sqrt{a^2 - r^2}}da \quad (4.57)$$

4.4.3 Recursive Inversion of TEC

Figure 4.8 shows a view of the recursive scheme of the Abel inversion corresponding to the case in ionosphere RO (Jin et al. 2014). Neglecting the discretization errors and with the same assumptions mentioned in the last section,

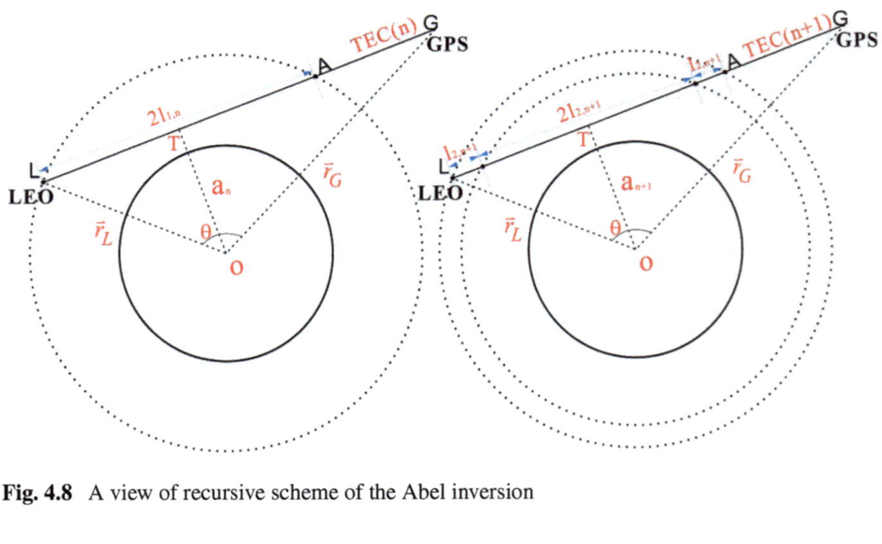

Fig. 4.8 A view of recursive scheme of the Abel inversion

$$TEC \approx 2\sum_{i=1}^{n} Ne(a_i) \cdot l_{i,n} \tag{4.58}$$

Obviously, each TEC observation can determine a layer in the vertical profile. Further, the sampling rate of receivers equipped on the LEO satellites for ionosphere sounding determines the vertical resolution of the profile.

$$Ne(a_n) = \frac{TEC(n) - 2 \cdot \sum_{i=1}^{n-1} Ne(a_i) \cdot l_{i,n}}{2 \cdot l_{n,n}} \tag{4.59}$$

Using Eq. 4.59, the electron density can be obtained layer by layer from the uppermost observation to lowermost observation.

4.4.4 Amplitude Inversion

As we know, not only phase observations but also amplitude variation information contains in GPS-RO observations. Although the precision of bending angle derived from amplitude measurement is not as good as phase measurement, it is applied in GPS-RO physical inversion method such as backpropagation algorithm (Gorbunov et al. 2000), diffraction method (Melbourne et al. 1994). The variation of the amplitude is mainly caused by focusing or defocusing, and it also associates with the atmospheric absorption and scattering effect. It may play an important role in optimizing phase inversion, monitoring multipath propagation, finding anomalies of electron density et al. Amplitude measurement also can be used for ionosphere inversion (Liou et al. 2002). Its computing formulae are shown as follows:

$$\frac{dNe}{dh} = \frac{f^2}{40.28\pi}\frac{1}{r}\int_{r}^{\infty} \frac{a\frac{d\alpha(a)}{da}}{\sqrt{a^2 - r^2}} da \tag{4.60}$$

$$\frac{d\alpha(a)}{da} = \left(1 - \frac{A_0^2}{A^2}\right)\left[\frac{1}{\sqrt{|\vec{r}_G|^2 - a^2}} + \frac{1}{\sqrt{|\vec{r}_L|^2 - a^2}}\right] \tag{4.61}$$

$$r = \mathrm{Re} + h \tag{4.62}$$

Re is the radius of Earth. A_0 and A are amplitudes before and after one occultation event. These parameters from space-borne GNSS radio occultation provide global coverage for study on climate change and space weather.

References

Aa, E., Huang, W., Yu, S., et al. (2015). A regional ionospheric TEC mapping technique over China and adjacent areas on the basis of data assimilation. *Journal of Geophysical Research: Space Physics, 120*(6), 5049–5061.

Austen, J. R., Franke, S. J., & Liu, C. H. (1988). Ionospheric imaging using computerized tomography. *Radio Science, 23*(3), 299–307.

Bassiri, S., & Hajj, G. A. (1992). *Modeling the global positioning system signal propagation through the ionosphere*.

Blewitt, G. (1990). An automatic editing algorithm for GPS data. *Geophysical Research Letters, 17*(3), 199–202.

Cai, C., Liu, Z., Xia, P., et al. (2013). Cycle slip detection and repair for undifferenced GPS observations under high ionospheric activity. *GPS Solutions, 17*(2), 247–260.

Chen, P. (2012). *GNSS ionospheric chromatography and ionospheric pre-earthquake anomalies*. Ph.D. thesis, Wuhan University.

Datta-Barua, S., Walter, T., Blanch, J., & Enge, P. (2008). Bounding higher-order ionosphere errors for the dual-frequency GPS user. *Radio Science, 43*(05), 1–15. https://doi.org/10.1029/2007rs003772.

Feltens J. (1998). *Chapman profile approach for 3-D global TEC representation*. IGS presentation.

Fleury, R., Clemente, M., Carvalho, F., et al. (2009). Modelling of ionospheric high-order errors for new generation GNSS. *Annals of Telecommunications—Annales des Télécommunications, 64*(9–10), 615–623.

Gorbunov, M. E., Gurvich, A. S., & Kornblueh, L. (2000). Comparative analysis of radioholographic methods of processing radio occultation data. *Radio Science, 35*(4), 1025–1034. https://doi.org/10.1029/1999RS002247.

Hajj, G. A., & Romans, L. J. (1998). Ionospheric electron density profiles obtained with the Global Positioning System: Results from the GPS/MET experiment. *Radio Science, 33*(1), 175–190.

Hocke, K. (1997). Inversion of GPS meteorology data. *Annales Geophysicae-Atmos Hydrospheres Space Science, 15*(4), 443–450.

Hofmann-Wellenhof, B., Collins, J., & Lichtenegger, H. (1992). *GPS: Theory and practice*. Springer.

Jin, S.G., Cardellach, E., & Xie, F. (2014). *GNSS remote sensing: Theory, methods and applications*, Springer, Netherlands, 276 pp, ISBN 978-94-007-7481-0.

Klobuchar, J. A. (1987). Ionospheric time-delay algorithm for single-frequency GPS users. *IEEE Transactions on Aerospace and Electronic Systems, 3*, 325–331.

Komjathy, A. (1997). *Global ionospheric total electron content mapping using the Global Positioning System*.

Kursinski, E. R., Hajj, G. A., Schofield, J. T., et al. (1997). Observing Earth's atmosphere with radio occultation measurements using the Global Positioning System. *Journal of Geophysical Research: Atmospheres, 102*(D19), 23429–23465.

Li, Z. H., & Huang, J. S. (2005). *GPS measuring and data processing*. Wuhan University Press.

Liou, Y. A., Pavelyev, A. G., Huang, C. Y., Igarashi, K., & Hocke, K. (2002). Simultaneous observation of the vertical gradients of refractivity in the atmosphere and electron density in the lower ionosphere by radio occultation amplitude method. *Geophysical Research Letters, 29*(19), 1937.

Liu, J. Y., Chen, C. H., Lin, C. H., et al. (2011). Ionospheric disturbances triggered by the 11 March 2011 M9.0 Tohoku earthquake. *Journal of Geophysical Research: Space Physics, 116*(A6).

Liu, L., Wan, W., Ning, B., et al. (2006). Solar activity variations of the ionospheric peak electron density. *Journal of Geophysical Research: Space Physics, 111*(A8).

Liu, J., Wang, Z., Wang, H., & Zhang, H. (2008). Modeling regional ionosphere using GPS measurements over China by spherical cap harmonic analysis methodology. *Geomatics and Information Science of Wuhan University, 33*(8), 792–795.

Manucci, A. J., Wilson, B. D., & Edwards, C. D. (1993). A new method for monitoring the Earth's ionospheric total electron content using the GPS global network. In *ION GPS-93* (pp. 1323–1332).

Mannucci, A. J., Wilson, B. D., Yuan, D. N., et al. (1998). A global mapping technique for GPS-derived ionospheric total electron content measurements. *Radio science, 33*(3), 565–582.
Melbourne, W. G., Davis, E. S., Duncan, C. B., et al. (1994). The application of spaceborne GPS to atmospheric limb sounding and global change monitoring.
Ou, J. K. (1996). Atmosphere and its effects on GPS surveying. *LGR-Series* (Vol. 14). Delft: Delft Geodetic Computing Centre.
Sardon, E., Rius, A., & Zarraoa, N. (1994). Estimation of the transmitter and receiver differential biases and the ionospheric total electron content from Global Positioning System observations. *Radio Science, 29*(3), 577–586.
Schaer, S. (1999). Mapping and predicting the Earth's ionosphere using the Global Positioning System. *Geodätisch–Geophysikalische Arbeiten in der Schweiz* (Vol. 59, p. 59).
Syndergaard, S. (1998). Modeling the impact of the Earth's oblateness on the retrieval of temperature and pressure profiles from limb sounding. *Journal of Atmospheric and Solar-Terrestrial Physics, 60*(2), 171–180.
Syndergaard, S. (2000). On the ionosphere calibration in GPS radio occultation measurements. *Radio Science, 35*(3), 865–883.
Wu, X. C., Hu, X., & Gong, X. Y. (2008). Three dimensional model constrained inversion method for ionospheric occultation. *Chinese Journal of Geophysics, 51*(3), 618–625. (in Chinese).
Yuan, Y., & Ou, J. (2004). A generalized trigonometric series function model for determining ionospheric delay. *Progress in Natural Science, 14*(11), 1010–1014.
Yuan, Y., & Ou, J. (2005). Generalized trigonometric function ionospheric delay model. *Progress in Natural Science, 15*(8), 1015–1019.
Yue, X., Wan, W., Liu, L., et al. (2007). Statistical analysis on spatial correlation of ionospheric day-to-day variability by using GPS and Incoherent Scatter Radar observations. *Annales Geophysicae, 25*(8), 1815–1825.
Zhang, H. (2006). *Research on ionospheric monitoring and delay correction in China rainy area based on ground-based GPS*. Ph.D. thesis, Chinese Academy of Sciences.

Chapter 5
Detection Methods for Ionospheric Disturbances

Due to the complex spatial distribution and variation characteristics of the ionospheric layer, it is difficult to separate or remove each anomaly. The ionosphere has regular variations (daily and seasonal scales) and anomalies due to the magnetic storm, solar flares, and scintillation as well as other hazard events. The non-earthquake ionospheric variation background in the GNSS ionospheric observation is difficult to eliminate for earthquake ionospheric detection. One arc of the GNSS satellite normally continues for only several hours of continuous GNSS observations. It normally observes large spatial-scale earthquake ionospheric disturbance from two-dimensional or three-dimensional GNSS ionospheric results. For small-scale earthquake ionospheric disturbances, the sparse GNSS observations cannot meet the spatial-temporal resolution for the ionospheric disturbance. How to determine whether the ionospheric disturbance is related to the earthquake is the primary problem for the study of seismo-ionospheric disturbances.

5.1 Detection Methods

Although the ionospheric has a complex spatial-temporal structure, its main driving forces, such as solar radiation, geomagnetic field, etc., are normally relatively steady in silent state, i.e., steady in a short period. When its distribution or spatial-temporal characteristics are no longer stable, e.g., during earthquakes, abnormal disturbances occur in the ionosphere. To obtain seismo-ionospheric abnormal disturbances, it is important to eliminate the ionospheric background variation. The following are several detection methods for GNSS seismic ionospheric disturbances.

S. Jin et al., *GNSS Atmospheric Seismology*,
https://doi.org/10.1007/978-981-10-3178-6_5

5.1.1 Statistical Probability Method

Under the silent state, for the same position in the month scale, the periodic weekly variation of the ionosphere is the most significant variation characteristic, and the ionospheric electron content in the same location at the same time is nearly a constant. Based on the variation characteristic, the ionospheric electron content at the same day at different epochs is subject to the Gaussian distribution, namely

$$F(x) = \frac{1}{\sqrt{2\pi}\sigma_{vTEC}} \int_{-\infty}^{x} e^{-\frac{(x-\mu_{vTEC})^2}{2\sigma_{vTEC}^2}} dx \tag{5.1}$$

In Eq. (5.1), μ_{vTEC} and σ_{vTEC} are the expectation and standard deviation of $vTEC$, respectively, characterizing the concentration position and the dispersion of the probability distribution of $vTEC$, and $F(x)$ is the probabilistic density of $vTEC$ at x. As shown in Fig. 5.1, the probabilities of $vTEC$ at $\mu \pm 3\sigma$, $\mu \pm 2\sigma$, and $\mu \pm \sigma$ are the areas of the regions surrounded by the Gaussian distribution curve, $x = \pm 3\sigma$, $x = \pm 2\sigma$, $x = \pm\sigma$, and the x-axis. By integrating the distribution curve, we can see that the corresponding probabilities are 0.999937, 0.9545 and 0.6827, respectively. If the value of $vTEC$ exceeds the scope of $\mu \pm 3\sigma$, i.e., an incident happens with the probability of 0.000063. However in the probability, normally an incident happens with a probability lower than 5% is called a small-probability incident. For an incident with a probability of 0.000063, it almost never happens. Then, we define the $vTEC$ at the epoch as abnormal variations and consider that there are other driving factors causing disturbance of the $vTEC$. Generally, in the GNSS seismic ionospheric anomaly detection, the statistical parameters are usually determined by the corresponding epochs and positions in a month or several days before and after the event. After determining the statistical parameters, we can estimate the variation limit of $vTEC$ so as to identify the abnormal disturbance using Eq. (5.2). Usually, the value of k can take 2–3, and the confidence is 95–99%.

$$\begin{aligned} vTEC_{up} &= \mu + k\sigma \\ vTEC_{up} &= \mu - k\sigma \end{aligned} \tag{5.2}$$

Divided by the standard deviation and the expectation estimation of the $vTEC$ variation limit, Liu et al. (2000, 2001, 2004) determined the abnormal criterion for detection of abnormal ionospheric disturbances using the quartiles, namely

$$\begin{aligned} vTEC_{up} &= \bar{x} + kIQR \\ vTEC_{up} &= \bar{x} - kIQR \end{aligned} \tag{5.3}$$

Here, k is generally taken as 1.5, equivalent to 1.34σ, and the corresponding confidence is 80–85% (Klotz and Johnson 1983).

Fig. 5.1 Schematic figure of Gaussian probability density distribution (3 sigma standard)

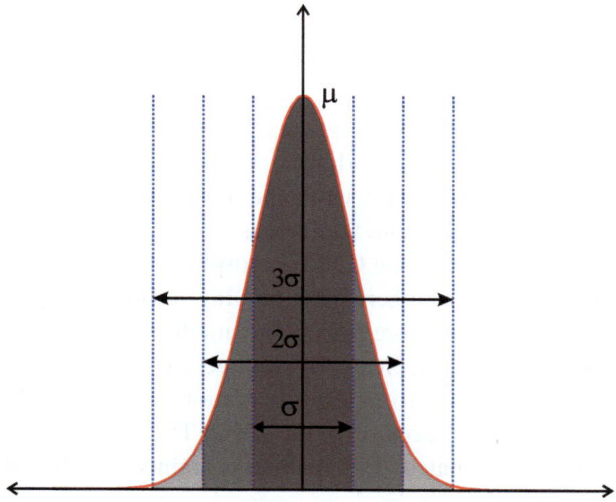

The detection method of envelope abnormal disturbances based on the quartiles distance and the above mentioned method based on the sequence expectation and standard deviation are essentially the same, and both are based on the probability under the premise of the stable daily ionospheric variation. These methods are widely used in seismic ionospheric abnormal disturbance detection due to its simpleness and effectiveness. The key step for these methods is the determination of statistical parameters and the threshold. Due to the existence of seasonal and other long-term scale TEC variations in the ionosphere, the number of days selected to determine the statistical parameters should not be too long, but we must ensure that enough samples are to accurately determine its statistical characteristics. For disturbances with small amplitudes due to the spatial-temporal smoothing effect on GNSS ionospheric modeling, the corresponding abnormal disturbance often cannot be detected by high confidence threshold. Then, we can properly degrade the confidence of the threshold to achieve the purpose of anomaly detection.

5.1.2 Polynomial Fitting Method

For relatively large ionospheric abnormal disturbance detection the envelope method based on probabilistic statistics is very effective, however, the detection ability for small disturbance is limited. The variation of the ionosphere not only has the daily variation that is not completely stable, but also the method cannot be directly applied to more accurate relative TEC sequence of arcs. Since the ionosphere changes relatively smoothly in a calm state, we can omit the physical background of the variation, and directly fit the background variation using a polynomial like Eq. (5.4), and then the residual is used to detect abnormal disturbances (Heki 2011).

$$sTEC = \frac{vTEC}{\cos(zen)} = \sum_{i=0}^{n} a_i t^i \tag{5.4}$$

Generally, the order n takes 3–4. On one way the order based on the polynomial fitting may cause rank defect, on the other way for the ionospheric variation in the scale of several hours, when the order reaches 3–4, the order of fitting residuals no longer decreases with the increase of the order. It should be pointed out that for the TEC sequence of the GNSS ionospheric oblique path, the polynomial fits the total effect of the ionospheric time-varying and horizontal gradient of the horizontal movement of the puncture point. This method can be directly applied to high-precision relative TEC sequences. By subtracting the TEC sequence simulated by the polynomial Eq. (5.4) from the observation sequence, the background variation of TEC along with the variation of the GNSS TEC sequence causing the drift of the puncture point is eliminated. This method is mainly used for ionospheric abnormal detection in short timescale. For longer period ionospheric variations, the polynomial cannot fit the variation characteristics well due to the complex structure and other influencing factors.

5.1.3 Moving Average Method

The moving average method can detect ionospheric anomalous disturbances by subtracting the moving average values around the epoch from the observed values. (Afraimovich et al. 2001; Jin et al. 2010, 2014, 2015, 2017, 2018).

$$TEC(t) = TEC(t) - \frac{\sum_{i=t-\frac{\Delta t}{2}}^{i=t+\frac{\Delta t}{2}} TEC(i)}{n} \tag{5.5}$$

In Eq. (5.5), Δt is the length of the moving window, and n is the number of epoch in the moving window. The method is based on the fact that the variation of the ionosphere is nearly linear during a short period. In fact, the moving average disturbance detection method is the special case of zero-order polynomial fitting. Other than the polynomial fitting, the residual sequence of the moving average method does not depend on the whole observation arc. Figure 5.2 shows an example of the effect of moving average processing on the original signal spectrum. As shown in Fig. 5.2, the components with period less than Δt can be eliminated by the moving average processing with a window length Δt, which is equivalent to high-pass filtering. The original sequence minus the moving average sequence is equivalent to a low-pass filtering. In actual ionospheric abnormal disturbance detection, the original signal sequence is usually first processed by sliding average with a window length Δt_1, and then the sliding average sequence of $\Delta t_2(>\Delta t_1)$ is subtracted to achieve the effect of bandpass filtering, so as to remove the ionospheric background variation and detect the ionospheric disturbance signal in a specific frequency. By comparing

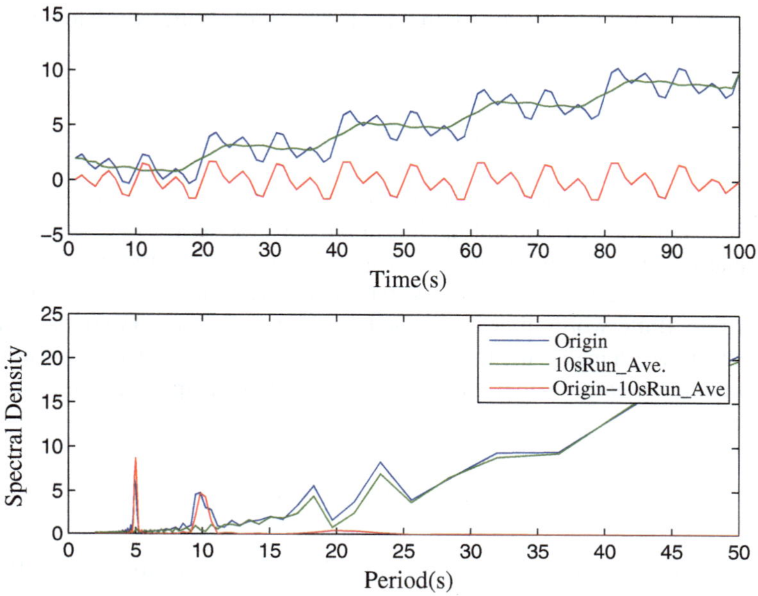

Fig. 5.2 Effect of the moving average processing on the original signal spectrum. The above graph shows the original signal (blue line) composed of three periodic components with periods of 5, 10, 20 s and a linear trend expressed as $y = 0.1x + \sin(\frac{pi}{5}x) + \sin(\frac{2pi}{5}x) + \sin(\frac{pi}{10}x)$, 10 s moving average signal (green line) and the difference between the original signal and the moving average signal (red line). The graph below shows the power spectral densities of the corresponding series

and analyzing the results of sliding average processing in different time windows, it will contribute to the discrimination of the ionospheric disturbance from different mechanisms and help us study the differences and similarities of these characteristics.

5.1.4 Butterworth Filtering Method

The Butterworth filter was firstly proposed by Butterworth (1930). The response frequency of the Butterworth filter can be expressed as

$$G(\omega) = \left|H(j\hat{\omega})\right| = \sqrt{\frac{1}{1 + \varepsilon^2(\frac{\omega}{\omega_p})^{2n}}} \tag{5.6}$$

where n is the order of Butterworth filter, ω is the angular frequency of the signal, $\hat{\omega} = \omega/\omega_p$ is the angular frequency of the normalized regularized signal (Normally choosing the half-power point frequency), $\left|H(j\hat{\omega})\right|$ is the propagation function, j is the complex unit, and ε is a constant parameter. According to Eq. (5.6), it can be known that the derivative of the response frequency of the Butterworth low-pass filter is

$$\frac{dG}{d\hat{\omega}} = -n\varepsilon^2 G^3 \hat{\omega}^{2n-1} \tag{5.7}$$

It can be concluded that the frequency response function of the Butterworth filter is monotonically decreasing and its first to $2n - 1$ order derivatives are all zero at $\omega = 0$, that is to say that the frequency response curve of the Butterworth filter has no ripple and is the flattest in the low-pass band. Therefore, the most significant feature of the Butterworth filter is that the frequency response curve in the passband is the flattest, which makes the Butterworth filter to retain the signal waveform in the passband to the greatest extent. For the high frequency component, the change rate of the gain is

$$n = -\lim_{\hat{\omega}\to\infty}\left[\frac{d\log(G)}{d\log(\hat{\omega})}\right] \tag{5.8}$$

For the n-order Butterworth filter, the high-frequency attenuation is $20n$ dB every 10 times frequency. Normally, the high-order Butterworth filter is decaying faster. The commonly used design indices for the low-pass Butterworth filter are cutoff frequency ω_p, bandpass attenuation α_p, bandpass starting frequency ω_s, and stop-band frequency α_s. The essence of constructing Butterworth filter satisfying the corresponding index is to determine the corresponding transfer function H using the given index.

$$\alpha(\hat{\omega}) = 10\log\frac{1}{|H(j\hat{\omega})|^2} = 10\log(1+\hat{\omega}^{2n}) \tag{5.9}$$

Then, we can derive the order of the filter using the attenuation index:

$$\begin{aligned} \varepsilon &= \sqrt{10^{\frac{\alpha_p}{10}} - 1} \\ n &= \frac{\log(10^{\frac{\alpha_s}{10}} - 1) - 2\log\varepsilon}{2\log\omega_s} \end{aligned} \tag{5.10}$$

Usually, the cutoff frequency of the passband is taken at the half-power point, i.e., $\alpha_p = 3\,\text{dB}$, and then ε is equal to 1. In the two-dimensional complex plane, the modulus of $s = j\omega$ is

$$H(s)H(-s) = |G(j\hat{\omega})|^2 = \frac{1}{1+(-1)^n s^{2n}} \tag{5.11}$$

The polar points can be derived as

$$s_k = e^{j\frac{2k+n-1}{2n}\pi}, k = 1, 2, \ldots, n \tag{5.12}$$

Therefore we can conclude that

$$H(s) = \frac{1}{(s - s_1)(s - s_2)\ldots(s - s_n)} \tag{5.13}$$

Combining two lower pass filters, it is easy to realize the Butterworth bandpass filter. In seismic ionospheric disturbance detection, normally the fourth-order band-pass Butterworth filter is used (Calais and Minster 1995; Calais et al. 1998). In order to eliminate the phase shift of the original signal caused by filtering, the zero-phase filtering method is usually adopted, namely that the first filtering result is reversed in the time domain and then the second filtering result is performed by the same filtering method, and the final filtering result is obtained by reversing the result again in the time domain (Gustafsson 1996). The Butterworth filter can not only eliminate the background variation in the GNSS TEC sequence but also maintain the characteristics of ionospheric disturbances in specific frequency.

5.2 Test of Ionospheric Anomaly Detection

In order to test whether above methods can accurately detect the abnormal disturbance in GNSS TEC sequence, the TEC sequences at the puncture points are derived based on the empirical ionospheric background IRI-2017 model, and then an artificial disturbance is added. By comparing the difference between the detection results and the artificial disturbance, the effectiveness of these detection methods can be obtained. Here, we mainly take into account of the relative TEC sequence based on the high-precision GNSS observation arcs. For the statistical abnormal detection mainly used for long-term TEC sequence, it is provided by the statistical parameters and the abnormal criteria. Assuming that there is a GNSS station located at 32°N, 132°W with tracking GPS satellite RPN01 as an example, the slant TEC sequence and the corresponding observation geometry based on IRI-2017 simulation with the precise ephemeris provided by IGS are shown in Fig. 5.3. The ionospheric disturbance can be described as:

$$dTEC(D, t) = \mathrm{sinc}[\omega_b(t - t_0) - \frac{D}{v}] \cdot cos(\omega_c(t - t_0) - \frac{D}{v}) \tag{5.14}$$

where D is the distance from the disturbance source, t_0 is the epoch when the disturbance occurs, v is the propagation velocity of the disturbance, ω_b and ω_c are the bandwidth and center frequency of the disturbance signal, respectively. Here, we suppose that the disturbance source is above the station, and the propagation velocity is 1 km/s. As shown in Fig. 5.3, a perturbation with a bandwidth of 0.5 mHz and center frequencies of 4 mHz and 7 mHz is added in at UT0300 and UT0400, respectively, and then we can simulate the ionospheric slant TEC sequence with anomalous disturbances. The polynomial fitting, moving average, and the Butterworth filtering are

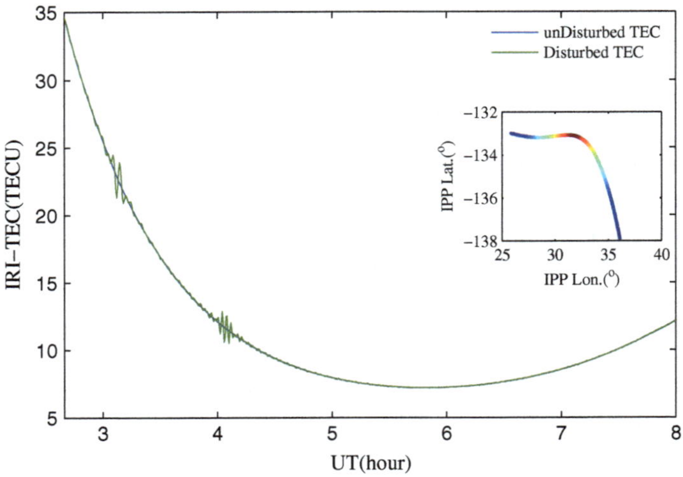

Fig. 5.3 The simulated slant TEC sequence of GPS satellite PRN01 on October 28, 2012 based on IRI-2017 model. Here we suppose that the station location is 32°N, 132°W, the satellite ephemeris is the precise ephemeris released by IGS on October 28, 2012, and the disturbance signal is simulated from Eq. (5.14). The subgraph shows the variations of the corresponding signal path puncture points with the elevation angle (the red represents the high elevation angle, and the blue represents the low elevation angle)

all used to detect abnormal disturbances, while the order of polynomial fitting is 6, the window length of the moving average is 166–500 s, and the order of Butterworth filter is 4, with a passband of 2–6 mHz. As shown in Fig. 5.4, all three methods can detect the disturbance signal very well.

The polynomial fitting method does not take into account the frequency of the disturbance signal, two artificially added disturbances are both reflected in the residual error sequence, and are in good agreement with the constructed disturbance signal. The time scale of co-seismic GNSS ionospheric disturbance is usually much smaller than a complete GNSS observation arc (5–6 h), and ionospheric disturbance with short period almost has no effect on the GNSS TEC observation sequence. Therefore, through selecting appropriate order of the polynomial fitting, we can precisely fit the mean TEC variation from the TEC sequence, and then the ionospheric abnormal disturbance can be detected. However, this method cannot distinguish the ionospheric anomalous disturbances of different frequencies.

In the residual of the moving average method, only the first disturbance is detected and the second signal is filtered out, depending on the period range of the sliding window. It should be noted that the residual amplitude of the moving average method is slightly smaller than the original disturbance signal. Comparatively speaking, the residual of the Butterworth filter can not only reflect the disturbance in the first passband but also filter the disturbance in the second non-passband. Since the characteristics of the ionospheric disturbances following earthquakes are relatively complex, the disturbances of different frequencies are generated from different physical

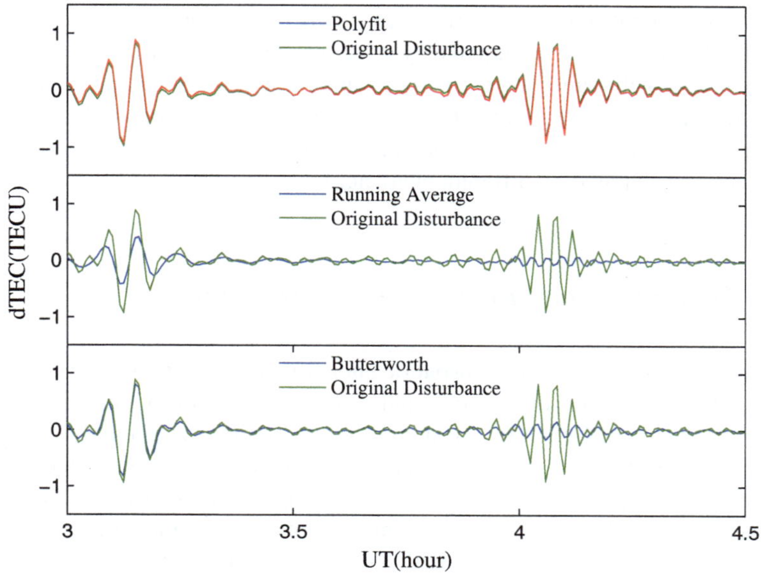

Fig. 5.4 Abnormal detection results of three anomaly detection methods. The order of polynomial fitting is 6, the window length of the moving average is 166–500 s, and the order of Butterworth filter is 4 with a passband of 2–6 mHz

mechanisms. In order to better analyze the characteristics of ionospheric disturbance generated from different physical mechanisms, it is usually necessary to separate the disturbances of different frequencies, so as to analyze the effect of different factors. The Butterworth filtering and the moving average method can both meet the requirement, while the former can maintain the disturbance phase characteristic better. Therefore, the method of Butterworth filtering is used to detect the co-seismic ionospheric anomalous disturbance. As a matter of fact, the GNSS TEC sequence based on the arcs contains not only the spatial-temporal information of the ionosphere but also the horizontal gradient information due to the movement of the puncture point caused by satellite motion. Because of the spatial correlation of the ionosphere, the ionospheric horizontal gradient information is usually neglected for GNSS TEC sequences with little variations in the distance of disturbance source and short disturbance duration.

5.3 Determination of Seismic Correlation

There are many factors that can cause the ionospheric disturbance. How to determine the earthquake-related or earthquake-induced abnormal disturbance is an important issue in the study of GNSS seismic ionospheric disturbances. On one way, we can discuss the temporal and spatial correlation between ionospheric disturbance and the epoch and location of the earthquake by selecting earthquakes when the ionosphere and the geomagnetic environment is quiet and without other natural or man-made events such as volcanic eruption and rocket launching, so as to avoid the influence of other factors. On the other way, the ionospheric disturbances caused by different factors can be separated by different characteristics.

5.3.1 Determination of Pre-seismic Correlation

Different from the co-seismic ionospheric anomaly, there is a relatively large uncertainty in the distribution of the time and spatial distribution of pre-seismic ionospheric disturbances. Generally speaking, the pre-earthquake anomaly is analyzed by GNSS TEC sequence from several days to month scale. This timescale is similar with the ionospheric disturbance caused by geomagnetic anomaly. The ionospheric disturbance caused by geomagnetic storms is global, while the earthquake-related ionospheric anomaly is correlated with the seismogenic zone. Therefore, we can distinguish the geomagnetic correlated disturbance and earthquake correlated disturbance by spatial cross-correlation analysis (Pulinets et al. 2004). In the seismognic zone and out of seismogenic zone we separately choose a GNSS observation station, the former one is used as an abnormal disturbance detection site, and the latter one is the reference site. In order to weaken the effect of the geomagnetic latitude and local time, the latitudes of the two stations should maintain coincidence, and the longitude is kept in a certain range. Then, the correlation coefficient of TEC variations at the two stations can be described as

$$
\begin{aligned}
C &= \frac{\sum_{i=0,k} (TEC_{1,i} - aTEC_1)(TEC_{2,i} - aTEC_2)}{k(\sigma_1\sigma_2)} \\
aTEC &= \frac{\sum_{i=0,k} TEC_i}{k+1} \\
\sigma^2 &= \frac{\sum_{i=0,k} (TEC_i - aTEC_i)^2}{k+1}
\end{aligned}
\tag{5.15}
$$

where the subscripts 1 and 2 refer to the stations in the seismogenic zone and outside the seismogenic zone, k is the number of epoch, $aTEC$ is the daily average total electron volume observed at the corresponding stations, and σ^2 is the standard deviation of the k epoch. In the quiet state or during the magnetic storm, for regions in the same latitude and longitude within a certain range, the ionospheric TEC is nearly the same, showing a strong spatial correlation, and the daily correlation coefficient is very big. The ionospheric disturbance related to earthquake mainly exists in the seismogenic zone, and there is no similar disturbance outside the seismogenic zone, so the daily correlation coefficient will reduce. Therefore, the daily correlation coefficient of the ionosphere inside and outside the seismogenic area can be used to distinguish the ionospheric anomalies associated with earthquakes and magnetic storms. Along with the increase of GNSS observation stations and the improvement of the precision and the resolution of the two-dimensional and three-dimensional ionospheric modeling, the global and regional GNSS ionospheric TEC model is used as important data for the pre-earthquake ionospheric disturbance research. Using the methods in the previous section, the spatial and temporal distribution of ionospheric disturbances at the corresponding epoch can be derived, and then the coincidence of the time and spatial distribution of the disturbance with the epicenter location and the seismogenic time can be used to determine whether the disturbance is related to the earthquake. Currently, the study of pre-earthquake ionospheric disturbance is post-event analysis, i.e., the earthquake occurring is known, and then looking for anomalous ionospheric disturbances before earthquakes. Due to the fact that the generating mechanism of pre-seismic ionospheric anomalies is not known well, currently it is hard to determine whether the ionospheric anomaly indicates the occurrence of earthquakes.

5.3.2 Determination of Co-seismic Correlation

The co-seismic ionospheric abnormal disturbance usually refers to the ionospheric anomalous disturbance within a few hours after the earthquake. Different from the pre-earthquake ionospheric disturbance, the co-seismic ionospheric disturbance is the spatial effect triggered by the earthquake. The determination of the correlation of the co-seismic ionospheric disturbances is, as a matter of fact, to determine whether the ionospheric disturbances in the corresponding space-time region are caused by earthquakes. Earthquakes start from the vibration caused by the rock layer rupture from the inner Earth, and the generated energy is propagated as the form of seismic wave. The epicenter, i.e., the surface projection point of the earthquake source is generally the region with the most obvious seismic effects. In real case, the area with the most obvious seismic effect usually has a certain range, which is usually called epicenter area. However, compared with the scale of seismic ionospheric disturbances, the epicentral area is much smaller, and therefore in the study of the characteristic of co-seismic ionospheric disturbances, it is usually regarded as point source disturbance, and the location of the point source is considered as the epicentral location. Based on the hypothesis of epicentral point source disturbance, we can judge the correlation of co-seismic ionospheric disturbance by the time delay and epicentral distance of the disturbance to know epicentral propagation characteristics.

There are three possible ways for the epicentral disturbance to propagate to the ionosphere (Heki and Ping 2005), namely, the vertical propagated acoustic wave, the tsunami generated gravity wave and the secondary acoustic wave generated by the Rayleigh wave. GNSS TEC observations can accurately monitor the ionospheric horizontal distribution and the time-varying information, however it is insensitive to the vertical disturbance of the ionosphere, so the GNSS co-seismic ionospheric disturbance is mainly focused on the horizontal propagation characteristic of the disturbance. For ionospheric disturbance near the epicenter, the point source hypothesis is no longer reasonable, but the disturbance is relatively strong, so whether it is caused by the earthquake can be directly determined by its correlation with the time and location of the earthquake along with the exclusive method. For the far-field co-seismic ionospheric disturbance, the time delay of the disturbance may take the peak of GNSS TEC sequence or the time interval of initial epoch of the disturbance and the earthquake occurrence time. According to the type of disturbance, the epicentral distance can take the geodesic distance D_{sip} from the epicenter to the ground projection point of the corresponding puncture point, the geometric distance D_{geo} from the epicenter to the puncture point or the geodetic distance D_{ipp} from the position at the ionospheric height over epicenter to the puncture point. For local regions, the geospatial parameters are nearly the same, and thus the disturbance can normally be considered as propagating with an approximately uniform velocity in the study of co-seismic ionospheric disturbance. The regression coefficient can be derived using the linear regression equation.

$$r = \frac{\sigma_{xy}^2}{\sigma_x \cdot \sigma_y} = \frac{\sum (x - \bar{x})(y - \bar{y})}{\sqrt{\sum (x - \bar{x})^2}\sqrt{\sum (y - \bar{y})^2}} \tag{5.16}$$

Here, x is the time delay of the disturbance, and y is the epicentral distance of the disturbance. Since the seismic ionospheric anomaly is point source disturbance with uniform velocity propagation, if the disturbance is caused by earthquake, the regression coefficient will be significantly higher than that in quiet state. However, due to the diversity of ionospheric disturbances, the propagation velocities of different types of disturbances vary greatly. It is obviously unreasonable to use all disturbance data indiscriminately to obtain the regression coefficient. On one way, we can distinguish different types of seismic ionospheric disturbances by selecting the appropriate filtering band, on the other way, the cross-correlation function Eq. (5.17) can be used to determine whether the disturbances detected by different GNSS TEC sequences are of the same type.

$$C_{12}(\tau) = \int_{-\infty}^{+\infty} f_1 * f_2(t+\tau)dt \tag{5.17}$$

where τ is the disturbance delay and * represents the convolution of two sequences. When the disturbances of different sequences are the same type and the observation areas are basically the same, the correlation of the two sequences is strong and vice versa.

5.4 Case Analysis and Validation

Although the magnitude of ground deformation caused by seismic wave is small, the frequency band of the energy released is at the scale of 10^{-3}–10^{-1} Hz, and the atmospheric noise is weak and close to the atmospheric eigenmode in this frequency band, which makes it easy to form effective coupling. Studies have shown that the effective frequency of disturbances caused by the coupling of the atmosphere and the solid Earth is at the level of several milli-Hertz (mHz) (Lognonné et al. 2006; Artru et al. 2005). Here, the 30-s sampling rate of the GPS observation data from GEONET is used, so the highest detectable frequency is 16.7 mHz according to the Nyquist Sampling Theorem. Therefore, for co-seismic ionospheric disturbances, the anomalous signal is extracted using 1–15 mHz zero-phase Butterworth filter. Figure 5.5 shows the frequency response of the second to eighth-order Butterworth filter with 1–15 mHz passband. Obviously, when the order of the filter reaches 4, the enhancement of the order cannot significantly improve the performance of the filter. The higher the order of the filter, the larger the corresponding time delay and the larger the calculation. Therefore, in the extraction of co-seismic ionospheric disturbances, the order of the Butterworth filter normally takes 4 so as to get a filter with optimal filter frequency response, shorter time delay and faster computation.

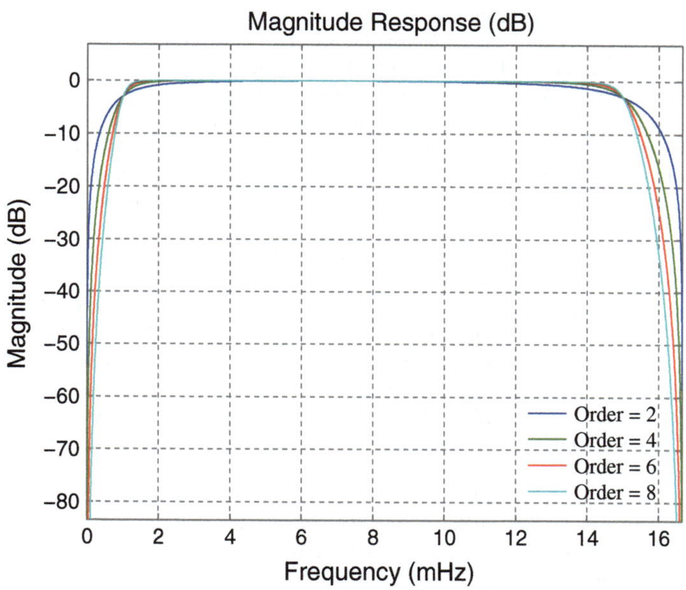

Fig. 5.5 Frequency response curves of the second-, fourth-, sixth-, and eighth-order Butterworth filters with passband frequencies ranging from 1 to 15 mHz

5.5 Summary

In this chapter, the current mainstream methods of the ionospheric TEC abnormal detection are introduced. Based on the TEC background provided by IRI-2017 model, an artificial disturbance signal is introduced, and the feasibility and advantages of several detection methods are validated. The results showed that the Butterworth filtering method is not only convenient for separating the abnormal ionospheric disturbance at different frequency ranges, but also fairly maintain the disturbance signal waveform, so the Butterworth filtering method is used for detecting co-seismic ionospheric disturbances in late chapters. Furthermore, the correlation of the disturbance and the earthquake is discussed for the pre-earthquake and co-seismic ionospheric disturbances. Since the coupling mechanism of seismic ionospheric disturbances is not known well, currently it is mainly determined based on the coincidence of the ionospheric disturbance with the striking time of the earthquake and the spatial correlation of the epicenter.

References

Afraimovich, E. L., Chernukhov, V. V., & Kiryushkin, V. V. (2001). Spatial-temporal characteristics of the ionospheric disturbance induced by shock acoustic waves due to rocket launching. *Journal of Communications Technology & Electronics, 46*(11), 1199–1206.

Artru, J., Ducic, V., Kanamori, H., et al. (2005). Ionospheric detection of gravity waves induced by tsunamis. *Geophysical Journal International, 160*(3), 840–848.

Butterworth, S. (1930). On the theory of filter amplifiers. Wireless Engineer.

Calais, E., Minster, J. B., Hofton, M., et al. (1998). Ionospheric signature of surface mine blasts from global positioning system measurements. *Geophysical Journal International, 132*(1), 191–202.

Calais, E., & Minster, J. B. (1995). GPS detection of ionospheric perturbations following. *Geophysical Research Letters, 22*(9), 1045–1048.

Gustafsson, F. (1996). Determining the initial states in forward-backward filtering. *IEEE Transactions on Signal Processing, 44*(4), 988–992.

Heki, K. (2011). Ionospheric electron enhancement preceding the 2011 Tohoku-Oki earthquake. *Geophysical Research Letters, 38,* L17312. https://doi.org/10.1029/2011GL047908.

Heki, K., & Ping, J. (2005). Directivity and apparent velocity of the coseismic ionospheric disturbances observed with a dense GPS array. *Earth and Planetary Science Letters, 236*(3), 845–855.

Jin, S. G., Jin, R., & Li, D. (2017). GPS detection of ionospheric Rayleigh wave and its source following the 2012 Haida Gwaii earthquake. *Journal of Geophysical Research: Space Physics, 122*(1), 1360–1372. https://doi.org/10.1002/2016JA023727.

Jin, S. G., Jin, R., & Li, J. H. (2014). Pattern and evolution of seismo-ionospheric disturbances following the 2011 Tohoku earthquakes from GPS observations. *Journal of Geophysical Research: Space Physics, 119*(9), 7914–7927. https://doi.org/10.1002/2014JA019825.

Jin, S. G., Occhipinti, G., & Jin, R. (2015). GNSS ionospheric seismology: Recent observation evidences and characteristics. *Earth-Science Reviews, 147,* 54–64. https://doi.org/10.1016/j.earscirev.2015.05.003.

Jin, S. G. (2018). Two-mode ionospheric disturbances following the 2005 Northern California offshore earthquake from GPS measurements. *Journal of Geophysical Research: Space Physics, 123*(10), 8587–8598. https://doi.org/10.1029/2017JA025001.

Jin, S. G., Zhu, W. Y., & Afraimovich, E. (2010). Co-seismic ionospheric and deformation signals on the 2008 magnitude 8.0 Wenchuan Earthquake from GPS observations. *International Journal of Remote Sensing, 31*(13), 3535–3543. https://doi.org/10.1080/01431161003727739.
Klotz, S., & Johnson, N. L. (1983). Encyclopedia of statistical sciences. Wiley.
Liu, J. Y., Chen, Y. I., Chuo, Y. J., et al. (2001). Variations of ionospheric total electron content during the Chi-Chi earthquake. *Geophysical Research Letters, 28*(7), 1383–1386.
Liu, J. Y., Chen, Y. I., Pulinets, S. A., et al. (2000). Seismo-ionospheric signatures prior to M $\geq$ 6.0 Taiwan earthquakes. *Geophysical Research Letters, 27*(19), 3113–3116.
Liu, J. Y., Chuo, Y. J., Shan, S. J., et al. (2004). Pre-earthquake ionospheric anomalies registered by continuous GPS TEC measurements. *Annales Geophysicae, 22*(5), 1585–1593.
Lognonné, P., Clévédé, E., & Kanamori, H. (2006). Computation of seismograms and atmospheric oscillations by normal-mode summation for a spherical earth model with realistic atmosphere. *Geophysical Journal International, 23*(1), 106–117.
Pulinets, S., Gaivoronska, T., Contreras, A. L., et al. (2004). Correlation analysis technique revealing ionospheric precursors of earthquakes. *Natural Hazards & Earth System Sciences, 4*(5), 697–702.

Part II
Seismic Atmospheric Disturbances

Chapter 6
Seismic Lower Atmospheric Anomalies

6.1 Background

The earthquake is related to various scales motions of the Earth's surface and interior as well as the microscopic processes, such as the electric field, chemical reactions, pressure, and temperature (e.g., Pulinets and Boyarchuk 2005). The first distinctive signals in the air pressure were detected before the 1964 Alaskan earthquake (Bolt 1964), which was the generation of the strong electric field around the epicenter and as a result of this electric field plasma drift from earthquake epicenter excited free ions motion in the atmosphere. The accumulated huge energy release and generated waves from the Earth's interior may propagate into the atmosphere through weak zones on the Earth's crust, which may cause anomalies in the temperature, the pressure, and electromagnetic fields in the atmosphere. Seismic activities generate electric field disturbances in the ionosphere with magnitudes up to 10 mV/m and may induce the temperature increase of Earth's surface. Thermal anomalies remain active around the earthquake preparation zone due to the release of uncontrolled energy and stress (Freund 2000), which spread out like electric dynamo from Earth's interior around earthquake preparation zone. Blackett et al. (2011) computed 6 years of MODIS Land surface temperature (LST) to study anomalous pattern before the Gujarat earthquake and spiky peaks have been observed near the earthquake day as thermal anomalies. It is practically proved in laboratory that porosity of the rocks near the epicenter is truncated. It could be possible that variations might be due to enormous energy discharges at some earthquake zones while the remaining non-active zones might not suffer large energy release (Freund et al. 1993).

Additionally, dense dual-frequency GNSS observations can estimate the tropospheric and ionospheric delays, which can monitor atmospheric variations (e.g., Jin et al. 2004, 2007, 2008; Jin and Park 2007). The GNSS atmospheric disturbances may reflect various changes of the atmospheric compositions or airwaves, e.g., acoustic wave propagation due to the movement of air particles. Numerous studies have found co-seismic or pre-seismic GNSS ionospheric anomalies and speculated the possible reason due to the acoustic coupling of the atmosphere and solid-Earth (Shah and

S. Jin et al., *GNSS Atmospheric Seismology*,
https://doi.org/10.1007/978-981-10-3178-6_6

Jin 2015). For example, Jin et al. (2010) and Afraimovich et al. (2010) have found significant co-seismic ionospheric total electron content (TEC) disturbances during the 2008 Wenchuan Earthquake with an intensive N-shape shock-acoustic wave. The co-seismic TEC disturbance is possibly due to the coupling of the ionosphere and solid-Earth with airwave propagation from the ground to the ionosphere. If the airwaves propagate from the ground to the upper atmosphere, it may also affect the lower atmosphere (i.e., the troposphere). However, such possible impacts on the troposphere are not clear. In this chapter, the possible low-atmospheric variations following the earthquake are investigated from GPS, MODIS, and OLR measurements.

6.2 MODIS, OLR, and GPS Observations

6.2.1 Surface Temperature and Radiation

The MODIS provided the daily Land Surface Temperature (LST), which are retrieved at 1 km pixels by the generalized split-window algorithm and at 6 km grids by the day/night algorithm. In the day/night algorithm, daytime and nighttime LSTs and surface emissivities are retrieved from pairs of day and night MODIS observations in seven TIR bands. MODIS (LST) data (1×1 km^2) are routinely published at http://modis.gsfc.nasa.gov/data, which can be used to investigate seismic temperature anomalies. Outgoing Long-wave Radiation (OLR) is a measurement of the amount of energy emitted to space by Earth's surface, oceans, and atmosphere. The spatial resolution is $2.5° \times 2.5°$ while the temporal resolution varies on a daily basis. The OLR data can be used to estimate the variations in atmospheric conditions related to earthquake activities, which are caused by imbalanced heat exchange between earthquake breeding zones and the atmosphere (Shah 2017).

6.2.2 ZTD from GNSS Observations

The national project "Crustal Movement Observation Network of China (CMONOC)" was initiated in August 1998 (Jin et al. 2010), including a nationwide fiducial network of 28 permanent GPS sites observed from August 1998 to now, and 56 survey mode sites with yearly operations for the period 1998–2018 as well as ~1000 regional campaign GPS stations operated by the State Bureau of Surveying and Mapping (SBSM) and the China Earthquake Administration (CEA) in 1999, 2001, 2004, 2007, 2008, etc. with at least 4-day observations in each session. Unfortunately, only 28 continuous GPS sites observations data are available around this event on May 12, 2008. Figure 6.1 shows the GPS stations distribution, where the square is continuous GPS stations, the triangle is the yearly observing GPS stations,

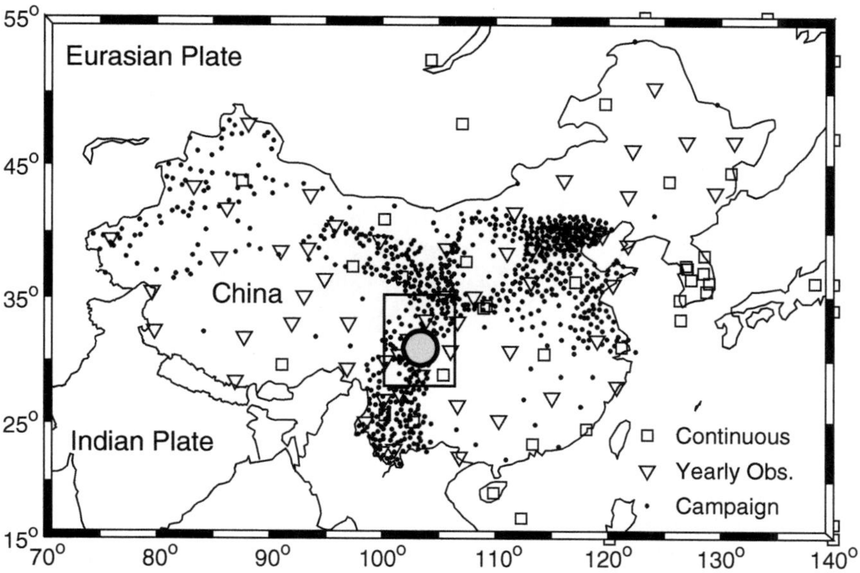

Fig. 6.1 Distribution of GPS observation stations and the epicenter location in China. The square is continuous GPS stations, the triangle is the yearly observing GPS stations, the dot is campaign GPS stations and the large solid circle is the epicenter of the main shock

the dot is campaign GPS stations, and the large solid circle is the epicenter of the main shock.

The GPS observation data are processed by the GAMIT software (King and Bock 2018), which solves the ZTD and other parameters using a constrained batch least squares inversion procedure. The International GNSS Service (IGS) final orbits, International Earth Rotation and Reference Systems Service (IERS) Earth orientation parameters, and azimuth- and elevation-dependent antenna phase center models recommended by the IGS are used. The cutoff elevation angle is 15° and the temporal resolution the ZTD estimate is 5 min. In order not to degrade the accuracy of ZTD estimates, a 12-h sliding window strategy is designed and the ZTD is extracted from the middle 4 h of the window and then forward 4 h of moving window (Jin et al. 2007). Finally, the ZTD time series in several days around the main shock is obtained at 28 continuous GPS sites in China with a temporal resolution of 5 min.

6.3 Seismic Thermal Anomalies

The Mw = 7.8 Gorkha earthquake (Nepal) occurred on April 25, 2015, at (28.2°N, 84.6°E) and hit the countryside near Kathmandu with causing severe destruction of lives and infrastructures. Due to its shallow depth and high intensity, it perturbed

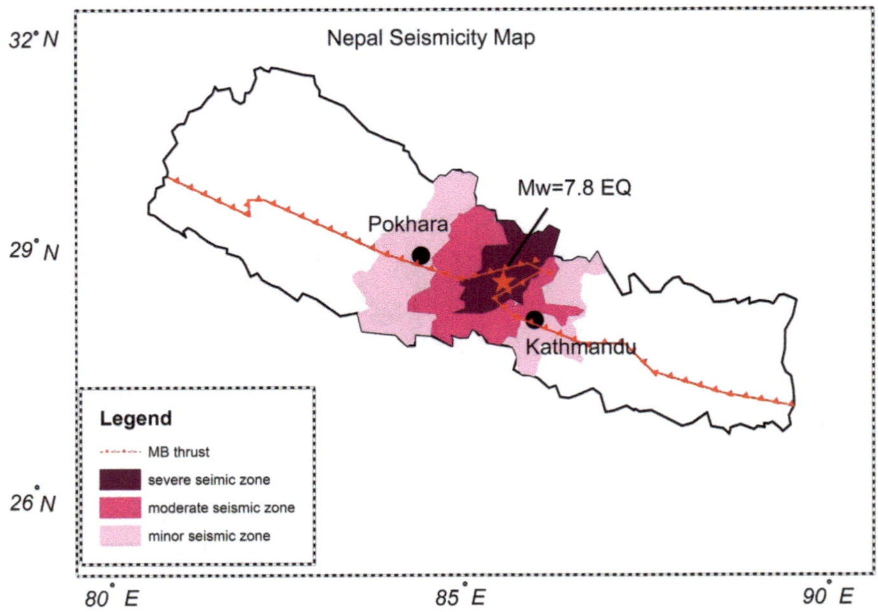

Fig. 6.2 The seismicity zone and tectonic setting of the study area. The red star is epicenter and two black squares are the Pokhara and Kathmandu Meteorological observatories

the nearby regions along the fault line known as Main Boundary thrust (MB). This earthquake was declared the most devastating event in a decade by the National Disaster and Management Authority of Nepal. The region is located in different zones of seismicity. The epicenter of Mw = 7.8 falls in the severe seismic zone according to USGS (United States Geological Survey) (Fig. 6.2). It is important to investigate some possible anomalies and processing related to this earthquake.

6.3.1 MODIS LST Anomalies

MODIS LST and OLR data are analyzed both spatially and temporally to quantify the perturbation in the lower atmosphere following this earthquake. Figure 6.3 shows MODIS LST and air temperature deviations of more than 3 °C on April 20 before the impending earthquake (Shah 2017). MODIS LST images within 26°–30° north and 80°–88° east are averaged to analyze the seismic region. The upper and lower bounds in the form of mean and standard deviation ($\mu \pm 2\sigma$) are calculated for normal distribution of LST to understand the instantaneous changes caused by the earthquake. After implementing the statistical window of the confidence interval, a very prominent thermal anomaly can be seen 5 days before in the air temperatures of Pokhara and Kathmandu. The anomaly is beyond the upper bound is 5 °C, which

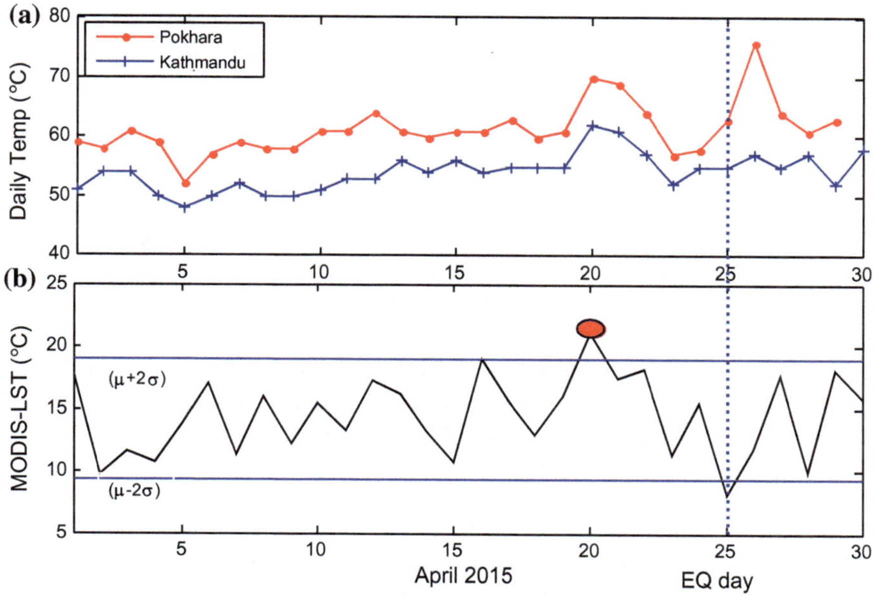

Fig. 6.3 MODIS LST temperature anomalies following the 2015 Mw = 7.8 Nepal earthquake. **a** Atmospheric temperature enhanced anomalies of nearby meteorological centers before the main shock, and **b** MODIS land surface temperature perturbation on 20 April before the 2015 Mw = 7.8 earthquake (Shah 2017)

is very unusual for the Land Surface Temperature. LST and air temperature have anomalies occurring at the same local time that could be related to the earthquake. According to Saradjian and Akhoondzadeh (2011), MODIS LST and air temperature anomalies may have the same source, but the mechanism of perturbation is different and not clear.

Pulinets and Ouzounov (2011) examined the characteristics of Radon emission in seismically active regions and found that the increase of the surface temperature may be due to vertical electric fields, ultimately causing thermal anomalies. An abnormal temperature variation is seen on April 19, 2015 along a line from Pokhara to Kathmandu in the middle of the epicenter (Shah 2017). Although the phenomenon is consistent with the eastward drift of atmospheric winds prior to earthquake, it still has many uncertainties. But a real cluster of high-temperature values covered the epicenter on April 20th, which is obvious. These clouds travel eastwards, which confirms the existence of an electro dynamo source, probably the massive earthquake. It has been proven by Freund (2000) that rock fractures could activate mobile positive holes, which eventually leads to thermal anomalies that make bulk clouds. The spreading of these positive holes often ionizes the atmosphere, which may absorb infrared radiation. Due to its high density, it forces the less dense atoms of the ambient atmosphere to move eastward. Some of them collide with atmospheric ions, transferring energy, which is seen as thermal anomalies. Previous findings showed the current density of these mobile ions, its consequences on the nighttime plasma bubbles, as well as the expansion of the equatorial drift prior to earthquakes (Tronin et al. 2004).

6.3.2 *OLR and AIRS Anomalies*

In order to be fully confident about the existence of atmospheric anomalies before the 2015 Mw = 7.8 Nepal earthquake, OLR (Outgoing Long-wave Radiation) data are analyzed in combination with MODIS LST data and air temperature. In this study, the daily OLR grid 2.5° × 2.5° (Longitude × Latitude) has been studied relative to the earthquake days. Implementing statistical methods, the abnormal value of OLR can be seen on April 21 as shown in Fig. 6.4 (Shah 2017), which coincides with the perturbation of MODIS LST data. This correlation between OLR and MODIS shows the energy accumulation in atmosphere caused by the earthquake. It can be seen clearly that a significant OLR anomaly covers the earthquake breeding zone with the maximum value. It is also evident that the OLR anomaly is moving along the fault line from April 19 to April 22. The main cause of this terrestrial enhancement is the drift of strong wind raised from the seismogenic region, which eventually forces the ions in the near atmosphere to move eastward.

According to previous studies, light ions and inert gases are observed in high concentration over the seismogenic region, but the real source of these gas particles

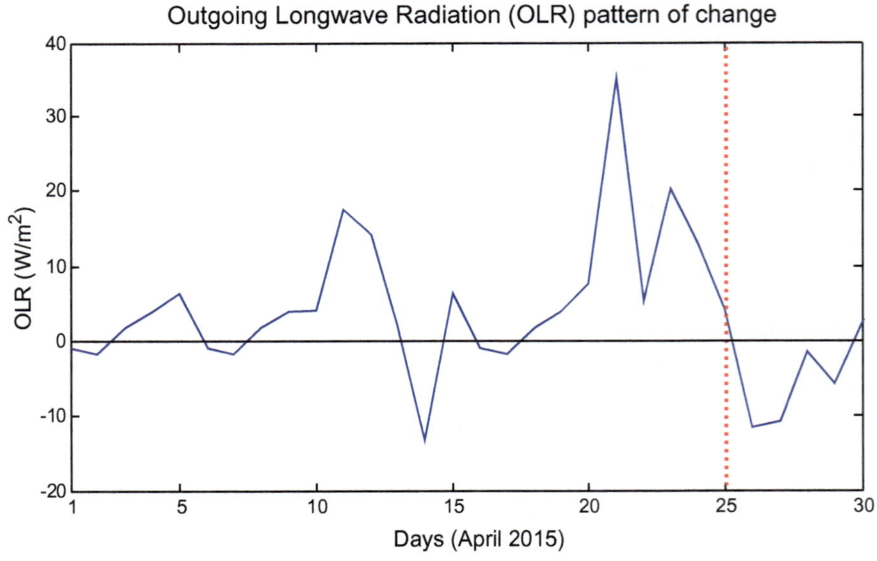

Fig. 6.4 Outgoing Long-wave Radiation (OLR) time series over the epicenter of the 2015 Mw = 7.8 earthquake (Nepal). The earthquake day is denoted by the red line, while OLR values are along the y-axis and days are along x-axis (Shah 2017)

is still controversial. The actual variation of different ions can be estimated from the Atmospheric Infrared Sounder (AIRS) of NASA. The percentage of nitrogen (N_2), oxygen (O_2), and carbon dioxide (CO_2) in air are 78.08, 20.95, and 0.035, respectively. Daily variation of atmospheric gases over Nepal is analyzed (Shah 2017). The variations in all ion measurements near the earthquake day are very prominent (Fig. 6.5). Nitrogen gas crosses the upper bound from the normal distribution is up to 3–5%. Oxygen gas has a negative anomaly on the 20 April and a positive perturbation on 22 April. Oxygen reacts with hydrogen to further increase the humidity level in the atmosphere (Shah 2017). This increment in humidity can increase the atmospheric temperature, which is finally converted into thermal anomalies. Another important component of the atmosphere is carbon, which is 3% up the normal mean distribution. All the air components are perturbed due probably to the accumulation of high energy in the system. It needs more works to further study the gas chemistry in the atmosphere to gain a better interpretation of thermal anomalies in the future.

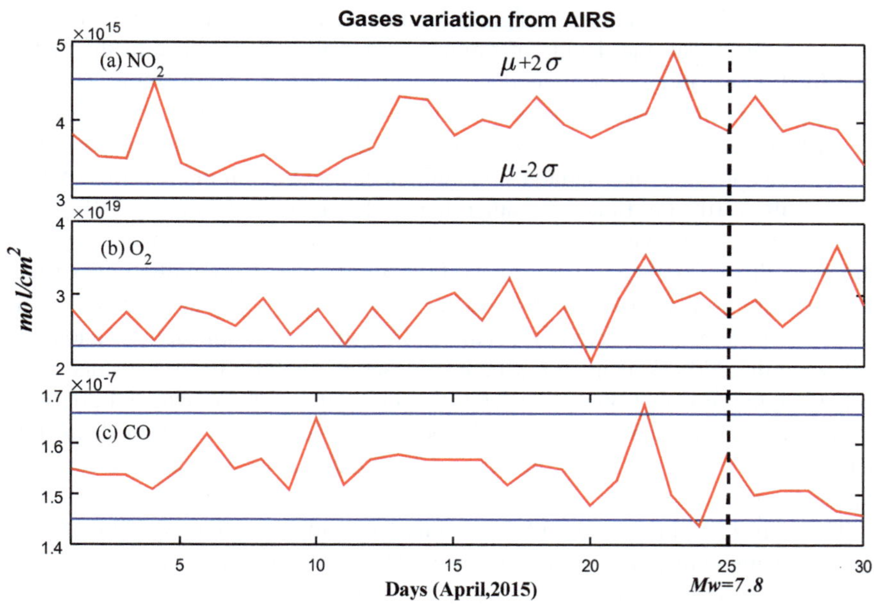

Fig. 6.5 Different gas molecules in the atmosphere related to the 2015 Mw = 7.8 Nepal earthquake. **a** Red line shows the NO_2 concentration for the month of April 2015. The upper and lower bounds are represented by two blue lines, **b** the variation in oxygen for the month of April, 2015, and **c** the significant enhancement in CO gas is seen on t 23rd April 2015 (Shah 2017)

6.4 Seismo-tropospheric Anomalies

6.4.1 2018 Wenchuan Earthquake

The Wenchuan earthquake with Mw = 8.0 occurred at 14:28:01.42 LT (06:28:01.42 UT) on May 12, 2008 in Sichuan province, West China. The extent of the earthquake and after shock-affected areas lies northeast, along the Longmen Shan fault, a thrust structure along the border of the Tibetan Plateau and the western Sichuan Basin. It was a deadly earthquake to hit China since the 1976 Tangshan earthquake, which killed ten thousands of lives in several cities along the western Sichuan basin in China (Fig. 6.1). This event ruptured about 300 km of the Longmen Shan fault zone and lasted close to 120 s, with the majority of energy released in the first 80 s (Burchfiel et al. 2008; Toda et al. 2008; Parsons et al. 2008). The rupture started from the Wenchuan County and propagated at an average speed of 3.1 km/s, 49° toward northeast, rupturing a total of about 300 km (http://www.csi.ac.cn). Although robust seismic signals around the globe could estimate the gross nature of this event, the details of rupture are usually obscure due to the lack of near-field observations. Local geodetic measurements can provide more details on the kinematic rupture and processes and the size of this continental event. For example, GPS observation results showed that co-seismic deformations move towards the earthquake epicenter and the largest magnitude is up to 2.3 m in the horizontal and 0.7 m in the vertical in Beichuan rather than in the epicenter (Jin et al. 2010).

Ionospheric disturbances following this large earthquake have been found recently (Jin et al. 2014, 2015, 2017; Jin 2018). In a similar way, the coupling also affects the lower atmosphere (i.e., the troposphere) as airwaves propagate from the ground to the ionosphere. For example, Jin et al. (2011) found lower atmospheric anomalies following the 2008 Wenchuan Earthquake observed by GPS measurements. In the following, the seismic tropospheric changes during the 2008 Wenchuan Earthquake are investigated by the total Zenith Tropospheric Delay (ZTD) from dense GPS network observation data collected from China national continuous and campaign GPS network.

The daily ZTD time series with 5-min interval from 28 continuous GPS measurements in China are analyzed around the 12 May 2008 Wenchuan earthquake. It is interesting to find significant Co-seismic Tropospheric Disturbance (CTD) in term of ZTD near the epicenter. For example, Fig. 6.6 shows the daily ZTD residual time series at 5-min interval (ZTD minus the daily mean) from 11 to 13 May 2008 at Kmin site, southern part of the epicenter. The vertical dash line shows the main shock time of the Wenchuan Earthquake and solid black line represents the ZTD residuals on May 12, 2008. The ZTD at the main shock time (06:28 UTC, May 12, 2008) has a significant increase with up to about 15 mm, and then decrease after 2 h, while the ZTD on the non-earthquake days before and after May 12 almost has no anomaly changes. The ZTD variations at Lhas site, west of the epicenter, also show the similar disturbances with about 10 mm increase and then decrease during the main shock, while the ZTDs on the non-earthquake days before and after the earthquake also have

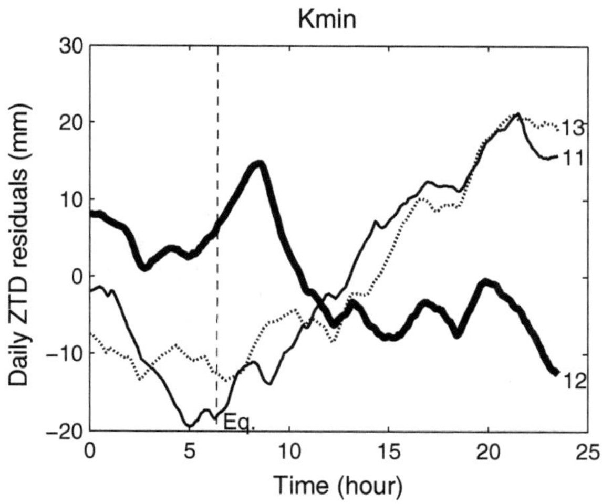

Fig. 6.6 Daily 5-min ZTD residual time series (ZTD minus daily mean) from day 11 to 13 May 2008 at Kmin site. The vertical dash line shows the main shock time of the Wenchuan earthquake and solid black line represents the ZTD residuals on May 12, 2008

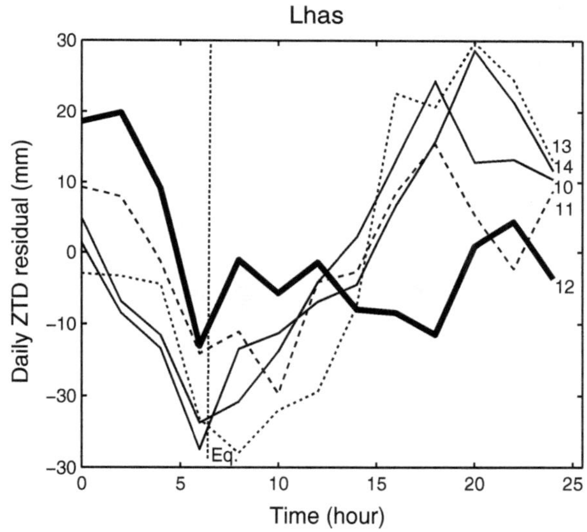

Fig. 6.7 Daily 5-min ZTD residual time series from day 10 to 14 May 2008 at Lhas site

no anomaly changes with a very consistent pattern, even 2 days ago or later (days 10 and 14 of May 2008) (Fig. 6.7).

In addition, the independent Bernese software (Dach et al. 2007) is employed to process the raw GPS data. The new ZTD time series at 2-h interval are obtained at the 28 continuous GPS sites. It has shown the similar co-seismic ZTD disturbances with the GAMIT processed results. For example, Fig. 6.8 is the daily ZTD residuals at 2-h interval from day 11 to 13 May 2008 at Kmin site. The co-seismic ZTD disturbances at 2-h interval have almost the same as the 5-min ZTD time series variations during this earthquake (Fig. 6.1). The co-seismic ZTD at northern parts of the epicenter has similar increase and then decrease during the main shock, e.g., at Xiaa (Fig. 6.9).

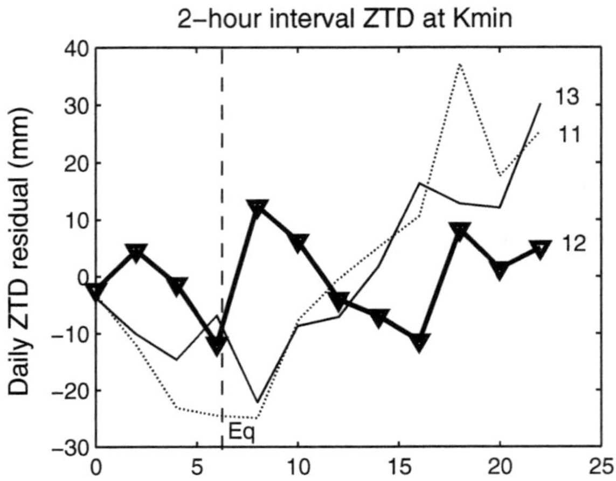

Fig. 6.8 Daily ZTD residual time series at 2-h interval from day 11 to 13 May 2008 at Kmin site. The vertical dash line shows the time of the Wenchuan earthquake main shock and solid black line represents the ZTD residuals on May 12, 2008

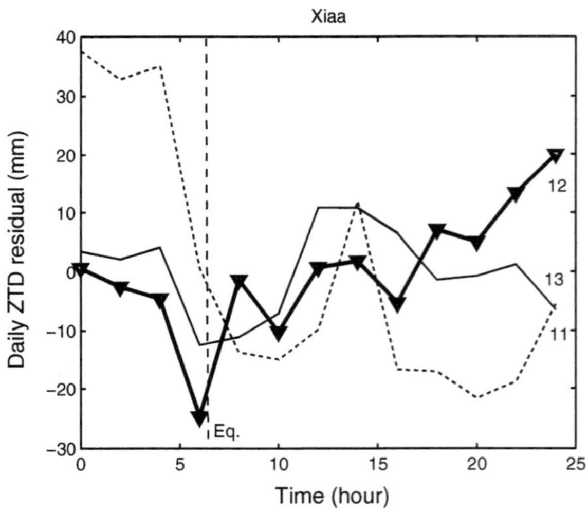

Fig. 6.9 Daily ZTD residual time series at 2-h interval from day 11 to 13 May 2008 at Xiaa site

Since it has no rain in these areas according to the Chinese Meteorological Administration in the daytime of May 12, 2008, the co-seismic ZTD anomalies during the event are mainly in the Zenith Hydrostatic Delay component (ZHD).

The available surface meteorological data at Lhas site are further analyzed and compared during this event. Figure 6.10 shows the daily temperature residuals from day 11 to 13 May 2008 at Lhas site. The co-seismic temperature anomalies are found during the main shock, which may be caused by the rapidly heating releasing and then cooling during the earthquake (Tranmutoli et al. 2005). Furthermore, the atmospheric pressure has shown the same co-seismic ZTD disturbance pattern with increase and then decrease during the main shock at Lhas site (Fig. 6.11) as the ZTD disturbance. Since the quick seismic surface waves propagating produces the radiated sound and infrasonic waves, these integrated interactions will generate

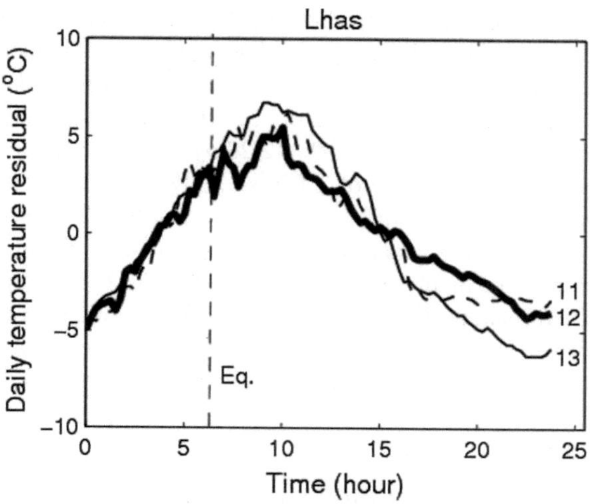

Fig. 6.10 Daily temperature residuals from day 11 to 13 May 2008 at Lhas site

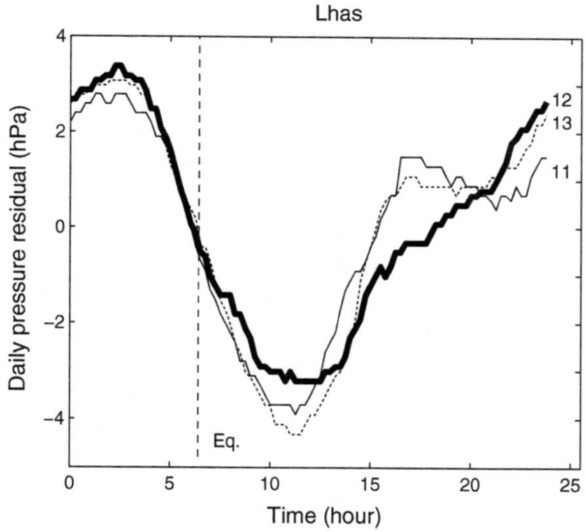

Fig. 6.11 Daily pressure residuals from day 11 to 13 May 2008 at Lhas site

locally ground-coupled airwaves. The airwaves propagate first upward or obliquely toward the upper atmosphere and after some time spread horizontally as acoustic-gravity waves in the lower atmosphere (Mikumo 1968). The acoustic-gravity waves from the ground vertical motion will drive the atmospheric 4pressure disturbances (Watada et al. 2006). For example, Lighthill (1978) showed a simple relationship between the ground velocity and the pressure change at the surface as $\Delta p = \rho c_s w$, where Δp, ρ, c_s, and w are the pressure change, air density, sound velocity, and velocity of fluid motion, respectively.

Under the assumption of hydrostatic equilibrium, the change in pressure profile is related to the total density at altitude h through the approximate relationship as

$dp = -\rho(h)g(h)dh$, where $\rho(h)$ and g(h) are the density and gravity at the altitude h, respectively. Ignoring the change of gravity acceleration g with respect to height, the zenith hydrostatic delay (ZHD) can be expressed as (Saastamoinent 1973), $ZHD = kp_0$, here k is a scale factor (2.28 mm/hPa) and p_0 is the pressure at height h_0 (Davis et al. 1985), namely $ZHD = 2.28\ p_0$. The hydrostatic component ZHD accounts for approximately 90% of ZTD, so ZTD is strongly correlated with surface pressure p_0 at the site. If surface pressure varies, the ZTD has a similar change with a scale factor of 2.28. Therefore, the consistent co-seismic atmospheric pressure changes during the main shock at Lhas site (Fig. 6.11) also support Co-seismic Tropospheric Disturbances (CTD), mainly in the hydrostatic delay component (ZHD), which are driven by the acoustic coupling of the atmosphere and solid-Earth with seismic waves propagating from the ground to the top atmosphere.

6.4.2 2010 Chile Earthquake

The 2010 Chile earthquake occurred off the coast of central Chile (35.91°S, 72.73°W) on Saturday, 27 February at 03:34:14 local time (06:34:14 UTC), with a magnitude of 8.8 on the moment magnitude (Mw) scale and lasting for about 3 min. This was the strongest earthquake affecting Chile since the 1960 Mw = 9.5 Valdivia earthquake, and worldwide since the 2004 Indian Ocean earthquake and until the 2011 Tōhoku earthquake. The earthquake was generated by the gently sloping fault that conveys the Nazca plate eastward and downward beneath the South American plate. The two plates are converging at 7 m per century. The fault rupture, largely offshore, exceeded 100 km in width and extended nearly 500 km parallel to the coast. The rupture began deep beneath the coast and spread westward, northward, and southward. As it spread, the fault slip generated earthquake shaking. The fault slip also warped the ocean floor, setting off the tsunami along the fault rupture area.

If the airwaves propagate from the ground to the top atmosphere, it may affect the lower atmosphere (i.e., the troposphere). However, such a possible impact on the troposphere is not clear. The possible tropospheric variations during the 2010 Chile Earthquake are investigated using GPS network observation data collected from SIRGAS and IGS network. It further compared the data with surface atmospheric pressure observations and temperature observations at co-located GPS site.

Here, GPS data from February 24 to February 28, 2010, are processed from GPS network, including IGS sites (SANT, CONZ, COPO, COYQ, LHCL, VALP, CFAG, and ANTC). The ZTD time series of several days are obtained around the main shock are obtained at 8 continuous GPS sites in Chile and Argentina with a temporal resolution of 10 min. It is interesting to find the same significant co-seismic tropospheric disturbance (CTD) as the above case with respect to the ZTD around the epicenter (Gurbuz and Jin 2017). As an example Fig. 6.12 shows the daily ZTD residual (ZTD minus daily mean) time series at 10 min interval from 25 to 27 Feb 2010 at CONZ station, which is directly at the epicenter of the earthquake. While vertical dash line points out the main shock time of the 2010 Chile earthquake, red

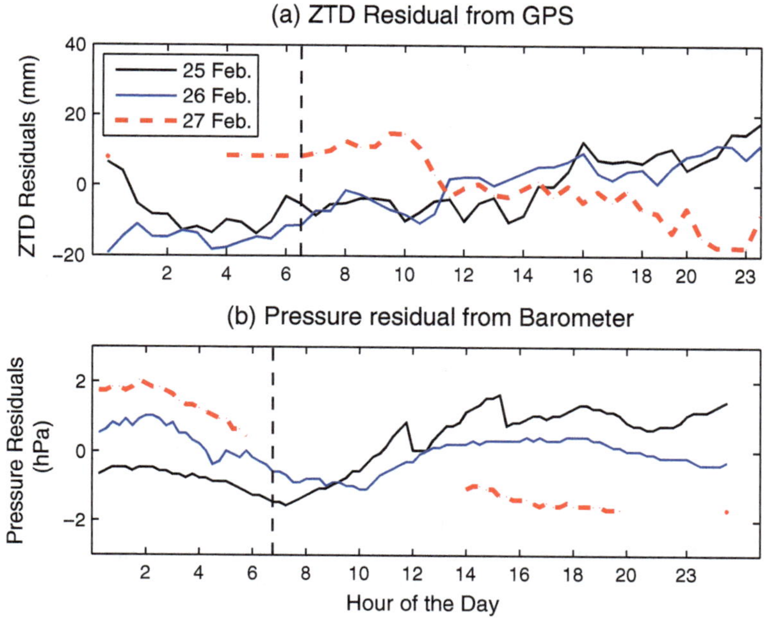

Fig. 6.12 Residual time series of GPS ZTD (**a**) and pressure (**b**) at CONZ station

dash line represents the ZTD residuals on February 27, 2010. As can be seen in Fig. 6.11, there are no anomalies on the ZTD before earthquake days (25 and 26 February 2010).

The ZTD at the main shock time has a little increase, after couple hours ZTD at CONZ station has been significantly decreased. Available surface meteorological data at CONZ station are analyzed and compared with non-earthquake days. Figure 6.12b shows the daily pressure residuals from 25 February to 27 February 2010 at CONZ station which has post-seismic pressure anomalies after the main shock. Unfortunately, there is no available meteorologica l data to compute ZTD or zenith hydrostatic delay (ZHD) on 28 February, which is the day of the earthquake at CONZ station. ZTD residuals at VALP station, north of the epicenter, also show similar anomalies with significantly decrease after the earthquake, while the ZTD residuals on the non-earthquake days (before and after the earthquake) also have no disturbances (Fig. 6.13) (Gurbuz and Jin 2017).

In addition, to check the accuracy of employed solutions to process the raw GPS data of CONZ station, JPL solutions at 26 February 2010 (DoY 057) are further used from the Crustal Dynamics Data Information System (CDDIS)'s FTP site which are processed by GIPSY and compared with our solutions in this study. Results show that the ZTD's agrees on millimeter level. The stations in south and east of the earthquake epicenter have less ZTD residual anomalies when compared to VALP or CONZ station (Gurbuz and Jin 2017).

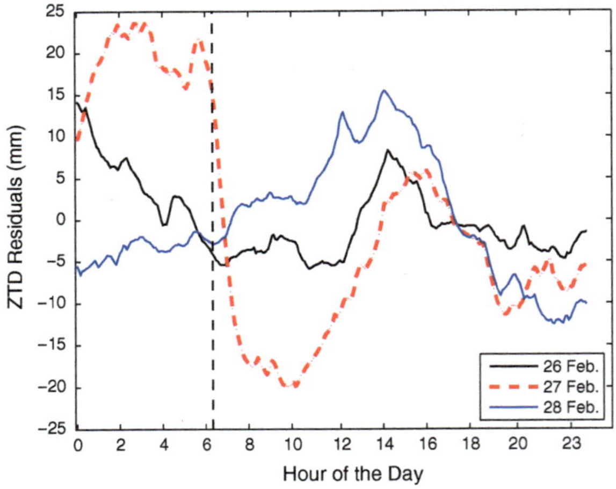

Fig. 6.13 Time series of GPS-estimated ZTD residual at VALP station

Since there was no precipitation in these areas according to the Chilean Met Office (Dirección Meteorológica de Chile) around February 27, 2010, the co-seismic and post-seismic ZTD anomalies during and after an earthquake are probably associated with this earthquake. However, further details should be investigated related with pressure and other atmospheric parameters.

6.5 Conclusion

The instantaneous enhancement in different atmospheric parameters prior to the 2015 Mw = 7.8 Nepal earthquake shows that these abnormalities are probably related to the impending earthquake. MODIS LST and OLR anomalies are strictly aligning the fault plane. It means that thermal anomalies can be seen along the seismologic zones. Another interesting thing is examined that the land surface temperature and OLR anomalies form dense clouds around the epicenter 10 days before the earthquake (Shah 2017). The gas particle in the atmosphere has 2–5% deviation from the rest of distribution 5 days before the earthquake. The significant enhancement of 0.1–0.5 mol/cm^2 of O_2, N_2, and CO concentration in the atmosphere can be due to the impending earthquake. On the contrary, the strong electric field around the earthquake epicenter is capable of driving these clouds towards the east. The induced E $\times$ B winds before the earthquake cause thermal anomalies. The atmospheric clouds of earthquake are susceptible to E $\times$ B drift that is execution of earthquake, while more cases are needed to study in the future with more temperature and radiation data as well as triggering process and mechanism.

Furthermore, significant Co-seismic Tropospheric Disturbances (CTD) are found from GPS observations following the 2008 Wenchuan earthquake and 2010 Chile earthquake. The ZTD at the main shock time has increased and then decreased, while it is also supported by the surface observed atmospheric pressure change at co-located GPS site. Since there was no rain in these areas in the daytime of earthquakes, the co-seismic ZTD anomalies are mainly in the zenith hydrostatic delay component (ZHD). The co-seismic lower atmospheric disturbances are probably driven by the ground-coupled airwaves from ground vertical motion of seismic waves propagating. Therefore, the Co-seismic Tropospheric delay Disturbances (CTD) show the acoustic coupling effect of the atmosphere and solid-Earth with airwave propagation from the ground to the ionosphere. In the future, it will further investigate the lower atmospheric anomalies, coupling processes, and mechanism between the lower atmosphere and solid-Earth with more cases and observation data.

References

Afraimovich, E., Feng, D., Kiryushkin, V., Astafyeva, E., Jin, S. G., & Sankov, V. (2010). TEC response to the 2008 Wenchuan earthquake in comparison with other strong earthquakes. *International Journal of Remote Sensing, 31*(13), 3601–3613. https://doi.org/10.1080/01431161003727747.

Blackett, M., Wooster, M. J., & Malamud, B. D. (2011). Exploring land surface temperature earthquake precursors: A focus on the Gujarat (India) earthquake of 2001. *Geophysical Research Letters, 38,* L15303. https://doi.org/10.1029/2011GL048282.

Bolt, B. A. (1964). Seismic air waves from the great 1964 Alaskan earthquake. *Nature, 202,* 1094–1095.

Burchfiel, B. C., Royden, L. H., van der Hilst, R. D., et al. (2008). A geological and geophysical context for the Wenchuan earthquake of 12 May 2008, Sichuan, People's Republic of China. *GSA Today, 18*(7), 4–11. http://www.gsajournals.org/perlserv/?request=get-abstract&doi=10.1130%2FGSATG18A.1.

Dach, R., Hugentobler, U., Fridez, P., & Meindl, M. (2007). The Bernese GPS Software Version 5.0, University of Bern, Bern, Switzerland.

Davis, J. L., Herring, T. A., Shapiro, I., Rogers, A., & Elgered, G. (1985). Geodesy by radio interferometry effects of atmospheric modeling errors on estimates of baseline length. *Radio Science, 20*(6), 1593–1607.

Freund, F. (2000). Time-resolved study of charge generation and propagation in igneous rocks. *Journal of Geophysical Research, 105*, 11,001–11,019. https://doi.org/10.1029/1999jb900423.

Freund, F., Freund, M. M., & Batllo, F. (1993). Critical review of electrical conductivity measurements and charge distribution analysis of magnesium oxide. *Journal Geophysical Research, 98,* 22209–22229.

Gurbuz, G., & Jin, S. G. GPS observations of tropospheric disturbances following the 2010 Mw = 8.8 Chile earthquake. In *Proceeding of IEEE International Geoscience and Remote Sensing Symposium (IGARSS), July 23–28, 2017* (pp. 4718–4721). Fort Worth, Texas, USA. https://doi.org/10.1109/igarss.2017.8128055.

Jin, S. G. (2018). Two-mode ionospheric disturbances following the 2005 Northern California offshore earthquake from GPS measurements. *Journal of Geophysical Research: Space Physics, 123*(10), 8587–8598. https://doi.org/10.1029/2017JA025001.

Jin, S. G., & Park, J. (2007). GPS ionospheric tomography: A comparison with the IRI-2001 model over South Korea. *Earth Planets Space, 59*(4), 287–292.

Jin, S. G., Wang, J., Zhang, H. P., & Zhu, W. Y. (2004). Real-time monitoring and prediction of the total ionospheric electron content by means of GPS observations. *Chinese Journal of Astronomy and Astrophysics, 28*(3), 331–337. https://doi.org/10.1016/j.chinastron.2004.07.008.

Jin, S. G., Park, J. U., Cho, J. H., & Park, P. H. (2007). Seasonal variability of GPS-derived Zenith tropospheric delay (1994–2006) and climate implications. *Journal Geophysical Research, 112,* D09110. https://doi.org/10.1029/2006JD007772.

Jin, S. G., Luo, O., & Park, P. (2008). GPS observations of the ionospheric F2-layer behavior during the 20th November 2003 geomagnetic storm over South Korea. *Journal of Geodesy, 82*(12), 883–892. https://doi.org/10.1007/s00190-008-0217-x.

Jin, S. G., Zhu, W., & Afraimovich, E. (2010). Co-seismic ionospheric and deformation signals on the 2008 magnitude 8.0 Wenchuan Earthquake from GPS observations. *International Journal of Remote Sensing, 31*(13), 3535–3543. https://doi.org/10.1080/01431161003727739.

Jin, S. G., Han, L., & Cho, J. (2011). Lower atmospheric anomalies following the 2008 Wenchuan Earthquake observed by GPS measurements. *The Journal of Atmospheric and Solar-Terrestrial Physics, 73*(7–8), 810–814. https://doi.org/10.1016/j.jastp.2011.01.023.

Jin, S. G., Jin, R., & Li, J. H. (2014). Pattern and evolution of seismo-ionospheric disturbances following the 2011 Tohoku earthquakes from GPS observations. *Journal of Geophysical Research: Space Physics, 119*(9), 7914–7927. https://doi.org/10.1002/2014JA019825.

Jin, S. G., Occhipinti, G., & Jin, R. (2015). GNSS ionospheric seismology: Recent observation evidences and characteristics. *Earth Science Reviews, 147,* 54–64. https://doi.org/10.1016/j.earscirev.2015.05.003.

Jin, S. G., Jin, R., & Li, D. (2017). GPS detection of ionospheric Rayleigh wave and its source following the 2012 Haida Gwaii earthquake. *Journal of Geophysical Research: Space Physics, 122*(1), 1360–1372. https://doi.org/10.1002/2016JA023727.

King, R. W., & Bock, Y. (2018). *Documentation for the GAMIT GPS analysis software*. Cambridge Mass: Mass. Inst. of Technol.

Lighthill, J. (1978). *Waves in fluids*. New York: Cambridge Univ. Press.

Mikumo, T. (1968). Atmospheric pressure waves and tectonic deformation associated with the Alaskan Earthquake of March 28, 1964. *Journal of Geophysical Research, 73*(6), 2009–2025. https://doi.org/10.1029/jb073i006p02009.

Parsons, T., Ji, C., & Kirby, E. (2008). Stress changes from the 2008 Wenchuan earthquake and increased hazard in the Sichuan basin. *Nature, 454,* 509–510.

Pulinets, S., & Boyarchuk, K. (2005). Ionospheric precursors of earthquakes. Springer Science & Business Media.

Pulinets, S., & Ouzounov, D. (2011). Lithosphere-atmosphere-ionosphere coupling (LAIC) model—An unified concept for earthquake precursors validation. *Journal of Asian Earth Sciences, 41,* 371–382. https://doi.org/10.1016/j.jseaes.2010.03.005.

Saastamoinent, J. (1973). Contribution to the theory of atmospheric refraction. *Bulletin Geodesique, 107,* 13–34.

Saradjian, M. R., & Akhoondzadeh, M. (2011). Thermal anomalies detection before strong earthquakes (Mw > 6.0) using interquartile, wavelet and Kalman filter methods. *Natural Hazards and Earth Systems Sciences, 11,* 1099–1108. https://doi.org/10.5194/nhess-11-1099-2011.

Shah M. (2017). Statistical analysis and modelling of seismic ionospheric anomalies from GNSS and other satellite observations, Ph.D. Thesis, University of Chinese Academy of Sciences.

Shah, M., & Jin, S. G. (2015). Statistical characteristics of seismo-ionospheric GPS TEC disturbances prior to global Mw $\geq$ 5.0 earthquakes (1998–2014). *Journal of Geodynamics, 92,* 42–49. https://doi.org/10.1016/j.jog.2015.10.002.

Toda, S., Lin, J., Meghraoui, M., & Stein, R. S. M. (2008). 12 May 2008 M = 7.9 Wenchuan, China, earthquake calculated to increase failure stress and seismicity rate on three major fault systems. *Geophysical Research Letters, 35,* L17305.

Tranmutoli, V., Cuomo, V., Filizzola, C., Pergola, N., & Pietrapertosa, C. (2005). Assessing the potential of thermal infrared satellite surveys for monitoring seismically active areas: The case of Kocaeli (I˙zmit) earthquake, August 17, 1999. *Remote Sensing of Environment, 96,* 409–426.

Tronin, A. A., Biagi, P. F., Molchanov, O. A., Khatkevich, Y. M., & Gordeev, E. I. (2004). Temperature variations related to earthquakes from simultaneous observation at the ground stations and by satellites in Kamchatka area. *Physics and Chemistry of the Earth, 29,* 501–504.

Watada, S., Kunugi, T., Hirata, K., Sugioka, H., Nishida, K., Sekiguchi, S., et al. (2006). Atmospheric pressure change associated with the 2003 Tokachi-Oki earthquake. *Geophysical Research Letters, 33,* L24306. https://doi.org/10.1029/2006GL027967.

Chapter 7
Pre-seismic Ionospheric Anomalies

7.1 Introduction

Earthquakes, as one of the important natural hazards, often threaten our lives and livelihoods, which often result in civilian casualties and huge damage to civilian property. Therefore, monitoring and forecasting the earthquake is one of the hot topics all the time. Ones try to find some possible pre-seismic anomalies through various methods and observations. However, it is still challenging to obtain convincing pre-earthquake anomalies or evidences up to now. The earthquake is a very complex and broad topic. Not only the motion of tectonic plates but also the microscopic processes, such as the generation of electric charge and chemical reactions, are possibly associated with earthquakes. Both Earth's crustal dynamics and the state of Earth's atmosphere will be affected by earthquakes. Currently, seismological studies can be divided into two parts: one is to study solid Earth such as curst deformation, material flow, and so on, and the other one is to monitor Earth's surface parameters variations, such as ionospheric perturbations (Davies and Baker 1965; Calais and Minster 1995) and anomalous changes in the column water vapor (Dey et al. 2004).

GPS TEC can provide some insight into seismo-ionospheric disturbances related to seismic events (Afraimovich et al. 2010; Jin et al. 2010). A number of publications report that Seismo-Ionospheric Disturbances (SID) are caused by a large amount of energy released into the ionosphere from earthquake preparation zones. Variations in TEC, foF2, and NmF2 have correlated with large amounts of energy released during the earthquake preparation period, which are probably signs of a forthcoming large magnitude earthquake (Bolt 1999; Freund 2000; Singh et al. 2010; Jin et al. 2013). The radon emission from the Earth's crust in the earthquake preparation zone causes changes in the atmosphere, especially, the ionosphere (Jin et al. 2011). The statistical analysis of seismo-ionospheric TEC disturbances related to earthquakes in Taiwan showed that the TEC anomaly appears frequently 5 days before the earthquake and an increased TEC has been observed in the afternoon period 3 days prior to the earthquake (Liu et al. 2000). Similarly, statistical analysis and case studies for earthquakes in Japan during 1998–2010 reveal that seismo-ionospheric distur-

S. Jin et al., *GNSS Atmospheric Seismology*,
https://doi.org/10.1007/978-981-10-3178-6_7

bances related to a shallow hypocenter cause disturbances in TEC as compared to deep hypocenter earthquakes (Liu et al. 2006; Kon et al. 2011; Lin 2013). Interpreted results of TEC in North America showed that seismo-ionospheric disturbances on the continent were due to forthcoming earthquakes and sharpness in TEC increased with increasing magnitude as well as precursory alarm of the coming earthquake (Sardon et al. 1994; Pulinets et al. 2003; Hayakawa and Molchanov 2008; Zolotov et al. 2012). The seismo-ionospheric disturbance significantly decreases after the main shock, and the aftereffect of seismo-ionospheric disturbances is studied in detail and modeled in different ways to explain the behavior of seismo-ionospheric anomalies related to seismic activities (Jin et al. 2014, 2015; Pradipta et al. 2014). Analyses for threshold and different types of faults also show that disturbances in TEC can be seen for strike-slip faults with some explanations of the abnormality related to larger Richter scale earthquakes (Fatkullin et al. 1989; Pulinets et al. 1991; Astafyeva et al. 2014).

Furthermore, Le et al. (2011) statistically studied ionospheric anomalies from the GIM-TEC prior to 736 $Mw \geq 6.0$ worldwide earthquakes during 2002–2010 and found that the frequency of increases and decreases in SIPs is proportional to the magnitude, but inversely related to depth and day of the forthcoming earthquake. Their results suggest the effect of latitudinal enhancement in TEC, which is due to the fact that large amount of energy was transmitted into the atmosphere before the main shock of a larger earthquake than a smaller earthquake. But for small magnitudes like $Mw \leq 6.0$, the seismo-ionospheric anomalies in TEC related to earthquake breeding zone are less obvious than the large magnitude earthquake. Similarly, Kon et al. (2011) conducted a statistical analysis and showed that dissimilarities in TEC in Japan during 1998–2010 tend to appear 1–5 days before earthquakes, and positive anomalies tend to be more prominent than negative anomalies all over Japan. The seismo-ionospheric precursory mechanism was discussed in a model, which showed a lithosphere–atmosphere–ionosphere coupling and explained the disturbance in the ionosphere caused by seismic events (Pulinets and Ouzounov 2011). The TEC has variability influenced by large magnitudes and shallow depth earthquakes, which is a strong contributor of disturbance in the sensitive F2 layer of the ionosphere. Seismo-ionospheric disturbances 40 days before (after) the 2008 $Mw = 7.9$ Wenchuan earthquake were observed statistically, and the suspected seismo-ionospheric disturbance related to the 2008 Wenchuan in TEC was found at UT = 04:00. This variation was also observed in CHAMP (N_e) and DEMETER satellite data for the same days (Ryu et al. 2014). However, seismo-ionospheric anomalies prior to earthquakes are still challenging and difficult to know the coupling mechanism, particularly for different earthquake magnitudes and types of fault (Jin et al. 2007). In the following, a case study and a statistical analysis of pre-earthquake ionospheric disturbances related to the global $Mw \geq 5.0$ earthquakes during 1998–2014 are investigated from GPS TEC, including different earthquake sources worldwide as well as diverse magnitudes and depths.

7.2 Immediate Pre-seismic Ionospheric Anomalies

7.2.1 Observation Data

GNSS Earth Observation Network System (GEONET), which aims to monitor crust motion with high precision, consists of more than 1200 continuous operation GNSS stations covering the whole Japan, as shown in Fig. 7.1. Abundant continuous GNSS observations from GEONET provide us an opportunity to study temporal and spatial variation of Earth's ionosphere. Figure 7.2 shows the Ionosphere Puncture Point (IPP) track corresponding to five GEONET stations. It shows a good coverage rate. Obviously, its coverage is enough to discuss Earth's ionospheric changes following earthquakes happening in Tohoku, 2011, if we use all GPS stations (more than 1200).

According to USGS's reports, a series of earthquake occurred off the Pacific coast of Tohoku in 2011. Here, several significant earthquakes happened in these areas are selected to analyze pre-seismic ionospheric anomalies. Brief information of these earthquakes is shown in Table 7.1 (http://www.usgs.gov).

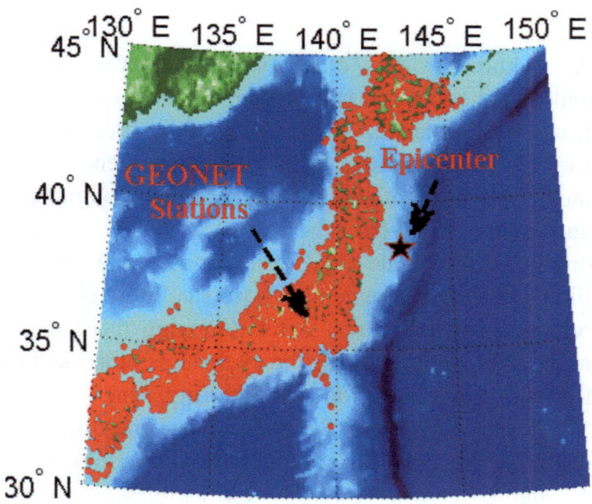

Fig. 7.1 GEONET stations' distribution. The red dot stands for GNSS stations of GEONET. The star is the epicenter of the earthquake occurred off the Pacific coast of Tohoku on March 9, 2011

Table 7.1 Brief information of several huge Tohoku earthquakes in 2011

Date	Onset (UTC)	Latitude	Longitude	Magnitude	Depth	Region
2011-03-09	02:45:20	38.44°N	142.84°E	7.3	32	Tohoku
2011-03-11	05:46:24	38.30°N	142.37°E	9.0	30	Tohoku
2011-07-10	00:57:12	38.04°N	143.29°E	7.0	23	Tohoku

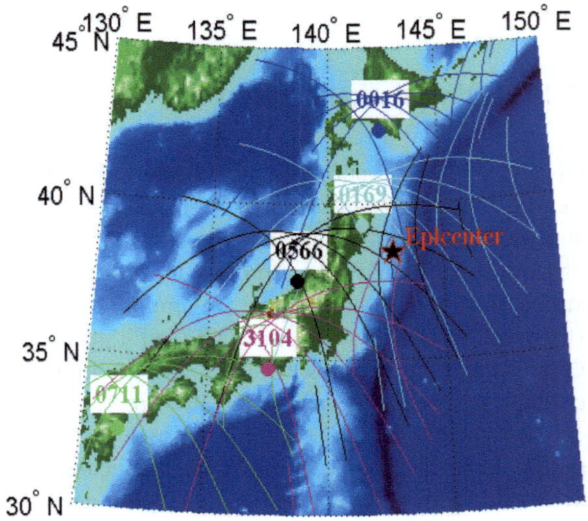

Fig. 7.2 IPP tracks for five GEONET stations (0711, 3104, 0566, 0169, and 0016). Each line stands for one IPP track relevant to one LOS during a whole and continuous arc

7.2.2 Data Preprocessing and Anomalies Detection

As we know, a variety of errors, which are associated with receivers, GNSS satellites and signal propagation, exist in GNSS raw data (Li and Huang 2005; Xu 2007; Liu 2008; Hernandez-Pajares et al. 1999; Jin et al. 2008). Preprocessing is necessary for TEC changes detection using dual-frequency GNSS observations. In this section, methods to deal with the raw data and TEC computation are presented in the first part, and in the second part, the approach of TEC anomalies detection is introduced. In order to get accurate STEC time series, cycle slips should be detected and repaired. First of all, measurements whose satellites' elevation angles are less than 30° are removed from the arcs. After removing these measurements with low signal-to-noise ratio, we can repair the cycle slips more accurately. Two algorithms are used for cycle slips detection and repairing. One is TE (TurboEdit) algorithm (Blewitt 1990) and the other is Second-order, Time-difference Phase Ionospheric Residual (STPIR) method (Cai et al. 2012). Combining these two methods cycle slips on L_1 and L_2 can be repaired well. Here, second-order difference is used to avoid confusing with dramatic changes of STEC in cycle slips repair. Clear arcs can be obtained by a successful cycle slip repair. STEC series can be derived from the clean dual-frequency GPS observations. Here, the thin shell assumption of the Earth's ionosphere has been generally adopted to estimate TEC (Klobuchar 1987; Smith et al. 2008). Earth's ionosphere ranges from 60 to 1000 km attitude. It is inconvenience for STEC integration and computation of GPS signal's ionospheric delay. Using a central ionosphere layer, whose attitude is usually set as 300–600 km (here the attitude is set as 450 km) instead of the real ionosphere, can simply the computation of GPS TEC and almost retain its property of variation. Mapping function should be applied in order to convert STEC to VTEC. In a small region and in short time, the variation of VTEC (vertical TEC: TEC of zenith direction in the relevant IPP) is smooth and steady. It shows a good agreement that using a polynomial expression to model the variation of VTEC during an arc in normal days. The four-order polynomial, which is shown in Eq. 7.1, is chosen for detecting TEC anomalies in this chapter.

$$VTEC = at^4 + bt^3 + ct^2 + dt + e \tag{7.1}$$

GPS ionospheric observations $L_4(= L_1 - L_2)$ are expressed in the following as a function of time and zenith distance:

$$L_4 = \frac{40.28(f_2^2 - f_1^2)}{f_1^2 f_2^2}\left[\frac{at^4}{\cos(z)} + \frac{bt^3}{\cos(z)} + \frac{ct^2}{\cos(z)} + \frac{dt}{\cos(z)} + \frac{e}{\cos(z)}\right] + f$$
$$f = \bar{N} - (B_1 - B_2) \tag{7.2}$$

Least-square (LS) method is used to estimate the coefficients $a \sim f$ in Eq. 7.2. The difference between the measured $VTEC$ and modeled $VTEC$(obtained from LS estimation) for thousands of GPS arcs in the Tohoku region gives us a good opportunity to analyze and investigate the TEC changes following the earthquakes. Obviously, the complex variation of GPS TEC cannot be represented accurately using a simple polynomial fitting, but it is enough for our residual analysis.

7.2.3 *Pre-seismic Ionospheric Anomalies*

Three earthquakes, shown in Table 7.1, are chosen for pre-seismic TEC anomalies analyses. All of their epicenters are located off the Pacific coast of Tohoku, which have denser GNSS data to study TEC anomalies prior to earthquakes. Figure 7.3 presents the TEC anomalies following the Tohoku earthquakes on March 9, March 11, and July 10 in 2011.

Due to the space limit and clear expression, one satellite and several GPS stations are picked up to show the TEC anomalies. All these four graphs in Fig. 7.3 do not show the vertical axis ticks, because they make no sense during studying TEC variations and residuals. Observed TEC values and modeled TEC values are added or subtracted a constant at the same time in order to put several arcs' TEC changes in the same axis for more directly expressing the relationship between the earthquakes and GPS TEC, which will not change the properties of TEC variations and residuals. Figure 7.3 just shows one small part of GPS TEC anomalies following the Tohoku earthquakes in 2011. TEC anomalies occurred around all of these three earthquakes. Both before and after the main shock time have TEC disturbances. Obviously, TEC disturbance amplitudes are related to the earthquake magnitudes. As shown in the upper right panel, the most dramatic disturbances happened following the Mw = 9.0 Tohoku earthquake occurred on March 11, 2011. Space weather and geomagnetic activities are also important impact factors. Large-scaled traveling ionospheric distur-

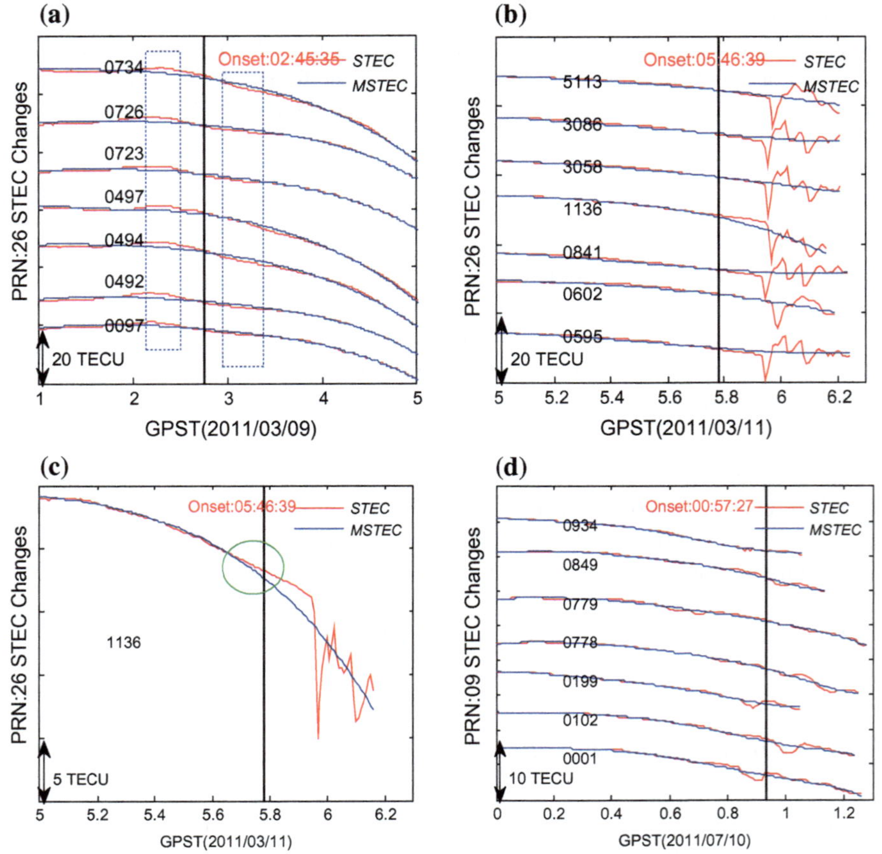

Fig. 7.3 GPS-TEC anomalies following the Tohoku earthquake in 2011. Here, the real TEC values are added or subtracted a constant, which do not change the residual and variation of GPS TEC, for better displaying in one figure

bances can be generated by high geomagnetic activities (Ding et al. 2008). However, fortunately, these days' space weather is relative quiet. It is also reported that strong sporadic E layer did not appear on March 11, when the Dst index comes up to -80nT (Heki 2011). Compared to co-/post-earthquake TEC anomalies, pre-earthquake TEC anomalies are slight for the earthquake on March 11, 2011.

7.2.4 *Pre-seismic Ionospheric Anomaly Amplitudes*

A tremendous amount of GPS measuring arcs around the onset of the 2011 Tohoku earthquakes in 2011 provided by GEONET is a good chance to monitor and study TEC anomalies following earthquakes. As we know, seismo-ionospheric distur-

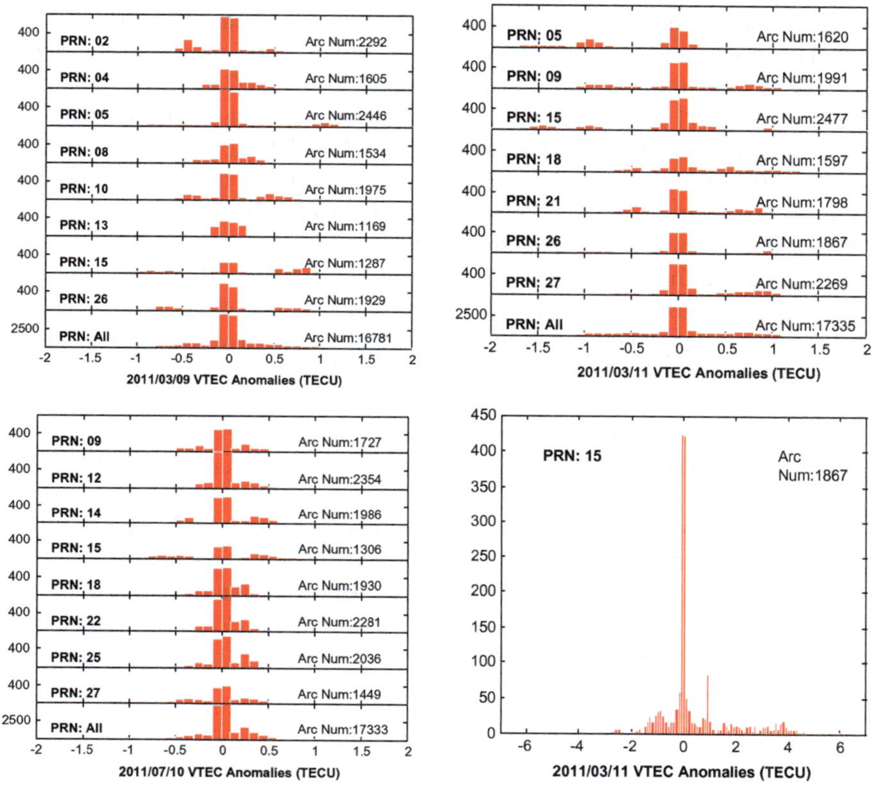

Fig. 7.4 Distribution of vertical TEC anomaly amplitudes following the Tohoku earthquake in 2011

bances have occurred a few hours or a few days before earthquakes (Pulinets and Boyarchuk 2005; Akhoondzadeh et al. 2010). In this chapter, we focus on the immediate TEC anomalies characteristics, so peak values of each arc during 2 h before and after the onsets are extracted for statistical analysis. Figure 7.4 shows the distribution of VTEC anomaly amplitudes corresponding to three earthquakes. In the four subgraphs of Fig. 7.4, "PRN: All" stands for all the visible GPS satellites in the area of Tohoku during relevant periods. Distributions for individual satellite, however, not all the visible satellites, are also presented in Fig. 7.4. Here, we chose several satellites which have enough observations to display. The sample numbers are large enough to demonstrate VTEC anomalies related to individual satellite. The distribution of TEC anomalies peak values for PRN 15, shown in the right bottom panel in Fig. 7.4, is picked out as an example to show the detail that cannot be distinguished in the right upper panel for the March 11 earthquake in 2011.

In normal days, VTEC derived from GPS dual-frequencies' measurements changes with the IPP variation smoothly. VTEC anomalies mainly vary between −0.1TECU and 0.1TECU (agreement with the noise of GPS TEC measurement).

However, the amount of VTEC anomaly amplitudes is even larger than 1TECU. Especially, for March 11, 2011, lots of arcs' peak TEC anomaly values are even up to 8 TECU. As shown in Fig. 7.4, the distribution of VTEC anomalies following the March 11, 2011 earthquake has the largest dispersion (range from about −5.5 to 8.2 TECU). The second one is the earthquake on March 9, 2011 (range from about −1.7 to 2.0 TECU), and the July 10 earthquake VTEC anomaly has the smallest dispersion (range from about −1.0 to 1.6 TECU). Generally, positive and negative VTEC anomalies almost have the same probability of occurrence for the three earthquakes, although for individual satellite, positive or negative VTEC anomalies may be overwhelming as shown in Fig. 7.4. However, the amplitudes of positive anomalies seem to be slightly larger than negative anomalies. The divergences and averages of arcs' peak TEC anomaly values are presented in Table 7.2. It is easy to see that the divergence of TEC anomaly peak amplitude values is positive correlation with the earthquake magnitude. However, the averages seem to not relate to the magnitude. For these three earthquakes, all of the averages of TEC anomalies range between 0 and 0.1 TECU.

Table 7.3 presents the comparison of pre-seismic and post-seismic arcs' peak TEC anomalies. For earthquakes on March 9 and July 10, the statistical properties of peak TEC anomalies do not show great differences. However, the disturbance of TEC enhances after the main shock for the earthquake on March 11. This enhancement is another direct evidence for that the ionospheric disturbance around the main shock is related to the earthquake.

Table 7.2 The divergences and averages of arcs' peak TEC anomaly values

Date	Arc number	Divergence	Average	Magnitude
2011/03/09	16,781	0.338	0.037	7.3
2011/03/11	17,335	0.732	0.064	9.0
2011/07/10	17,333	0.226	0.038	7.0

Table 7.3 Comparison of pre-seismic and post-seismic arcs' peak TEC anomalies

	Average	Divergence	Maximum	Minimum
Pre-seismic	0.047	0.346	1.993	−1.737
Post-seismic	0.026	0.329	1.279	−1.310
Pre-seismic	−0.034	0.502	2.049	−3.099
Post-seismic	0.134	0.855	8.297	−5.516
Pre-seismic	0.022	0.225	0.961	−0.991
Post-seismic	0.058	0.225	1.636	−0.735

7.2.5 *Pre-seismic Ionospheric Anomaly Time*

The properties of pre-seismic ionospheric anomalies with the maximum magnitude are further analyzed for the March 11, 2011 earthquake. Figure 7.5 shows the distribution of VTEC anomaly time offsets corresponding to onset of the Tohoku earthquake occurred on March 11, 2011. Here, we get rid of arcs whose peak anomalous values less than 0.1 TEC to avoid the normal disturbance in GPS TEC observations, which is caused by measuring noise. Zero in horizontal axis is corresponding to the main shock time of the earthquake. As shown in Fig. 7.5, lots of disturbances began to occur about 4 h before the main shock. This phenomenon may be one of the precursors of the strong earthquake. Comparing to pre-seismic effects, post-seismic is more apparent. The first peak appears 10 min after the main shock. Disturbances attenuate as time going, and sudden increase of anomalous arc occurs about 80 and 150 min after onset. Figure 7.5 shows three types of disturbances after the strong earthquake happened on March 11. Lots of precursors think that post-seismic ionospheric disturbances are caused by atmospheric waves going through the Earth ionosphere. Whether these three peaks are corresponding three types of atmospheric waves, which are direct acoustic wave from the focal area, gravity wave propagating obliquely, and secondary acoustic wave in far field, still need more independent evidences.

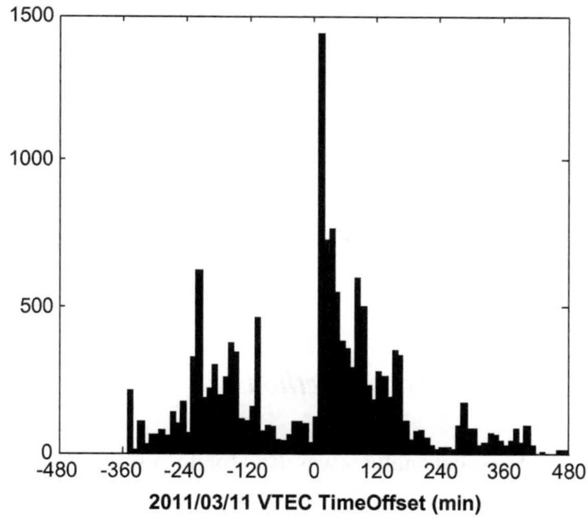

Fig. 7.5 Distribution of vertical TEC anomaly time series corresponding to onset of the Tohoku earthquake occurred on March 11, 2011

7.3 Statistics Analysis of Pre-seismic Ionospheric Anomalies

7.3.1 *TEC Time Series and Earthquakes*

Now, dual-frequency GPS observations can estimate the ionospheric TEC and its variations (e.g., Jin et al. 2007, 2008). We used the TEC data of GPS Global Ionosphere Maps (GIM) (ftp://cddisa.gsfc.nasa.gov/pub/gps/products/ionex). TEC data from GPS ionex have a 2-h temporal resolution and a spatial resolution of 2.5° × 5° in latitude and longitude (Schaer and Krankowski 2009). In our study, TEC data from IGS GIM are used for statistical analysis and investigating spatial variations of seismo-ionospheric disturbances. We extracted the TEC data from GPS ionex during 1998–2014 for global earthquakes of Mw $\geq$ 5.0 and investigated the seismo-ionospheric variations in TEC prior to the earthquake. TEC is expressed in TEC Unit (TECU) where 1 TECU = 10^{16} electron/m^2. More GIM-TEC information and recent improvements were given and discussed (Hernandez-Pajares et al. 1999). The earthquake database and fault catalogues were obtained from the USGS earthquakes archive (http://earthquake.usgs.gov/earthquakes/search/). The solar wind condition and geomagnetic indices data were obtained from NASA satellites (http://omniweb.gsfc.nasa.gov) and Kiyoto University (http://wdc.kugi.kyoto-u.ac.jp/kp).

Figure 7.6 shows the location of earthquakes with geographic latitude and longitude. The earthquake catalogues of USGS are used to accurately study worldwide earthquake occurrences (Shah and Jin 2015). The magnitude is denoted by the circle size. The earthquakes shown in Fig. 7.6 were not severely affected by the geomagnetic storm. 1,492 M $\geq$ 5.0 earthquakes during 1998–2014 were studied excluding the earthquakes whose 10 days fall near the geomagnetic storm of a higher scale (Dst ≤ -50 nT). According to Gutenberg and Richter (1956), large amounts of energy are released with a higher magnitude and seismic energy and the difference between two magnitudes of earthquake is 10 times.

7.3.2 *Statistical Methods*

The statistical analysis was used for detecting seismo-ionospheric disturbance related to earthquakes. To observe deviation of abnormal signals, the 10-day before (after) successive mean <TEC> and standard deviation (σ) for every 2 h of TEC for the period 1998–2014 were calculated (Shah and Jin 2015). An increment (decline) on the 11th day and the preceding 10 days was noted. Further to confirm the 95% deviation of TEC, we calculated the 95% confidence bound for TEC in the desired period. We omitted other earthquakes in these 10 days as well as the days following the magnetic storm. To calculate the Standard Error (SE), the standard deviation is divided by the square root of the number of 2-h TEC in successive 10 days from the time of study. Since the number of 2-h TEC in successive 10 days was 120,

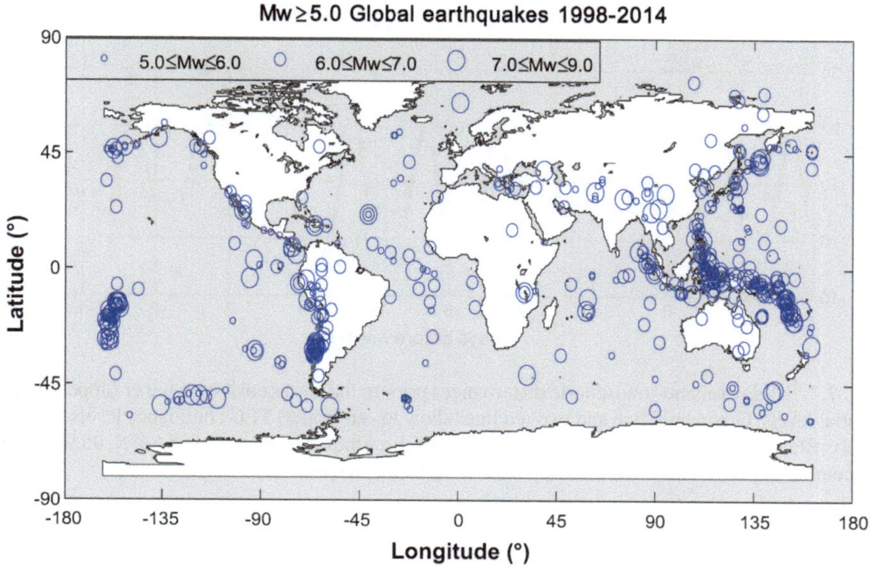

Fig. 7.6 Distribution of global Mw ≥ 5.0 earthquakes during 1998–2014. The size circle represents the magnitude

the value of z-percentile at $\alpha = 0.05$ significance level is 1.96. We computed the $Z_{1-\alpha/2}$ percentile for degree of freedom (d_f) of the number of 2-h TEC in 10 days. This percentile was used to calculate the upper and lower confidence intervals. For the normal distribution of mean <TEC> and standard deviation (σ), the concept of *t*-distribution for TEC was developed.

Since the number of 2-h TEC in successive 10 days from the day of study was 120 h, the deviation from normal distribution of <TEC> and σ at $1 - \alpha/2$ for t-distribution was significantly small. However, the deviation for z-distribution was large. Therefore, we used z-percentile instead of t-percentile for larger deviation from the normal distribution. In this study, we used z-distribution for $1 - \alpha/2$ percentile to observe a clear TEC anomaly. The 95% confidence bound for $n = 120$ with $d_f = n - 1$ in z-distribution is 1.96. Detailed methods canrefer to Shah and Jin (2015). Thus, when an observed TEC on some day increases (decreases) than previous 10 days <TEC> $\pm\, z_{df,\, 1-\alpha/2}$ *SE by more than value of threshold, which is defined as seismo-ionospheric disturbance related to the earthquake. We set the threshold value here (Mw ≥ 6) because the seismo-ionospheric disturbances are sharper for Mw ≥ 6.0. Now the TEC of 6 or more hours for a single day trip out from upper (lower) confidence level would be an abnormal signal. The earthquake days followed by Dst ≤ -50 nT and Kp > 3 are not included in this study as both effects cause changes in the ionosphere. This criterion was used to interrelate GPS TEC anomalies related to global Mw ≥ 5.0 earthquakes from 1998 to 2014 (Shah and Jin 2015).

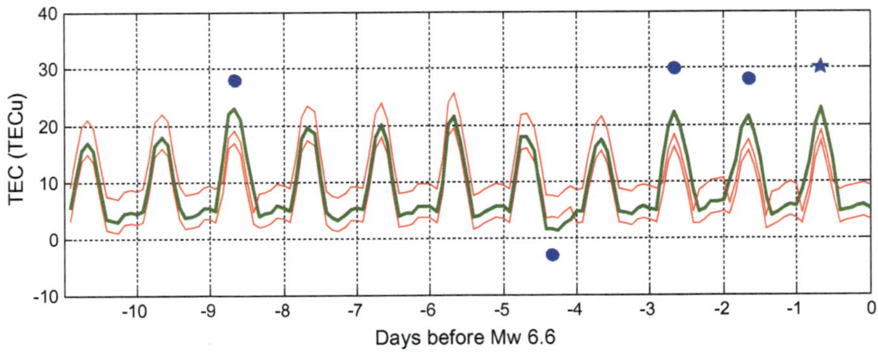

Fig. 7.7 10-day seismo-ionospheric disturbances prior to the earthquake with lower (upper) confidential levels. Green bold line and two red lines show lower (upper) TEC confidence levels, respectively. Blue star and dot symbols denote the Mw = 6.6 earthquake epicenter (56.84°N, 95.91°E) on December 27, 2011 and seismo-ionospheric disturbance, respectively

7.3.3 *Case Analysis*

To define the seismo-ionospheric disturbance related to earthquakes, the TEC values trip out the upper and lower confidence intervals, which are defined as seismo-ionospheric disturbances related to earthquakes by amount equal to the threshold of Mw = 6.0 (Marques de Sa 2007). The TEC value of more than 6 h for a single day of upper (lower) confidence level has been defined as a seismo-ionospheric disturbance. Figure 7.7 shows seismo-ionospheric disturbances related to the Mw = 6.6 earthquake epicenter (56.842°N, 95.911°E) in association with the lower (upper) confidential level and epicenter as well as pre-earthquake ionospheric anomalies. We have removed the days near the earthquake with Dst ≤ -50nT and Kp > 3. The TEC variations prior to the Mw = 6.6 earthquake revealed two strong positive anomalies on 2 days −1 and −2 and a negative anomaly on −4 day as shown in Fig. 7.7 (Shah and Jin 2015). Seismo-ionospheric anomalies in Fig. 7.7 within 5 days before the Mw = 6.6 earthquake are interpreted in terms of electric field generation and ground motion of earthquake breeding zone.

7.3.4 *Global Large Earthquakes Statistical Analysis*

In order to investigate the possible pre-earthquake ionospheric anomalies of global earthquakes, the same statistical process has been performed for global Mw $\geq$ 5.0 earthquakes during the period of 1998–2014. The TEC variations within 10 days of the main shock are large. The sudden enhancement in TEC crest is due to forthcoming earthquake (Zakharenkova et al. 2007; Tojiev et al. 2013). Figure 7.8 shows the cumulative counts in percentages which are obtained by dividing the number of anomalous

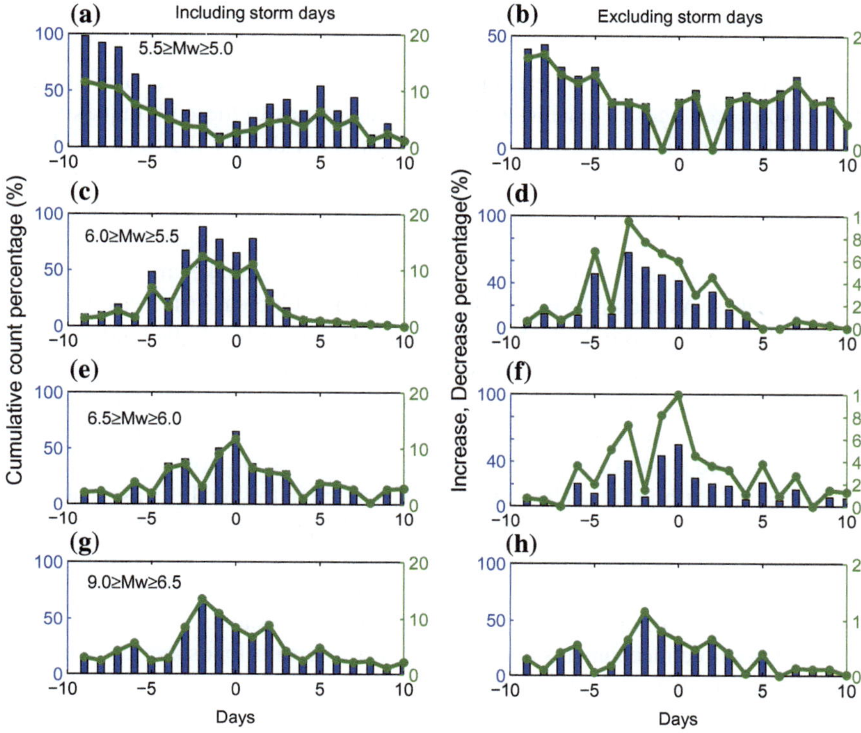

Fig. 7.8 Cumulative count in percentages (%) and increase (decrease) percentage (%) count for seismo-ionospheric disturbances related to global Mw $\geq$ 5.0 earthquakes during 1998–2004

days related to earthquakes by the total number of days (Shah and Jin, 2015). We further removed the magnetic storm anomalous days to get all seismo-ionospheric disturbances possibly due to earthquakes. The increase (decrease) percentage was calculated from the cumulative counts by dividing the cumulative count of a single day to sum of cumulative count days. Statistical results used above deviation from normal distribution of TEC due to using Z-percentile instead of t-percentile at $\alpha = 0.05$. Figure 7.8 shows an increase in tendency of TEC anomalies with Mw > 6.0, while sharp and regular spikes of anomalous TEC are significant for earthquakes of Mw > 6.5. Days with spikes and regular seismo-ionospheric anomalies are dominant from 4 to 5 days before earthquakes of Mw > 6.0. It has been observed that circulation of energy in the atmosphere and eastward dynamo effect on TEC is due to the seismic activity. Nevertheless, Mw > 6.0 and focal depth <50 km are key elements in causing these primary and short-term variations in the ionosphere.

7.3.4.1 Different Faults Earthquakes

We collected fault classes of earthquakes to check the phenomenon of increase (decrease) percentages of seismo-ionospheric anomalies to these earthquakes. The number of seismo-ionospheric disturbance days is divided by the total number of days multiplied by 100 to get increase (decrease) percentages. Here, we performed a statistical analysis for Mw $\geq$ 6.0 and observed an anomalous pattern with fault-related earthquakes from 1998 to 2014 (Shah and Jin 2015). Figure 7.9a shows no possible enhancement in TEC caused by normal fault earthquakes in 5 days prior to the earthquakes. As shown in Fig. 7.9b, the variation in increase (decrease) percentage of TEC anomalies is only for Mw $\geq$ 6.5 earthquakes with oblique fault, while the variation in TEC with the other earthquakes is not obvious. It can be seen in Fig. 7.9c, d that prominent abnormal TEC might be due to strike-slip fault and thrusting fault earthquakes in all magnitudes for the period of 1998–2014. Seismo-ionospheric disturbances related to strike-slip and thrusting earthquakes are noticeable for Mw = 6.0–Mw = 7.0 earthquakes. The Mw $\geq$ 6.5 was noted in this analysis as an anomaly in GPS TEC for Mw $\geq$ 6.5 in Fig. 7.9b–d is sharper than other magnitudes of earthquakes. It is obvious that pre-earthquake ionospheric anomaly for Mw $\geq$ 6.5 is sharper and more regular than any other magnitude in the statistical analysis carried out for fault classes of different magnitude earthquakes. Astafyeva et al. (2011) have presented possible mechanisms and discussions on seismo-ionospheric disturbance prior to major fault earthquakes around the globe.

7.3.4.2 Different Focal Depth Earthquakes

In addition, we performed a statistical analysis for Mw $\geq$ 6.0 earthquakes with different focal depths during 1998–2014 to ensure the variation in TEC related to the focal depth of earthquake because the focal depth is a vital feature of earthquake preparation period (Shah and Jin 2015). Earthquakes with a focal depth of less than 60 km and Mw $\geq$ 6 are presumably the root of deviation in ionospheric TEC because such earthquake preparation zones at shallower depth have a gigantic amount of energy. The peak of TEC 5 days before earthquakes in exists with lower focal depth, but for large focal depth, there is no obvious variation in TEC with 5 days before the earthquakes. Horizontal but not steep peak of TEC has been noticed for Mw $\geq$ 6.0 and focal depth of less than 120–150 km. Excluding the magnitude, the focal depth of an earthquake is responsible for discrepancies in TEC. The nighttime variation analysis method concluded that for shallow earthquakes (depth $<$ 40 km), the normalized variation and scattering of TEC exhibited a significant increase before earthquakes (Su et al. 2013). These statistical results suggest that propagation anomalies in the ionosphere and shallow earthquakes are directly related. The crest of TEC anomaly can be observed up to $\pm 40^\circ$ in latitude and $\pm 40^\circ$ in longitude, showing association with the TEC anomaly in the checked area (Dautermann et al. 2007). For describing the status and mechanism of seismo-ionospheric disturbance (Molchanov and

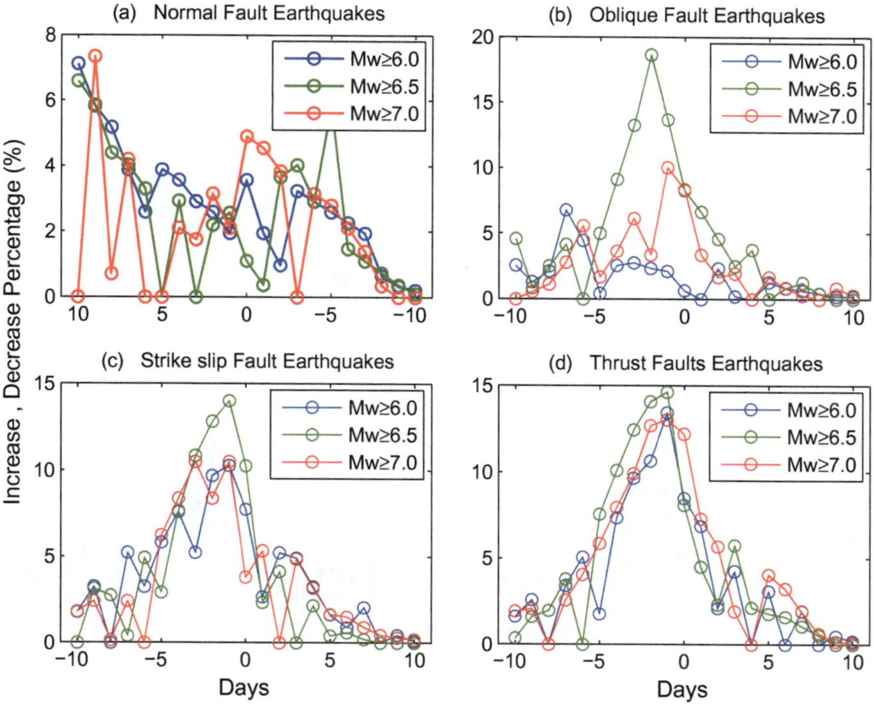

Fig. 7.9 **a** Increase (decrease) percentages of seismo-ionospheric disturbances for normal faults earthquakes of Mw ≥ 6.0. **b, c, d** are the increase (decrease) percentages of seismo-ionospheric disturbance for oblique fault, strike-slip fault, and thrust fault earthquakes during 1998–2014, respectively

Hayakawa 2002), the enhanced TEC may be due to the raised electric field in the eastward direction 1–5 days before earthquakes in Japan.

Figure 7.10 shows cumulative percentages and relative values of positive and negative anomalies related to global earthquakes of Mw ≥ 5.0 (Shah and Jin 2015). Cumulative percentages have been obtained by dividing the single-day TEC anomalies by overall TEC anomalies. The relative values of positive and negative TEC anomalies are expressed as

$$\text{Relative values} = \frac{TEC\ \text{Anomaly}_{\text{single day}} - TEC\ \text{Anomaly}_{mean}}{TEC\ \text{Anomaly}_{mean}} \tag{7.3}$$

where TEC $\text{Anomaly}_{\text{single day}}$ is the anomaly on a single day and TEC $\text{Anomaly}_{\text{mean}}$ is the mean value of 21 days (10 days before (after) earthquake) positive or negative anomalies. The relative values in Fig. 7.10 reveal a high ratio (up to 2) of positive anomalies 5 days prior to the global earthquakes, with a tendency similar to cumulative percentages of positive anomalies. The case of negative anomalies is adverse

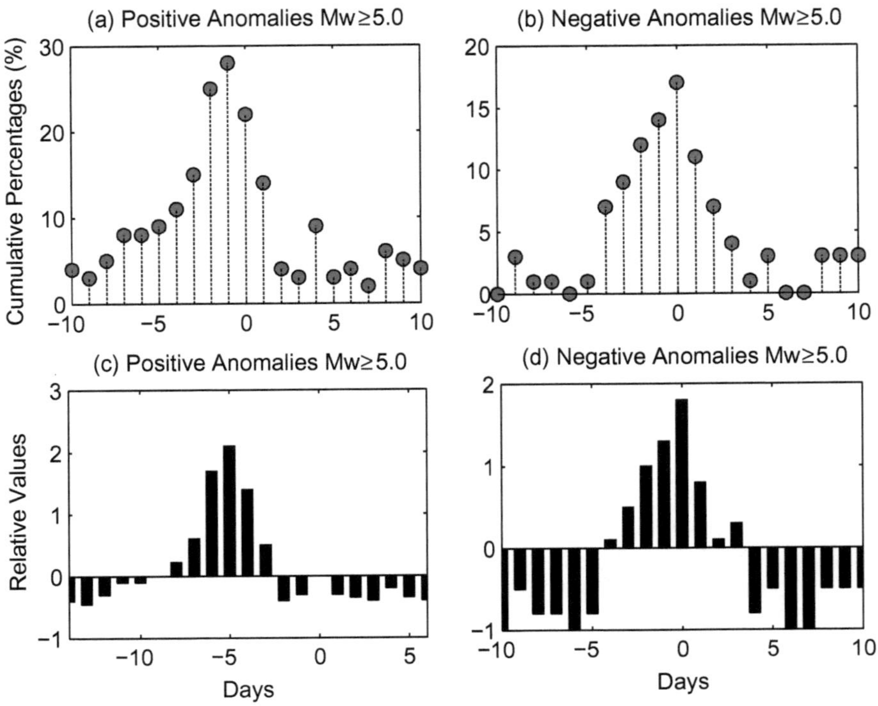

Fig. 7.10 Cumulative percentages and relative values for global Mw $\geq$ 5.0 earthquakes during 1998–2014

with low ratios (up to -0.5) in relative values. It is interesting to note that positive anomaly days with consistently more than negative anomalies 5 days before global earthquakes suggest that seismo-ionospheric anomalies are generated by earthquakes rather magnetic storms.

7.4 Concluding Remarks

A statistical analysis of seismo-ionospheric anomalies prior to global Mw $\geq$ 5.0 earthquakes is carried out with a 95% confidence level. Since ionospheric activities with Dst (magnetic storm) and Kp index affect TEC variations, we have removed days with Dst ≤ -50(nT) and Kp > 3. The cumulative count in percentages for seismo-ionospheric disturbances related to earthquakes shows that TEC anomalies for Mw = 6.0 and higher become evident within 5 days prior to earthquakes. The strike-slip fault and thrust fault earthquakes regularly cause disturbances in TEC. The anomalous pattern in TEC is regular and sharper for strike-slip earthquakes when compared to other classes of fault earthquakes. Oblique fault earthquakes with

Mw $\geq$ 6.5 cause seismo-ionospheric disturbancesionospheric disturbances and the enhancement in TEC is sharp as strike-slip earthquakes. Statistical analysis for Mw $\geq$ 6.0 and different focal depths show that pre-earthquake TEC anomalies with the focal depth of <50 km are obvious since TEC is susceptible to this large amount of breeding energy. With an increase of focal depth from 60 km, the seismo-ionospheric disturbances related to earthquakes are not clear. In addition, it is understandable that the relative percentage of positive anomalies in TEC is somehow sharper than the relative percentage of negative anomalies. Based on our results, the earthquake preparation zone most likely gives birth to seismo-ionospheric anomalies.

Although the percentage of positive anomalies is dominant, the percentages of negative anomalies sporadically appear 5 days before Mw $\geq$ 5.0 earthquakes all over the globe. Earthquakes are responsible for the generation of pre-seismo-ionospheric disturbance due to sensitivity of the ionosphere to energy evolved by earthquake preparation zone. Here, it just shows initial statistical results, which still needs more investigations to confirm with more observations and theoretical modeling in the future, particularly the mechanism of pre-earthquake ionospheric anomalies.

References

Afraimovich, E., Ding, F., Kiryushkin, V., Astafyeva, E., Jin, S. G., & Sankov, V. (2010). TEC response to the 2008 Wenchuan earthquake in comparison with other strong earthquakes. *International Journal of Remote Sensing, 31*(13), 3601–3613. https://doi.org/10.1080/01431161003727747.

Akhoondzadeh, M., Parrot, M., & Saradjian, M. R. (2010). Electron and ion density variations before strong earthquakes (M > 6.0) using DEMETER and GPS data. *Natural Hazards and Earth System Sciences, 10*(1), 7–18.

Astafyeva, E., Lognonné, P., & Rolland, L. (2011). First ionospheric images of the seismic fault slip on the example of the Tohoku-oki earthquake. *Geophysical Research Letters, 38,* L22104. https://doi.org/10.1029/2011GL049623.

Astafyeva, E., Rolland, L. M., & Sladen, A. (2014). Strike-slip earthquakes can also be seen in the ionosphere. *Earth and Planetary Science Letters, 405,* 180–193. https://doi.org/10.1016/j.epsl.2014.08.024.

Blewitt, G. (1990). An automatic editing algorithm for GPS data. *Geophysical Research Letters, 17*(3), 199–202.

Bolt, B. A. (1999). *Earthquake and its after effects* (4th ed.). Freeman, New York: W. H.

Cai, C., Liu, Z., Xia, P., & Dai, W. (2012). Cycle slip detection and repair for undifferenced GPS observations under high ionospheric activity. GPS Solutions.

Calais, E., & Minster, J. B. (1995). GPS detection of ionospheric perturbations following the January 17, 1994, Northridge Earthquake. *Geophysical Research Letters, 22*(9), 1045–1048.

Davies, K., & Baker, D. M. (1965). Ionospheric effects observed around the time of the Alaskan earthquake of March 28, 1964. *Journal Geophysical Research, 70*(9), 2251–2253.

Dautermann, T., Calais, E., Haase, J., & Garrison, J. (2007). Investigation of ionospheric electron content variations before earthquakes in southern California, 2003–2004. *Journal Geophysical Research, 112,* B02106. https://doi.org/10.1029/2006JB004447.

Dey, S., Sarkar, S., & Singh, R. P. (2004). Anomalous changes in column water vapor after Gujarat earthquake. *Advances in Space Research, 33*(3), 274–278.

Ding, F., Wan, W., Liu, L., Afraimovich, E. L., Voeykov, S. V., & Perevalova, N. P. (2008). A statistical study of large-scale traveling ionospheric disturbances observed by GPS TEC during major magnetic storms over the years 2003–2005. *Journal of Geophysical Research, 113*(A3), A00A01.

Fatkullin, M. N., Zelenova, T. I., & Legen'ka, A. D. (1989). On the ionospheric effects of asthenospheric earthquakes. *Physics of the Earth and Planetary Interiors, 57,* 82–85.

Freund, F. (2000). Time-resolved study of charge generation and propagation in igneous rocks. *Journal of Geophysical Research, 105,* 11001–11019.

Gutenberg, B., & Richter, C. F. (1956). Magnitude and energy of earthquakes. *Annali di Geofisica, 9,* 1–15.

Hayakawa, M., & Molchanov, O. A. (2008). Seismo electromagnetics: lithosphere-atmosphere-ionosphere coupling (pp. 477). Terra Sci., Tokyo.

Heki, K. (2011). Ionospheric electron enhancement preceding the 2011 Tohoku-Oki earthquake. *Geophysical Research Letters, 38*(17), L17312.

Hernandez-Pajares, M., Juan, J. M., & Sanz, J. (1999). New approaches in global ionospheric determination using ground GPS data. *Journal of Atmospheric and Solar Terrestrial Physics, 61,* 1237–1247.

Jin, S. G., Park, P. H., & Zhu, W. Y. (2007). Micro-plate tectonics and kinematics in Northeast Asia inferred from a dense set of GPS observations. *Earth and Planetary Science Letters, 257*(3–4), 486–496. https://doi.org/10.1016/j.epsl.2007.03.011.

Jin, S. G., Luo, O. F., & Park, P. H. (2008). GPS observations of the ionospheric F2-layer behavior during the 20th November 2003 geomagnetic storm over South Korea. *Journal of Geodesy, 82*(12), 883–892. https://doi.org/10.1007/s00190-008-0217-x.

Jin, S. G., Zhu, W. Y., & Afraimovich, E. (2010). Co-seismic ionospheric and deformation signals on the 2008 magnitude 8.0 Wenchuan Earthquake from GPS observations. International *Journal of Remote Sensing, 31*(13), 3535–3543. https://doi.org/10.1080/01431161003727739.

Jin, S. G., Han, L., & Cho, J. (2011). Lower atmospheric anomalies following the 2008 Wenchuan Earthquake observed by GPS measurements. *The Journal of Atmospheric and Solar-Terrestrial Physics, 73*(7–8), 810–814. https://doi.org/10.1016/j.jastp.2011.01.023.

Jin, S. G., van Dam, T., & Wdowinski, S. (2013). Observing and understanding the Earth system variations from space geodesy. *Journal of Geodynamics, 72,* 1–10. https://doi.org/10.1016/j.jog.2013.08.001.

Jin, S., Jin, R., & Li, J. H. (2014). Pattern and evolution of seismo-ionospheric disturbances following the 2011 Tohoku earthquakes from GPS observations. *Journal of Geophysical Research: Space Physics, 119.* https://doi.org/10.1002/2014ja019825.

Jin, S. G., Occhipinti, G., & Jin, R. (2015). GNSS ionospheric seismology: Recent observation evidences and characteristics. *Earth Science Reviews, 147,* 54–64. https://doi.org/10.1016/j.earscirev.2015.05.003.

Klobuchar, J. A. (1987). Ionospheric time-delay algorithm for single-frequency GPS users. *IEEE Transactions on Aerospace and Electronic Systems, 23*(3), 325–331.

Kon, S., Nishihashi, M., & Hattori, K. (2011). Ionospheric anomalies possibly associated with $M \geq 6.0$ earthquakes in the Japan area during 1998–2010: Case studies and statistical study. *Journal of Asian Earth Sciences, 41,* 410–420.

Le, H., Liu, J. Y., Liu, L. (2011). A statistical analysis of ionospheric anomalies before 736 M6.0+ earthquakes during 2002–2010. *Journal of Geophysical Research, 116*, A02303. https://doi.org/10.1029/2010ja015781.

Li, Z., & Huang, J. (2005). *GPS surveying and data processing*. China: Wuhan, Wuhan University Press.

Lin, J. W. (2013). Is it possible to detect earlier ionospheric precursors before large earthquakes using principal component analysis (PCA). *Arabian Journal of Geosciences, 6,* 1091–1100.

Liu, J. Y., Chen, Y. I., Chuo, Y. J., & Chen, C. S. (2006). A statistical investigation of pre-earthquake ionospheric anomaly. *Journal of Geophysical Research, 111,* A05304. https://doi.org/10.1029/2005JA011333.

Liu, J. Y., Chen, Y. I., Pulinets, S. A., Tsai, Y. B., & Chuo, Y. J. (2000). Seismo-ionospheric signatures prior to M6.0 Taiwan earthquakes. *Geophysical Research Letters, 27,* 3113–3116.

Liu, J. (2008). *The Principle and method of GPS navigation and positioning*. Beijing: China, Science Press.

Marques de Sa, J. P. (2007). Applied statistics using SPSS, STATISTICA, MATLAB and R (pp. 85–90). Springer, ISBN 978-3-540-71971-7.

Molchanov, O. A., & Hayakawa, M. (2002). Seismo electromagnetics and related phenomena: History and latest results (pp. 189). TERRAPUB, Tokyo.

Pradipta, R., Valladares, C. E., & Doherty, P. H. (2014). GPS observation of continent-size traveling TEC pulsations at the start of geomagnetic storms. *Journal of Geophysical Research: Space Physics, 119,* 6913–6924. https://doi.org/10.1002/2014JA020177.

Pulinets, S., & Ouzounov, D. (2011). Lithosphere-atmosphere-ionosphere coupling (LAIC) model—an unified concept for earthquake precursors validation. *Journal of Asian Earth Sciences, 41,* 371–382. https://doi.org/10.1016/j.jseaes.2010.03.005.

Pulinets, S. A., Legen'ka, A. D., Karpachev, A. T., Kochenova, N. A., Migulin, V. V., Orvevski, V. N., & Fligel, M. D. (1991). On the possibility of earthquakes prediction on the basis of topside satellite sounding, Preprint IZIMIRAN (Russia), 34a, 981.

Pulinets, S. A., Legen'ka, A. D., Gaivoronskaya, T. V., & Depuev, V. K. (2003). Main phenomenological features of ionospheric precursors of strong earthquakes. *Journal of Atmospheric and Solar-Terrestrial Physics, 65*(16–18), 1337–1347.

Pulinets, S., & Boyarchuk, K. (2005). *Ionospheric Precursors of Earthquakes*. Berlin Heidelberg: Springer.

Ryu, K., Parrot, M., Kim, S. G., Jeong, K. S., Chae, J. S., Pulinets, S., & Oyama, K.-I. (2014). Suspected seismo-ionospheric coupling observed by satellite measurements and TEC related to the M7.9 Wenchuan earthquake of 12 May 2008. *Journal of* Geophysical *Research:* Space Physics, *119.* https://doi.org/10.1002/2014ja020613.

Sardon, E., Rius, A., & Zarraoa, N. (1994). Estimation of the transmitter and receiver differential biases and the ionospheric total electron content from global positioning system observations. *Radio Science, 29,* 577–586.

Schaer, & Krankowski, A. (2009). The IGS VTEC maps: A reliable source of ionospheric information since 1998. *The Journal of Geodesy, 83*, 263–275. https://doi.org/10.1007/s00190-008-0266-1.

Shah, M., & Jin, S. G. (2015). Statistical characteristics of seismo-ionospheric GPS TEC disturbances prior to global Mw $\geq$ 5.0 earthquakes (1998–2014). *Journal of Geodynamics, 92,* 42–49. https://doi.org/10.1016/j.jog.2015.10.002.

Smith, D. A., Araujo-Pradere, E. A., Minter, C., & Fuller-Rowell, T. (2008). A comprehensive evaluation of the errors inherent in the use of a two-dimensional shell for modeling the ionosphere. *Radio Science, 43*(6), RS6008.

Singh, R. P., Mehdi, W., Gautam, R., Senthil Kumar, J., Zlotnick, J., & Kafatos, M. (2010). Precursory signals using satellite and ground data associated with the Wenchuan earthquake of 12 May 2008. *International Journal of Remote Sensing, 31*(13), 3341–3354.

Su, Y. C., Liu, J. Y., Chen, S. P., Tsai, H. F., & Chen, M. Q. (2013). Temporal and spatial precursors in ionospheric total electron content of the 16 October 1999Mw7.1 Hector Mine earthquake. *Journal of Geophysical Research: Space Physics, 118,* 6511–6517.

Tojiev, S. R., Ahmedov, B. J., Tillayev, Y. A., & Eshkuvatov, H. E. (2013). Ionospheric anomalies of local earthquakes detected by TEC measurements using data from Tashkent and Kitab stations. *Advances in Space Research, 52,* 1146–1154.

Xu, G. (2007). *GPS, theory, algorithms and applications*. Potsdam, Germany, Berlin, Heidlberg, New York: Springer.

Zakharenkova, I., Shagimuratov, I., Krankowski, A., & Lagovsky, A. (2007). Precursory phenomena observed in the total electron content measurements before great Hokkaido earthquake of September 25, 2003 (M = 8.3). *Studia Geophysica et Geodaetica, 51*(2), 267–278.

Zolotov, O. V., Namgaladze, A. A., Zakharenkova, I. E., Martynenko, O. V., & Shagimuratov, I. I. (2012). Physical interpretation and mathematical simulation of ionospheric precursors of earthquakes at midlatitudes. *Geomagnetizm i Aeronomiya, 52,* 390–397.

Chapter 8
Co-/Post-seismic Ionospheric Disturbances

8.1 Introduction

Giant earthquakes often result in great loss of economics and lives as well as geohazards, such as tsunami and landslide. Due to the lack of measurement and theoretical understanding for the Earth's interior, it is hard to predict the short-term earthquake's occurrence, location, and magnitude up to now. Getting more insights on earthquake processes will contribute to understand and predict earthquakes. Recently, seismo-ionospheric disturbances can be observed from Global Positioning System (GPS) (Klobuchar 1987; Calais et al. 1995; Jin et al. 2008, 2010; Afraimovich et al. 2010). The energy released by earthquake affects the Earth's atmosphere in the form of atmospheric wave from seismic vertical deformation and earthquake-generated tsunami. The seismic ionospheric disturbance will be amplified dramatically during its upward propagation from the ground to the ionosphere due to the attenuation of atmospheric density (Artru et al. 2001). With more and more continuous GPS stations, especially in Japan, it provides a great chance to study more details about the seismic atmospheric characteristics and coupling between the solid Earth and atmosphere further.

The seismic ionospheric disturbance was first found from GPS total electron content (TEC) in 1995 (Calais et al. 1995). Later, numerous studies have been attempted to seek pre- or co-seismic anomalies and its physical mechanism (Liu et al. 2004; Dautermann et al. 2007; Heki et al. 2013). Nowadays, the method to extract pre-seismic ionospheric effect is mostly using previous several normal days' TEC to estimate a predicted TEC value based on its statistical behavior for the earthquake days. By comparing the predicted value and the GPS TEC observations, the anomalous TEC variations are regarded as probable pre-seismic TEC disturbances, while its mechanism is still open debated. Different filter methods are applied to the observed GPS TEC time series to obtain the co-seismic ionospheric disturbances signals, waves could be induced and spread out after the fault rupture (Afraimovich et al. 2001; Heki et al. 2005, 2006; Lognonné et al. 2006; Rolland et al. 2011). However, the Earth's ionosphere has complex and dynamic spatial structure and is affected by amount

S. Jin et al., *GNSS Atmospheric Seismology*,
https://doi.org/10.1007/978-981-10-3178-6_8

of factors, such as geomagnetic field and solar flare. It is difficult to derive accurate pre- and co-seismic ionospheric anomalies for regional or global scales. Co-seismic ionospheric disturbances sources are normally thought to be near the epicenter, while few GPS stations are located near the epicenter of giant earthquakes, which is difficult to get accurate details about the pre-/co-seismic ionospheric effects and characteristics.

A giant earthquake with magnitude of up to 9 occurred in Tohoku near the east coast of Honshu, Japan at 05:46:24 UTC, March 11, 2011, which is the largest earthquake in the last 7 years. It was associating a number of aftershocks and huge tsunami (Fig. 8.1). The Tohoku earthquake's epicenter is located in 38.297°N, 142.372°E estimated by US Geological Survey (USGS). It is one of the shallow earthquakes whose focal depth is just 30 km. A dense GPS array of the GEONET with more than 1200 continuously operating stations within 2500 km away from the epicenter provides us a good chance to study the detailed seismic ionospheric effects and propagation characteristics.

A number of seismic ionospheric disturbances for the Tohoku earthquake have been investigated (Tsugawa et al. 2011; Liu et al. 2011). TEC disturbances spread out to almost all directions from the epicenter with the speed of several hundreds of meters to several kilometers per second. The ionospheric disturbances detected by GPS observations following the Tohoku earthquake showed three modes with different propagation velocities, which may be related to the seismic Rayleigh waves (2–3 km/s), acoustic waves (0.3–1.5 km/s), and tsunami-generated

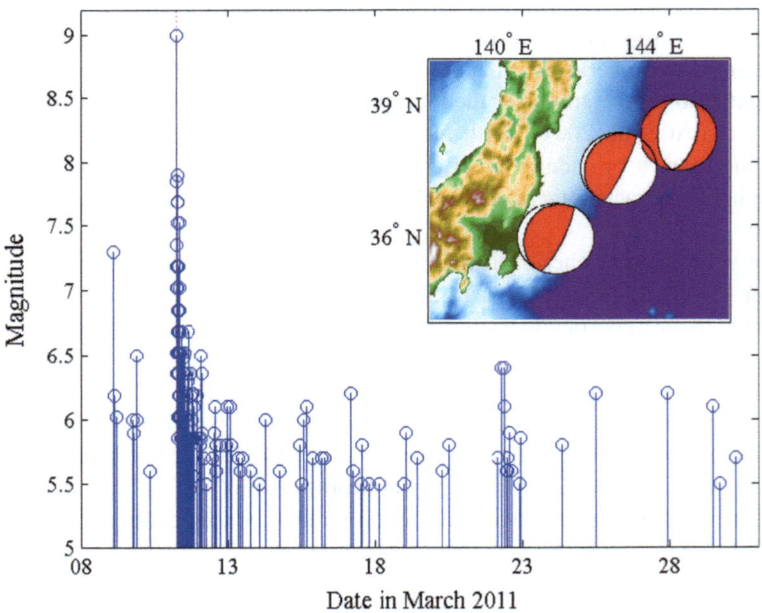

Fig. 8.1 Earthquake sequence in Tohoku, Japan, March 2011

gravity waves (0.1–0.3 km/s). Simulated results showed the southeastward ionospheric disturbances, which may be due to the coupling between tsunami and atmosphere (Galvan et al. 2012; Occhipinti et al. 2013). Two-dimensional simulation of near-field seismic ionospheric effects reproduced the ionospheric hole after the onset of Tohoku earthquake, indicating that the TEC decrease above the epicenter was caused by atmospheric expansion to upper layers derived by the sudden uplift of sea surface during the earthquake (Satio et al. 2011; Kakinami et al. 2012; Shinagawa et al. 2013). However, these results and simulations were mainly focused on the propagation velocities and the coupling of the tsunami or ground motion and atmosphere. Up to now, the pattern and evolution of seismic TEC disturbances following the Tohoku earthquake are not clear in detail. In this Chapter, the pattern and evolution of the disturbance amplitudes, period, propagation speed, and directions' features following the Tohoku earthquake are investigated comprehensively using dense GPS observations in Japan. Some discussions on seismic ionospheric characteristics are presented.

8.2 GPS Observations

As shown in Fig. 8.2, the GEONET with more than 1200 continuously operating GPS stations was established in 1996 by Geospatial Information Authority of Japan with the aim of monitoring crustal deformation. GPS ionosphere monitoring is one of the significant applications. The average distance between GEONET stations is about 25–30 km. Therefore, the dense GEONET provides us a good opportunity to monitor the Earth's ionosphere with high precision and resolution. In Fig. 8.2, the colorful dots on the assumed ionospheric thin shell show the sub-ionospheric pierce points (SIP) distribution from GEONET on UTC 05:46, March 11 in 2011, where the different colorful dots are corresponding to the SIP distribution for different GPS satellites. The white star is the location of the Tohoku earthquake's epicenter, the red dots on the ground terrain stand for the GEONET station, and the green rectangle is the rupture region estimated by the Tectonics Observatory, California Institute of Technology (http://www.tectonics.caltech.edu). Since the dense SIPs focus around the epicenter, the detailed seismic ionospheric disturbances can be obtained following the Tohoku earthquake. Here, continuous observations of more than 1200 stations are collected with the sampling interval of 30 s. Using the method mentioned above, the precise vertical TEC series at SIPs around the epicenter are obtained with 30 s interval. The precision of GPS carrier phase measurement is less than one millimeter. According to the error propagation law, the measurement error of TEC is about 0.01 TECU. The detrended TEC in quite condition is nearly the random noise, which is dominated by the measurement error. In quiet condition, the detrended TEC value will usually fall into the three-sigma range of about −0.03 to 0.03 TEC. When the detrended TEC values are out of this range, the TEC anomalies could be detected. The high precision and high temporal–spatial resolution ionospheric TEC from GEONET provides the opportunity to discuss the detailed seismic ionospheric pattern and evolution. The

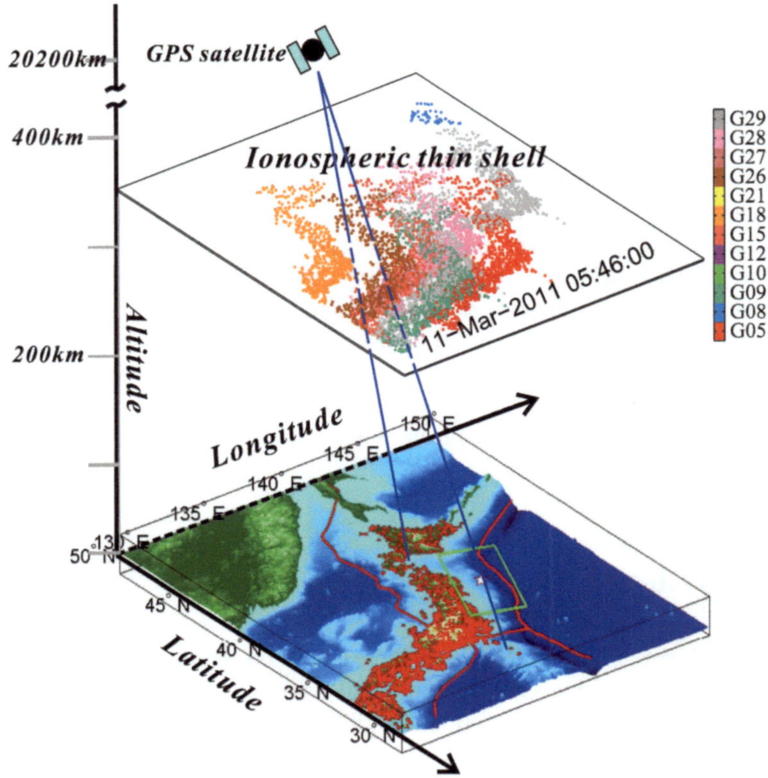

Fig. 8.2 Ground-based ionosphere sounding from dual-frequency GPS in Japan. The colorful dots on the assumed ionospheric thin shell show the SIP distribution on UTC 05:46, March 11 in 2011. The white star is the epicenter location of the 2011 Tohoku earthquake, the red dots on the ground show the distribution of GEONET stations, and the green rectangle is the rupture region estimated by the Tectonics Observatory, California Institute of Technology (http://www.tectonics.caltech.edu)

detailed TEC disturbances following the earthquake are investigated and discussed in the next section.

8.3 Co-/Post-seismic Ionospheric Disturbances

In this section, the ionospheric disturbances following the Tohoku earthquake with GEONET data are analyzed in detail, including the pattern and evolution of the disturbance amplitudes, spectrograms, propagation speed, and direction (Jin et al. 2014). The times in GEONET data are corrected by adding the difference between UTC and GPS time with −15 s on March 11, 2011 according to International Earth Rotation and References Systems Service (IERS) Bulletins (http://hpiers.obspm.fr/eop-pc). In order to degrade the multipath effects and the error of mapping function, measurements with low satellite elevation angles (less than 10 degrees if not specified in following) are not used. In order to remove the TEC trend and high frequencies fluctuations and other noise, the detrended TEC series are filtered with four-order zero-phase shift Butterworth filter that is designed with a most flat frequency response in the passband without phase shift.

8.3.1 TEC Anomalies

Figure 8.3 shows the detrended TEC maps derived from GEONET data with 15 min interval following the Tohoku earthquake. The topography is plotted in this region from US National Geophysical Data Center as the background, in order to show the topography influence on TEC disturbances. Here, the TEC series obtained by GEONET dual frequencies GPS measurements are filtered with a 1–15 mHz window for detrending. The trend is mainly caused by SIP's motion and ionospheric background changes. Here, we chose the 15 mHz as the high cutoff frequency in consideration of the 30 s interval of the GEONET ionospheric monitoring, avoiding the aliasing in signal processing. Dramatic negative and positive disturbances appeared near the epicenter around UTC 6:00, which were mainly on the two sides of the rupture as shown in Fig. 8.3. The Tohoku earthquake was resulted from a thrust between the Pacific and North America plates. The disturbance in epicenter region seems to be related to sea level change caused by the thrust directly and the source of seismic ionospheric disturbances is located in the rupture region. The seismic ionospheric disturbance following the 2011 Tohoku earthquake derived by Butterworth filter agrees well with the result based on moving average detrend (Tsugawa et al. 2011).

The ionospheric disturbance spreads out from the epicenter as a quasi-circular propagation pattern with the time. The large amplitudes are up to 3–4 TECU as shown in Fig. 8.4. The strong disturbances last nearly 2 h from UTC 6:00–8:00, while in most areas the disturbance amplitudes are 0.1–0.4 TECU, which are much larger than the normal level. The disturbances with large amplitudes are occurred above the Sea of Japan, except the epicenter region. Above the Sea of Japan, the relatively strong disturbances are found at UTC 07:16 and UTC 07:46. Till UTC 08:16, the TEC recovers to the state before the main shock as shown by the first and last subfigures in Fig. 8.3. The TEC disturbance has a second enhancement around UTC 06:15–06:30 as shown by the series of TEC disturbance maps. More details about this point will be discussed in the next section.

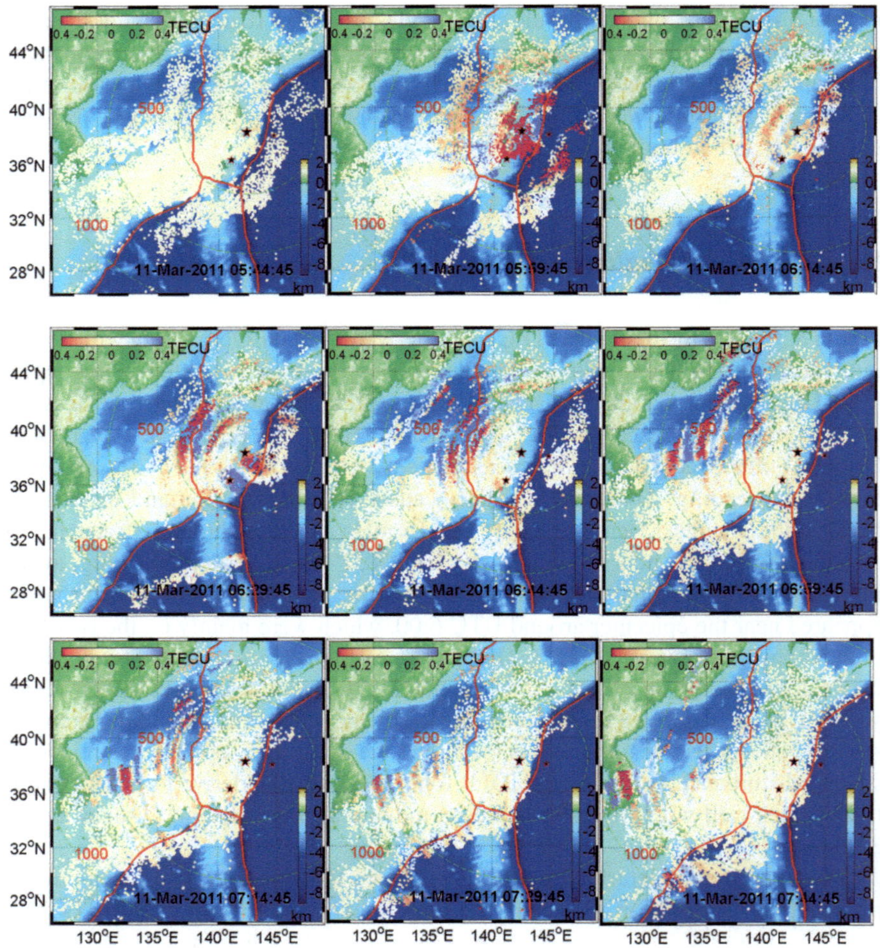

Fig. 8.3 Detrended TEC variations following the main shock in Japan with 15 min interval. The amplitude is shown in color. The three stars (the symbols' size is proportional to magnitude) stand for the location of the main shock epicenter and the two aftershocks with Mw > 7 during the 2011 Tohoku earthquake. The topographic data are provided by US National Geophysical Data Center (NGDC). Here, the 0 km is corresponding to the mean sea level. The red line is the plate boundary (Jin et al. 2007)

8.3.2 *TEC Disturbance Amplitudes*

Figure 8.4 is an example of TEC disturbance amplitudes following the Tohoku earthquake detected by the line-of-sight (LOS) between the station 2106 and GPS satellite 26. The left panel is the filtered TEC with different bandpass windows, and the right panel shows the observation geometry. Obvious disturbance appears several minutes after the main shock in all the four filtered TEC series (Jin et al.

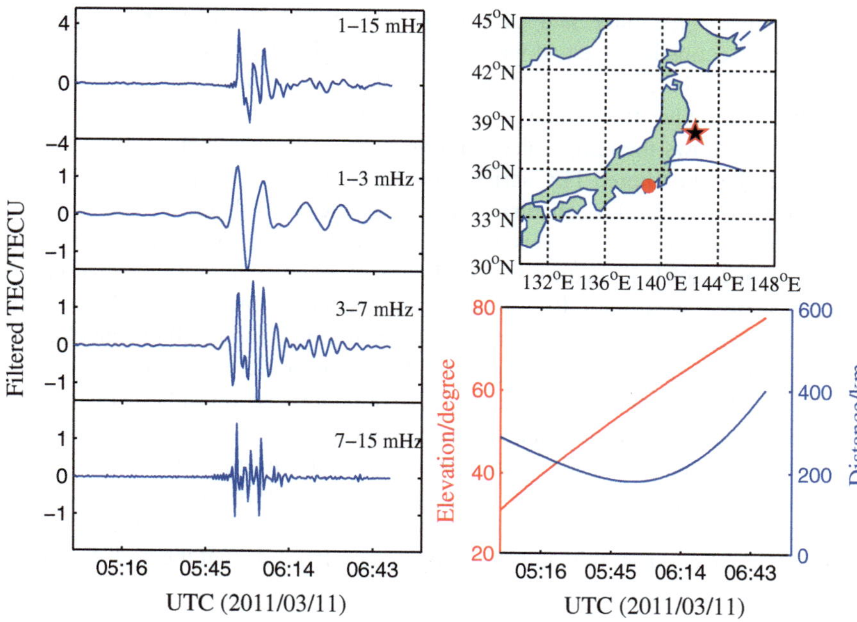

Fig. 8.4 TEC disturbances observed by the station 2106 with PRN 26 on March 11, 2013. The left panel is the filtered vertical TEC and the right panel describes the observation geometry. The blue line in the right-top panel is the SIP trajectory. The star and circle show the location of epicenter and the station 2016. The right-bottom panel shows the satellites' elevation angle (red line) and the geodetic distance between the SIP and epicenter (blue line)

2014). For the 1–15 mHz, it is the detectable frequency range of 30 s interval GPS measurement for ionospheric disturbance. Here, we chose 1 mHz as a threshold to remove the background TEC variation. As shown in the right panel of Fig. 8.4, the arc has a good view of the ionosphere following the Tohoku earthquake in the near field within 400 km away from the epicenter. The disturbance shows the dramatic characteristic like impulsion wave. The disturbance amplitude decreases rapidly with the increase of the time and the distance of SIP's location and the epicenter. The maximum amplitude is up to about 4 TECU for 1–15 mHz. Rolland et al. (2011) pointed out the LOS slant TEC with low elevation angle is more sensible to the seismic ionospheric disturbances. However, here for the vicinity of epicenter, the TEC series seem not to be positive correlation with the satellite elevation angle as shown in the right-bottom panel of Fig. 8.4. The distance and time variation should be the main factors for the disturbance amplitude and the disturbance even last until to UTC 8:00. The disturbance amplitudes are even larger than 1 TECU not only in 3–7 mHz but also the two other bands in the first half hour after the main shock. The ionospheric disturbance in 1–3 mHz has larger amplitude than the others about the second half hour, although the 3–7 mHzfiltered TEC series appear an second enhancement. The multi-frequencies' disturbances are induced after the earthquake in vicinity of the epicenter, and GPS ionospheric measurement near the epicenter can detect the acoustic wave and wave related to the Rayleigh wave (3–7 mHz) and tsunami-generated gravity wave (<3 mHz) (Matsumura et al. 2011; Occhipinti et al. 2013), but also higher frequencies signal (7–15 mHz).

Figure 8.5 shows the temporal variations of the mean value and the RMS (root mean square) of the filtered TEC series from the vicinity of the main shock epicenter to the far field. The focal mechanism solutions of the main shock and the two aftershocks with Mw > 7, whose onsets are UTC 05:46, UTC 06:15, and UTC 06:25, are marked by the beach balls in the upper subfigure. Their epicenters' location is marked in Fig. 8.3. Here, we do not discuss the signal between 7 and 15 mHz, because the magnitude of its mean amplitude variation in one region is too small in consideration of the GPS measurement noise. In general, the ionospheric variation in this area is recovered to the normal level at around UTC 08:00. Seismic ionospheric disturbance with 1–3 mHz has larger mean amplitude comparing to the one with 3–7 mHz. The average disturbance amplitude decreases dramatically during the ionospheric disturbance away from the epicenter. Compared to the disturbance in 1–3 mHz, the mean amplitude of disturbance in 3–7 mHz nearly recovers to zero for the region out of 300 km away from the main shock epicenter. It seems that the high-frequency

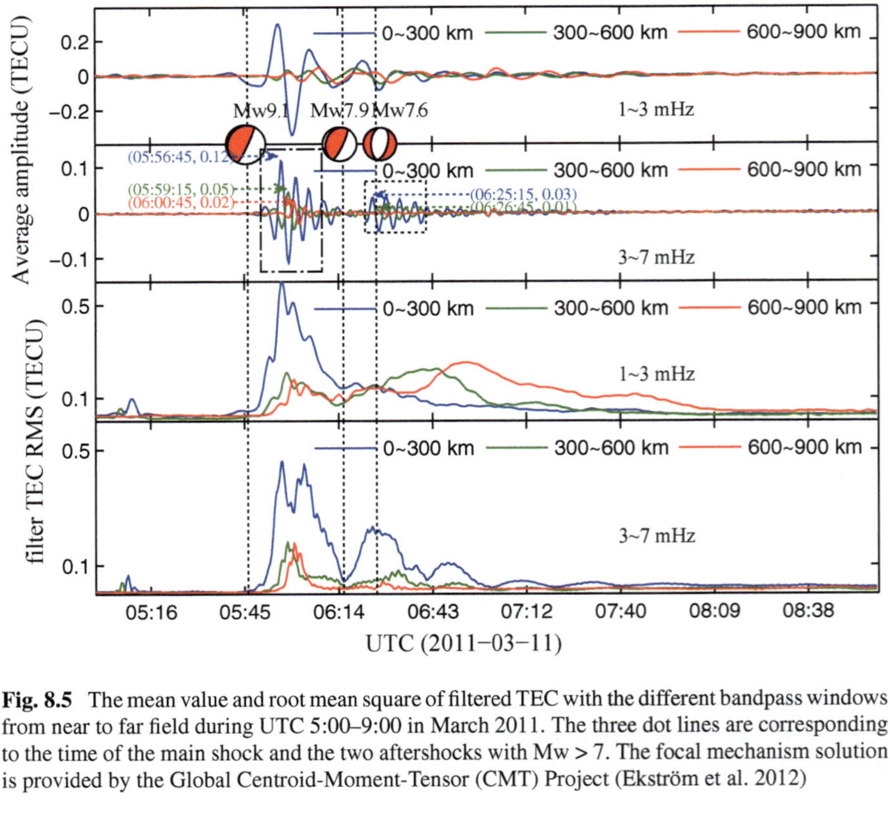

Fig. 8.5 The mean value and root mean square of filtered TEC with the different bandpass windows from near to far field during UTC 5:00–9:00 in March 2011. The three dot lines are corresponding to the time of the main shock and the two aftershocks with Mw > 7. The focal mechanism solution is provided by the Global Centroid-Moment-Tensor (CMT) Project (Ekström et al. 2012)

disturbance attenuates faster as its spreading out. It has been found that there is a second ionospheric disturbance with 3–7 mHz within 300 km away from the epicenter (almost covering the rupture region), which appears several minutes after the first giant aftershock (Mw7.9) and reaches the max in 10 min that is similar to the situation of the main shock ionospheric anomalies variation, as shown in the mean amplitude and RMS of the filtered TEC series.

As we know, ionospheric disturbances related to several types of wave, such as Rayleigh wave, acoustic wave, tsunami-generated gravity wave, etc., are probably induced following the giant earthquake (Ducic et al. 2003; Artru et al. 2004; Occhipinti et al. 2013). The Rayleigh and acoustic wave propagate with the speed of about 2–3 km/s and around 1 km/s (Heki et al. 2006). The propagation distance is far more than 300 km till to UTC 6:30. Figure 8.6 is the sea level change in vicinity of the epicenter recorded by Bottom Pressure Records (BPR). Two obvious disturbances exist in all three bottom ocean pressure residual series. The first one with a speed of 3071 m/s should be related to the seismic Rayleigh wave, while the second disturbance with a speed of 234 m/s should be related to the tsunami

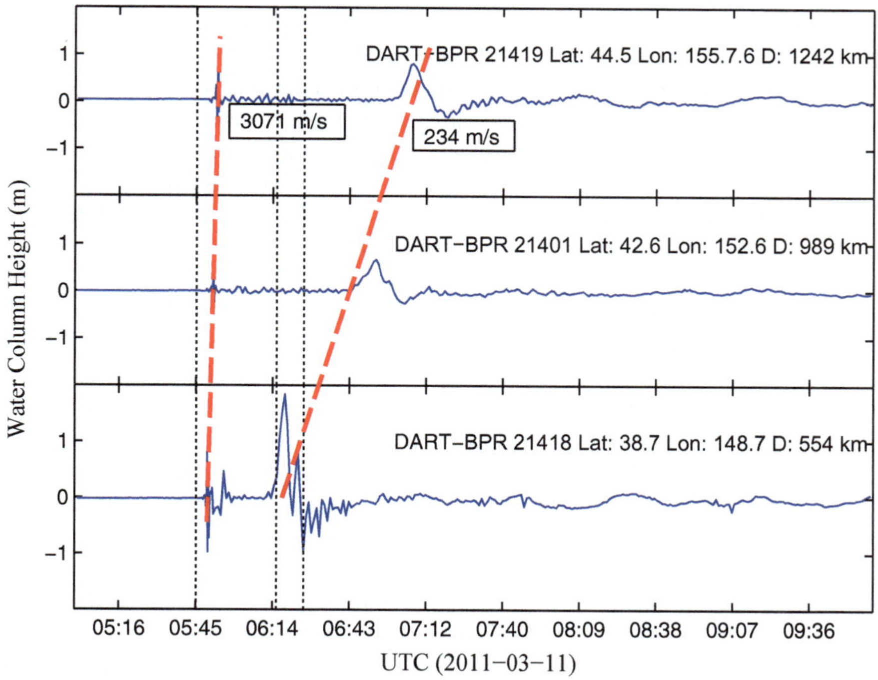

Fig. 8.6 The sea level variation recorded by three near-field Bottom Pressure Records (BPR).The data are provided by Deep-ocean Assessment and Reporting of Tsunami (DART) buoys (http://www.ndbc.noaa.gov/). The slops of the two red dash lines show the horizontal velocities (3071 and 234 m/s, based on linear fitting) of two disturbances. The three dot lines are same with the one in Fig. 8.5

propagation, which is the source of the tsunami-generated gravity wave in the open sea. The tsunami has been propagating to more than 500 km away from the epicenter. Although the tsunami-generated gravity wave along the tsunami propagates slowly, the tsunami signal can be visible in the ionosphere at least 500 km away from the epicenter according to Occhipinti et al. simulation (Occhipinti et al. 2013). Furthermore, from the perspective of TEC disturbance frequency, the tsunami-generated gravity wave is less than 3 mHz, while the Brünt-Vaïsalla frequency is in order of 3 mHz. The disturbances shown in the dotted rectangle, as shown in the second axis, should not be related to the three probable waves induced by the main shock directly. Considering the time shifts corresponding to peaks of 3–7 mHzregional average TEC disturbances series, the waves spread out from the near field to the far field with speeds of several kilometers per second, which are more agreement with the Rayleigh wave. The disturbances not only in the dash-dotted rectangle but also in the dotted rectangle propagate much faster than the tsunami gravity wave. When we assume the aftershocks' epicenter as the source for the second ionospheric disturbance, the situation is similar to propagating much faster than the tsunami gravity wave. It is difficult to distinguish the ionospheric effect induced by the two giant earthquakes because their onset interval is just 10 min. Note that the decease trends of the average amplitudes corresponding to the first and second disturbances presented in the blue line in the second subfigure of Fig. 8.5 show some subtle difference. The amplitude variation of the second one does not attenuate immediately but remain the high level till to almost 10 min (shown as the first to the third positive peak in the second ionospheric disturbance mentioned above) after the second giant aftershock (Mw7.6). Similar phenomenon is found in the filtered TEC RMS series for 0–300 km in 3–7 mHz. The second giant aftershock (Mw7.6) should be also one probably source of the ionospheric disturbance marked by the dotted rectangle in Fig. 8.5. Therefore, the second ionospheric disturbance marked by a dotted rectangle in Fig. 8.5 is related to the giant aftershocks, especially for the earthquake with Mw7.9, which has the similar focal mechanism solution, while there is no such feature for 1–3 mHz. Besides the two biggest earthquakes, there is the amount of aftershocks following the main shock with Mw < 7, but the TEC disturbance amplitudes have no significant variation. The RMS of the filtered TEC amplitude becomes bigger in far field after UTC 6:30 as shown in Fig. 8.5. The feature may be related to two factors. One is that the ionosphere above the epicenter recovers to the quiet condition gradually with the time. The other one is that the directional diversity becomes dramatic for far field.

Another slight but visible anomalies show in the amplitude average and RMS series of 0–300 km in 1–3 mHz around UTC 05:10, which is similar to Heki et al. (2013) with an enhanced TEC in dozens of minutes before the giant earthquake main shock. However, this anomaly does not last to the onset of the earthquake main shock, but just exists with several minutes. Their results were based on the polynomial fitting residual of TEC series and our results are based on the filtering TEC series. The polynomial fitting residual will be affected by the whole arc's measurement. The post-seismic disturbances have a great impact on the polynomial fitting result, while our statistical results have no such effects. The pre-seismic TEC disturbances of the Tohoku earthquake are also found significantly, but are weak when compared to the post-seismic disturbances. However, the geomagnetic is not quiet on March 11 registered in Dst, Ap, AE index series as shown in Fig. 8.12. The relationship among the ionospheric disturbance, geomagnetic variation, and the giant earthquake still needs to be studied in the near future.

Figure 8.7 shows the relationship between the time and distances between the SIPs' location and epicenter (here the satellite elevation angle threshold is set as 30 degrees). The blue line is onset of the Tohoku earthquake in the Distance–UTC plane. The right bottom is the number of SIP of corresponding time and regions. The right-bottom panel in Fig. 8.7 shows SIP numbers at special epoch. Here, we use the absolute value of the filtered TEC to get a view of the attenuation of seismic ionospheric disturbance. For most areas during this period, ionosphere monitoring from GEONET is available, except a few epochs near the epicenter and 1500 km far from the epicenter. The strongest disturbances mainly appeared in the near field. The seismic ionospheric disturbance amplitudes decreased dramatically with the SIPs' location far away from the epicenter, especially in the near field. Obviously, disturbances are found till to UTC 09:00 and 1500 km far away of the epicenter, while the average filtered TEC for the region with 50 km interval is even more than 1 TECU and decreases to about 0.2 TECU as it spreads out of 500 km away from the epicenter. It is interesting to see that the amplitude attenuation becomes slower when the ionospheric disturbance spreads out of 500 km. From 500 to 1000 km, the average amplitudes show no dramatic decrease but just have the time delay.

In general, the seismic ionospheric disturbances following the Tohoku earthquake are attenuated with the increase of the time and the distances between SIP and epicenter. The amplitudes can reach up to 1–4 TECU. Near the epicenter, the disturbances are triggered by not only the main shock but also the two strongest aftershocks. The average disturbance amplitudes decrease rapidly from epicenter to 500 km away, while the decrease becomes much more slowly at 500–1500 km region.

8.3.3 TEC Disturbances Speed

The propagation characteristics of post-seismic ionospheric disturbances are further investigated following the Tohoku earthquake (Jin et al. 2014). Since the TEC disturbances in the southeast part and the epicenter area have been discussed (Galvan et al. 2012; Occhipinti et al. 2013; Shinagawa et al. 2013), we will focus on the northwest part with dense observations and the strongest disturbances except near the epicenter. The SIPs with azimuth angles of 260–350° and satellite elevation angles of 20–90° are used. Figure 8.8 shows the propagation patterns corresponding to Rayleigh wave at 3.18 km/s, acoustic wave at 0.86 km/s, and gravity wave at 0.1–0.3 km/s. The disturbance amplitude is described by the scaled color. The disturbances of gravity wave have the longest duration. When the disturbances spread out by 200–400 km away, the disturbances are enhanced obviously. Moreover, the propagation velocities have a slight decrease in the area of 500–600 km as shown by waves a–e. The corresponding disturbance region just appears nearby the west coast of Japan as shown in

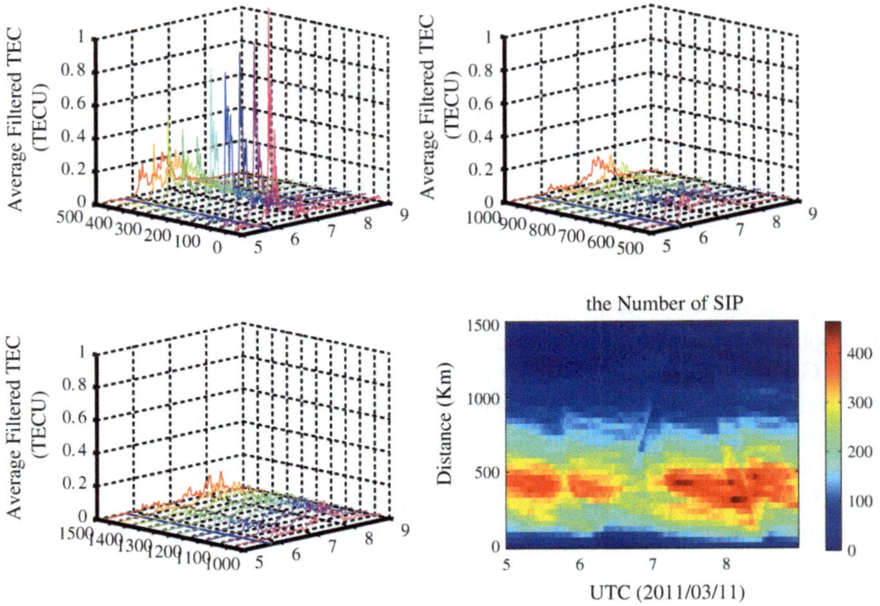

Fig. 8.7 Average amplitudes of the absolute filtered TEC variations with the distance to the epicenter during UTC 05:00–09:00, 0–500 km (top left); 500–1000 (top right); and 1000–1500 km (bottom left). The blue line is the onset of the Tohoku earthquake in the Distance–UTC plane. In the right bottom, the color scale is the number of SIP at corresponding time and regions

the bottom-right figure. The disturbances in this area are caused by the persisting of tsunami-generated gravity waves beyond the east coast of Japan (Galvan et al. 2012). According to the Occhipinti's (2013) simulation, the horizontal velocity of tsunami-generated wave has a deceleration. Here, the occurrence time of slowing horizontal velocity (50–100 min) and the region (400–600 km away from the epicenter) just are corresponding to the time and region of the tsunami-generated wave propagating to the altitude of around 125 km (the isothermal hypothesis ceases need to be valid), where the horizontal velocity of tsunami-generated wave has an strong deceleration. The slowing propagation of the TEC disturbance detected by the GPS ionospheric measurements in the northwest direction should be related to the effect of gravity wave upward propagation. An amplitude enhancement of the ionospheric disturbance appears around 200–400 km away from the epicenter. The amplitude variations along one stripe, which should be induced by the same disturbance source, do not simply attenuate as its spreading out, especially for the stripe b, whose amplitudes show no obvious attenuation even at 1400 km far away. As we know, the electron density increases dramatically from the ground to 300 km altitude. Although the variation of atmospheric mass density will attenuate, the electron density variation could be still dramatically because the electron density increases as the wave up propagation. Note that the tsunami-generated wave in deeper ocean has a higher propagation speed (Occhipinti et al. 2013). From wave a–e, which are corresponding to different

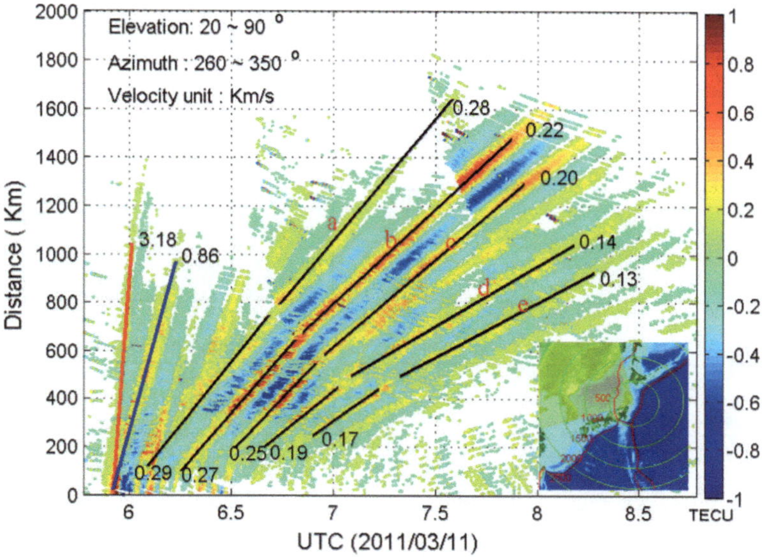

Fig. 8.8 Travel time diagram of post-seismic TEC disturbances following Tohoku earthquake on March 11, 2011. The SIPs with azimuth angles of 260–350° and satellite elevation angles of 20–90° are used. The disturbance amplitude is described by the scaled color. The bottom-right figure shows the SIPs location area with a yellow sector patch. The red circles are equidistance line to the epicenter corresponding to 500–2500 km with 500 km interval

tsunami waves, the horizontal velocities decrease gradually should be related to the tsunami's propagation from the source to the coast.

8.3.4 *TEC Disturbances Spectrogram*

Figure 8.9 presents TEC disturbances time series and spectrograms around the epicenter from UTC 05:00 to UTC 09:00. Here, the one-sided normalized Power Spectral Density (PSD) is computed using the short-time Fourier transform. The length of the window is set as 30 min. In the left-upper panel, the star is the location of epicenter and the circle is the location of corresponding GEONET station. The black line is SIP's tracks and the triangle is the SIP location at the onset of the Tohoku earthquake. The paralleled horizontal dash lines are the reference lines between 0.1 and −0.1 TECU. Since the interval of GEONET observation is 30 s, we could detect the signal with a frequency of less than 16.7 mHz according to Nyquist theorem. In general, the frequency with the largest power spectral density is around 1.3–1.8 mHz (about 10 min period) for all the TEC time series corresponding to different directions of the epicenter, and the signal duration is about dozens of minutes, although their propagation velocities are different. Another two signals with 3.83 and 5.83 mHz appeared

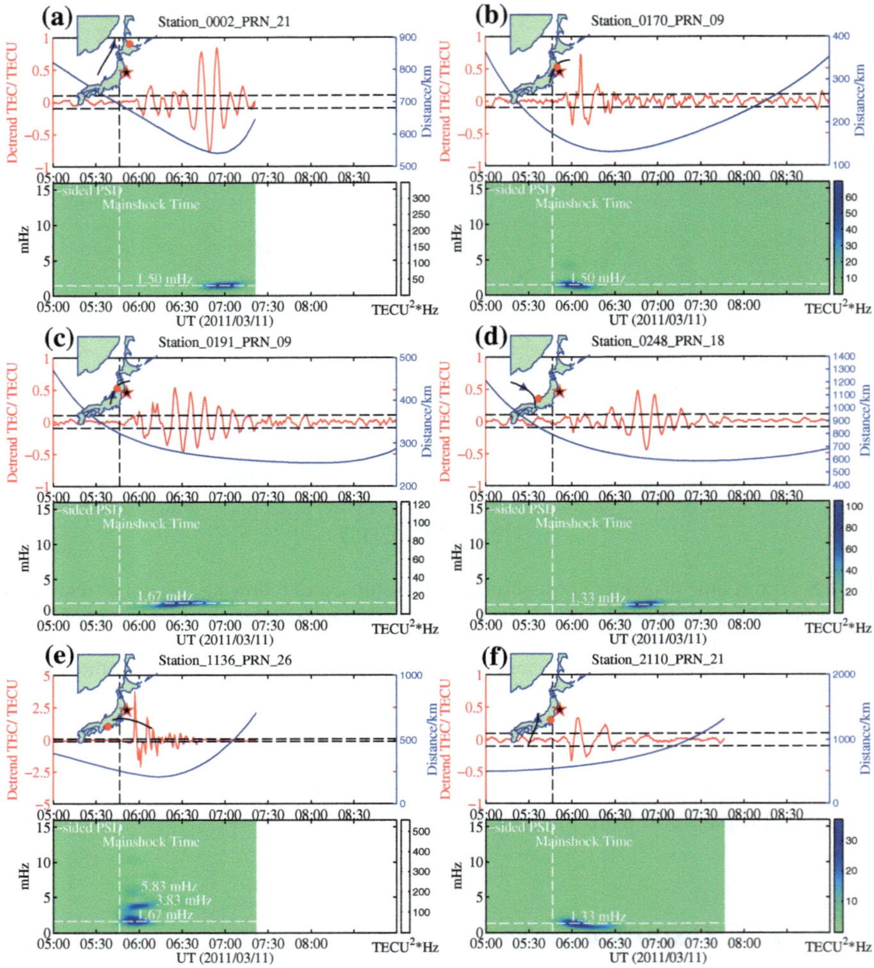

Fig. 8.9 TEC disturbances and spectrograms around the epicenter in several hours after the Tohoku earthquake. Here, the one-sided normalized Power Spectral Density (PSD) is computed using the short-time Fourier transform. The length of the window is set as 30 min. In the left-upper panel, the star is the location of epicenter and the circle is the location of corresponding GEONET station. The black line is the SIP's tracks and the triangle is the SIP location at the onset of the Tohoku earthquake. The paralleled horizontal dash lines are the reference lines between 0.1 and −0.1 TECU

in the time series, which are the TEC disturbances near the epicenter. While in the far field, there are no signals with these two frequencies. Therefore, the high-frequency disturbances are attenuated with its propagation.

8.3.5 TEC Disturbances Directivity

Figures 8.10 and 8.11 show the average amplitudes of the ionospheric disturbances following the Tohoku earthquake in different directions with respect to the epicenter at 30 min' interval from UTC 05:46 to UTC 07:46. The radius of the circles stands for the average amplitude. In the near field (0–500 km), stronger disturbances appear above the open ocean in the first half hour, while in the next hour, the largest amplitude is in the northwest direction. The average amplitudes decrease to about 0.05 TECU at UTC 7.3–7.8. The disturbances in southeast are related to the earthquake-generated tsunami (Galvan et al. 2012). The amplitudes in southeast decrease sharply at half an hour later, which may be related to the tsunami spreading out of 500 km in the open ocean, while no tsunami drives after about half an hour in the near field as shown in Fig. 8.6. In the northwest direction, the amplitudes reach up to the maximum value in the second half hour. Tsunami horizontal velocities decrease with the increase of the oceanic depth (Occhipinti et al. 2013). Unlike the southeastward tsunami, the

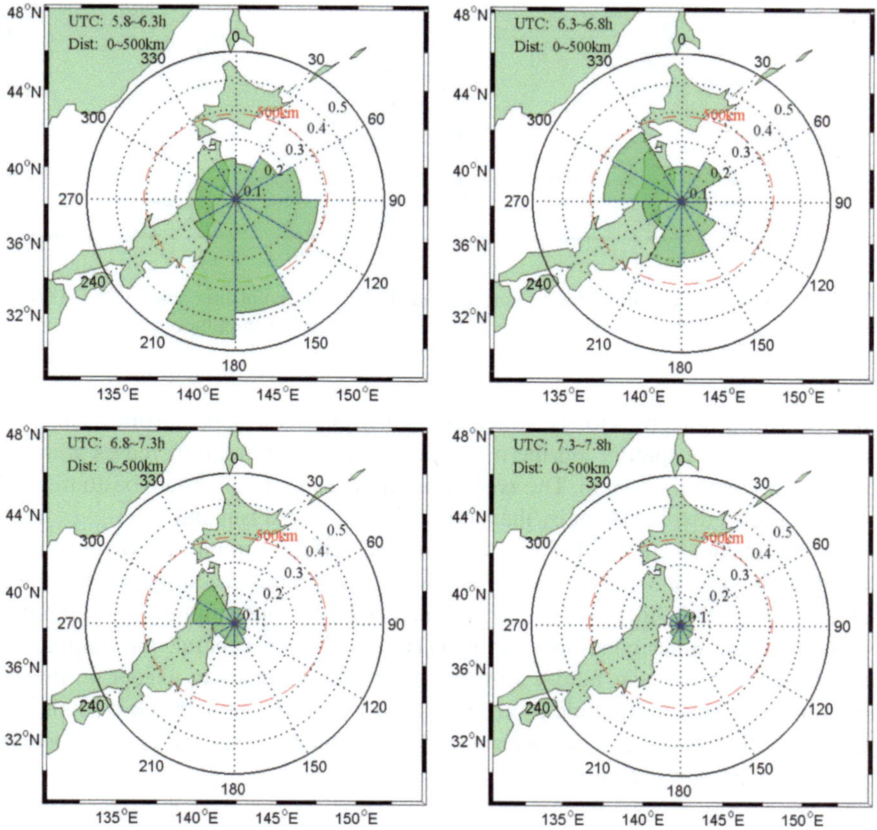

Fig. 8.10 Average amplitude of the absolute detrended TEC in different directions in 0–500 km from the epicenter. Here, the radius of the sector patch stands for the average amplitude

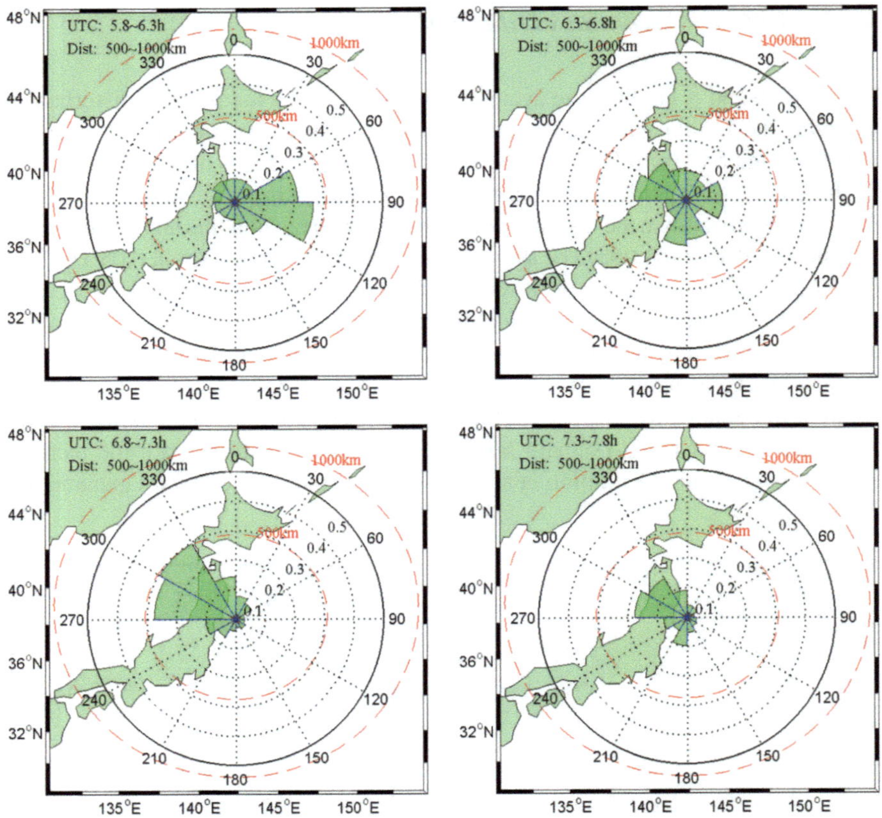

Fig. 8.11 Average amplitude of the absolute detrended TEC in different directions in 500–1000 km from the epicenter. The radius of the sector patch shows the average amplitude

northwestward tsunami velocity, which meets just half of the opposite direction, goes down with its propagation. Furthermore, the tsunami amplitude increases with the decrease of the oceanic depth. The average amplitude variations should be dominated by the sea level displacement by the earthquake-generated tsunami.

For the far-field area with 500–1000 km, the variation trend of most disturbed directions is similar with the near field, while amplitudes reach up to maximum value in UTC 6.8–7.3 in northwest direction, and the average amplitude is still much higher than the quiet condition during UTC 7.3–7.8. In general, the amplitudes are smaller than the near field with the propagation attenuation, except northwest direction. The large amplitude in northwest direction for 500–1000 km should be persisted obliquely due to upward propagation of tsunami-generated gravity wave as discussed in previous section.

8.4 Effects and Discussions

As we know, the geomagnetic and solar activities affect the Earth's ionosphere ionization and dynamics. Figure 8.12 shows the geomagnetic index and solar radio flux variations in March 2011. Here, the Dst (a) and AE (b) index series in March 2011. The Dst index is derived with the H components of geomagnetic filed observed by four magnetic observatories located at tropics nearby where are sufficiently far from the auroral and equatorial electrojets. The AE index is geomagnetic H components derived with 12 stations located in auroral zone, which is the indicator of electric currents flowing in the high-latitude ionosphere. The Ap is the linear scale of Kp index and monitors geomagnetic disturbance at subauroral region. Solar radio flux of 10.7 cm reflects the solar activities in corresponding days. The Dst, AE, and Ap index have some disturbances during the Tohoku earthquake. Heki et al. (2013) checked the TEC series and found no disturbances at hourly or minutely level on March 11, 2011. All Dst, AE, and Ap showed similar disturbances around the earthquake

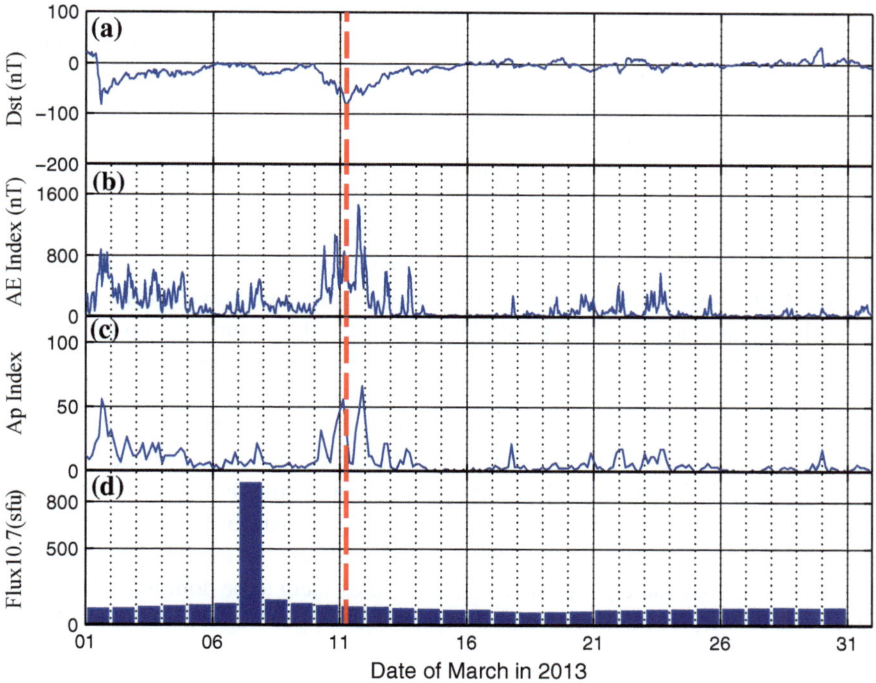

Fig. 8.12 Geomagnetic index and solar radio flux at the wavelength of 10.7 cm variation in March 2013. The Dst (**a**) and AE (**b**) index series in March 2011 are provided by World Data Centre (WDC) for Geomagnetism, Kyoto (http://wdc.kugi.kyoto-u.ac.jp), The Ap (**c**) index is from GeoForchungsZentrum (GFZ) (http://www.gfz-potsdam.de/), and solar radio flux (**d**) at the wavelength of 10.7 cm is provided by Space Physics Interactive Data Resource (SPIDR) (http://spidr.ngdc.noaa.gov)

days, indicating that the geomagnetic disturbances are not regional effects. However, almost all TEC anomalies are found at hourly or minutely level, which are related to the 2011 Tohoku earthquake. As for the relationship between the geomagnetic disturbances and the Tohoku earthquake will be further discussed in the future. The solar activity is quiet on the earthquake day and the TEC response to the solar activities is instantaneous (Mansoori et al. 2013).

8.5 Summary

In this chapter, we investigate the seismic ionospheric disturbance characteristics following the Tohoku earthquake in detail using the GPS data from GEONET, including disturbance amplitude, propagation speed and direction, spectrogram, and evolution. Except near the epicenter, the most disturbed region is located in the northwest, mainly above the Sea of Japan. The disturbed TEC almost recover to the pre-earthquake state in 2 h after the main shock. The post-seismic ionospheric effects are related to not only the main shock but also the two strongest aftershocks. This point may be meaningful for us to seek a more realistic source function of the seismic ionospheric effect that not only considers the effect of the main shock but also the aftershocks. A slight disturbance appeared before the onset of Tohoku earthquake in the near field, which may be a precursor of the giant earthquake, but still need further validate. Disturbance amplitude decreases rapidly in 0–500 km far from the epicenter. The ionospheric effects following the Tohoku earthquake are attenuated with the increase of the time and distance between the SIP and the epicenter, especially in the near field. The disturbance spreads out with three patterns, which are related to the Rayleigh wave, acoustic wave, and tsunami-generated wave. In the northwest area, the disturbances are enhanced obviously when the disturbances propagated to 200–400 km, and the propagation velocities decrease in the area of 500–600 km, which are related to the decrease of its horizontal velocities as upward propagation from the ground to the ionosphere. The frequency of detrended TEC perturbation with the largest PSD is 1.3–1.8 mHz. Signals with higher frequencies (3.83 and 5.83 mHz) are existed in TEC variation at the epicenter region, while do not appear in the far field, which may be caused by the attenuation with the disturbance propagation. The most disturbed direction is the northwest and southeast, which is perpendicular to the rupture approximately. The directional features seem to be dominated by the sea level vertical displacement caused by the earthquake-generated tsunami in the near field.

References

Afraimovich, E. L., Perevalova, N. P., Plotnikov, A. V., & Uralov, A. M. (2001). The shock-acoustic waves generated by earthquakes. *Annales Geophysicae, 19,* 395–409.

Afraimovich, E. L., Ding, F., Kiryushkin, V. V., Astafyeva, E. I., Jin, S., & Sankov, V. A. (2010). TEC response to the 2008 Wenchuan earthquake in comparison with other strong earthquakes. *International Journal of Remote Sensing, 31*(13), 3601–3613.

Artru, J., Lognonné, P., & Blanc, E. (2001). Normal modes modelling of post-seismic ionospheric oscillations. *Geophysical Research Letters, 28*(4), 697–700.

Artru, J., Farges, T., & Lognonné, P. (2004). Acoustic waves generated from seismic surface waves: Propagation properties determined from Doppler sounding observations and normal-mode modelling. *Geophysical Journal International, 158*(3), 1067–1077.

Calais, E., & Minster, J. B. (1995). GPS detection of ionospheric perturbations following the January 17, 1994, Northridge earthquake. *Geophysical Research Letters, 22*(9), 1045–1048.

Dautermann, T., Calais, E., Haase, J., & Garrison, J. (2007). Investigation of ionospheric electron content variations before earthquakes in southern California, 2003–2004. *Journal of Geophysical Research: Solid Earth, 112*(B2).

Ducic, V., Artru, J., & Lognonné, P. (2003). Ionospheric remote sensing of the Denali Earthquake Rayleigh surface waves. *Geophysical Research Letters, 30*(18).

Ekström, G., Nettles, M., & Dziewonski, A. M. (2012). The global CMT project 2004–2010: Centroid-moment tensors for 13,017 earthquakes. *Physics of the Earth and Planetary Interiors, 200–201,* 1–9. https://doi.org/10.1016/j.pepi.2012.04.002.

Galvan, D. A., Komjathy, A., Hickey, M. P., Stephens, P., Snively, J., Tony Song, Y., et al. (2012), Ionospheric signatures of Tohoku-Oki tsunami of March 11, 2011: Model comparisons near the epicenter. *Radio Science, 47,* RS4003. https://doi.org/10.1029/2012rs005023.

Heki, K., & Enomoto, Y. (2013). Preseismic ionospheric electron enhancements revisited. *Journal of Geophysical Research: Space Physics, 118*(10), 6618–6626.

Heki, K., & Ping, J. (2005). Directivity and apparent velocity of the coseismic ionospheric disturbances observed with a dense GPS array. *Earth and Planetary Science Letters, 236*(3), 845–855.

Heki, K., Otsuka, Y., Choosakul, N., et al. (2006). Detection of ruptures of Andaman fault segments in the 2004 great Sumatra earthquake with coseismic ionospheric disturbances. *Journal of Geophysical Research: Solid Earth* (1978–2012), *111*(B9).

Jin, S. G., Park, P. H., & Zhu, W. Y. (2007). Micro-plate tectonics and kinematics in Northeast Asia inferred from a dense set of GPS observations. *Earth and Planetary Science Letters, 257*(3–4), 486–496. https://doi.org/10.1016/j.epsl.2007.03.011.

Jin, S., Luo, O. F., & Park, P. (2008). GPS observations of the ionospheric F2-layer behavior during the 20th November 2003 geomagnetic storm over South Korea. *Journal of Geodesy, 82*(12), 883–892.

Jin, S. G., Zhu, W. Y., Afraimovich, E. (2010). Co-seismic ionospheric and deformation signals on the 2008 magnitude 8.0 Wenchuan Earthquake from GPS observations. *International Journal of Remote Sensing, 31*(13), 3535–3543. https://doi.org/10.1080/01431161003727739.

Jin, S., Jin, R., Li, J. H. (2014). Pattern and evolution of seismo-ionospheric disturbances following the 2011 Tohoku earthquakes from GPS observations. *Journal of Geophysical Research: Space Physics, 119*(9), 7914–7927.

Kakinami, Y., Kamogawa, M., Tanioka, Y., Watanabe, S., Gusman, A. R., Liu, J. Y., et al. (2012). Tsunamigenic ionospheric hole. *Geophysical Research Letters, 39*(13).

Klobuchar, J. A. (1987). Ionospheric time-delay algorithm for single-frequency GPS users. *IEEE Transactions on Aerospace and Electronic Systems, 3,* 325–331.

Liu, J. Y., Chuo, Y. J., Shan, S. J., Tsai, Y. B., Chen, Y. I., Pulinets, S. A., et al. (2004). Pre-earthquake ionospheric anomalies registered by continuous GPS TEC measurements. *Annales Geophysicae, 22*(5), 1585–1593.

Liu, J. Y., Chen, C. H., Lin, C. H., Tsai, H. F., Chen, C. H., & Kamogawa, M. (2011). Ionospheric disturbances triggered by the 11 March 2011 M9.0 Tohoku earthquake. *Journal of Geophysical Research: Space Physics, 116*(A6).

Lognonné, P., Artru, J., Garcia, R., Crespon, F., Ducic, V., Jeansou, E., et al. (2006). Ground-based GPS imaging of ionospheric post-seismic signal. *Planetary and Space Science, 54*(5), 528–540.

Mansoori, A. A., Khan, P. A., Bhawre, P., Gwal, A. K., & Purohit, P. K. (2013, July). Variability of TEC at mid latitude with solar activity during the solar cycle 23 and 24. In *Proceeding of 2013 IEEE International Conference on Space Science and Communication* (pp. 83–87).

Matsumura, M., Saito, A., Iyemori, T., et al. (2011). Numerical simulations of atmospheric waves excited by the 2011 off the Pacific coast of Tohoku Earthquake. *Earth, Planets and Space, 63*(7), 885–889.

Occhipinti, G., Rolland, L., Lognonné, P., & Watada, S. (2013). From Sumatra 2004 to Tohoku–Oki 2011: The systematic GPS detection of the ionospheric signature induced by tsunamigenic earthquakes. *Journal of Geophysical Research: Space Physics*.

Rolland, L. M., Lognonné, P., & Munekane, H. (2011). Detection and modeling of Rayleigh wave induced patterns in the ionosphere. *Journal of Geophysical Research: Space Physics* (1978–2012), *116*(A5).

Saito, A., Tsugawa, T., Otsuka, Y., Nishioka, M., Iyemori, T., Matsumura, M., et al. (2011). Acoustic resonance and plasma depletion detected by GPS total electron content observation after the 2011 off the Pacific coast of Tohoku Earthquake. *Earth, Planets and Space, 63*(7), 863–867.

Shinagawa, H., Tsugawa, T., Matsumura, M., Iyemori, T., Saito, A., Maruyama, T., et al. (2013). Two-dimensional simulation of ionospheric variations in the vicinity of the epicenter of the Tohoku-oki earthquake on 11 March 2011. *Geophysical Research Letters, 40*(19), 5009–5013.

Tsugawa, T., Saito, A., Otsuka, Y., Nishioka, M., Maruyama, T., Kato, H., et al. (2011). Ionospheric disturbances detected by GPS total electron content observation after the 2011 off the Pacific coast of Tohoku Earthquake. *Earth, Planets and Space, 63*(7), 875–879.

Chapter 9
Two-Mode Seismo-ionospheric Disturbances

9.1 Introduction

Ionospheric disturbances could be excited by various natural or artificial events from the Earth's interior to the top of the atmosphere, such as volcanoes, earthquakes, geomagnetic storms, etc (Jin et al. 2007; Jin and Park 2007; Afraimovich et al. 2010, 2013; Aoyama et al. 2016; Dautermann et al. 2009; Grawe and Makela 2016). These natural events excite acoustic resonance between the Earth's surface and the atmosphere. Some of the resonance waves leak upward into the ionosphere and then trigger ionospheric disturbances (Shinagawa et al. 2007). Nowadays, Global Positioning System (GPS) can estimate atmospheric delays (Jin et al. 2004, 2006, 2016), which can be used to monitor atmospheric and ionospheric disturbances (Catherine et al. 2017; Jin et al. 2011). Seismic ionospheric disturbance (SID) is one of the most important ways to study and understand the interaction and coupling of solid Earth and the ionosphere. Although it is still very challenging to understand the mechanism and electrodynamic-atmospheric interactions in different Earth's layers following the earthquake, the earthquake-induced ionospheric anomalies can be monitored by GPS TEC (e.g., Calais and Minster 1995; Rolland et al. 2013; Jin et al. 2015, 2017a).

The ionospheric disturbance triggered by great earthquakes can be observed as ionospheric TEC anomalies (Heki et al. 2006; Liu et al. 2006). GPS is a powerful tool to observe the atmospheric or ionospheric response to the earthquakes, particularly for areas with dense continuous operating GPS stations. For example, the TEC anomalies following the earthquake were found and known as an N-shaped disturbance (Heki et al. 2006). Later, the TEC disturbances following earthquakes were found to be more complicated and have different modes with different features. For example, the seismic ionospheric perturbations following the great 2004 Sumatra earthquake were observed with the direct acoustic wave from the epicenter area, the secondary acoustic wave caused by the far-field Rayleigh wave, and the gravity wave induced by tsunami (Heki et al. 2006; Jin et al. 2015). The results of the 2011 Tohoku earthquake in Japan also confirmed seismic ionospheric perturbations with three different propagation speeds, namely, the acoustic waves (0.3–1.5 km/s),

S. Jin et al., *GNSS Atmospheric Seismology*,
https://doi.org/10.1007/978-981-10-3178-6_9

seismic Rayleigh waves (2–3 km/s), and tsunami-generated gravity waves (0.1–0.3 km/s) (Jin et al. 2015; Rolland et al. 2011; Saito et al. 2011).

The features of TEC anomalies were closely associated with the intensity and type of the earthquake (Astafyeva et al. 2014). Despite the amplitude, the waveform of the variations was more or less depended on the focal mechanisms. For example, Afraimovich et al. (2010) found strong N-shaped acoustic waves following the 2008 Wenchuan earthquake with a plane waveform. The features of seismo-ionospheric disturbances including velocities, propagating direction, and amplitude of the ionospheric anomalies triggered by great earthquakes were further studied (Jin et al. 2014; Vukovic and Kos 2017). However, the disturbance mode and the coupling between the earthquake and the ionosphere are still not clear. Studing disturbance characteristics and the coupling process will contribute to a comprehensive understanding of the procedure and the mechanism of earthquakes, and help human to mitigate the damage. Furthermore, the characteristics and behaviors of seismic ionospheric perturbations vary with the earthquake fault type, depth, and focal mechanism.

The 2005 Mw = 7.2 earthquake occurred off the coast of Northern California with strike-slip faulting and 10 km in depth at 02:50:54 (UT), June 15, 2005, about 146 km West of Crescent City, CA in the middle of the Gorda Plate to the west of the Cascadia Subduction Zone. This quake was widely felt along the northern California–southern Oregon coastline. The motion with a NE striking strike-slip fault is similar to other sequences that have occurred in this region in the past. This strike-slip faulting earthquake with dense GPS observations operated by the University NAVSTAR Consortium (UNAVCO) may provide a new chance to monitor and understand seismic ionospheric disturbance characteristics and modes. In this chapter, the seismic ionospheric disturbances with two modes are estimated following the 2005 Northern California offshore earthquake from the dense GPS measurements, and the features of the ionospheric perturbation mode are further studied and discussed as well as possible mechanisms.

9.2 Methods and Data

Using dual-frequency GPS observations, the precise ionospheric TEC could be extracted from combining code-phase measurements with ignoring the high-order ionospheric effects (Brunini and Azpilicueta 2009; Jin et al. 2014, 2017a). In order to estimate the temporal-spatial distribution of seismo-ionospheric disturbances, the slant TEC along the GPS Line Of Sight (LOS) is converted into the vertical TEC ($vTEC$) using the mapping function as

$$vTEC = TEC * \cos\left[\arcsin\left(\frac{R\sin z}{R+H}\right)\right] \tag{9.1}$$

where H is the thin shell height of the ionosphere (here H is set as 300 km), R is the Earth's radius, and z is the zenith distance of the Line Of Sight (LOS) from the receiver to GPS satellites. The points at the thin shell are referred to as Ionospheric Piercing Points (IPPs). The precise *vTEC* can be obtained using code-smoothing phase measurements, while ambiguity, differential code biases, and noises are estimated as constant (e.g., Jin et al. 2017b). For seismic ionospheric disturbances, it is important to obtain the relative ionospheric variations following the earthquake rather than the magnitude of absolute TEC, so here the relative TEC time series are obtained from high-precision GPS carrier phase observations. The constant terms are removed by a low-frequency filtering, e.g., instrument deviations and ambiguities of GPS carrier phase observations. In addition, cycle slips of GPS measurements are detected and repaired using the ionospheric residual method of time-difference phase observations and TurboEdit algorithm (Blewitt 1990). In order to get the obvious seismic ionospheric perturbations, the background TEC variations are removed using the Butterworth filter of a fourth-order zero-phase finite impulse. Since the acoustic cutoff frequency is 3 mHz at above 150 km altitudes and the Nyquist frequency is about 7 mHz for GPS observations with the sampling interval of normally less than 60 s, the passband is set from 3 to 7 mHz (less than the Nyquist frequency).

Here, 504 stations of continuous GPS measurements are collected from the University NAVSTAR Consortium (UNAVCO) (http://www.unavco.org/data/gps-gnss/gps-gnss.html) and used for studying seismic ionospheric perturbations following the 2005 Northern California Earthquake (see Fig. 9.1), which are all located in 1500 km far from the epicenter (41.2°N, 126.0°W). The sampling interval for all GPS stations observations is 15 or 30 s. Using these dense GPS observations, the highly temporal-spatial seismic ionospheric perturbations can be estimated from the filtered TEC series. In the following sections, the seismic ionospheric disturbance pattern and modes are investigated and discussed.

9.3 Two-Mode Seismo-ionospheric Disturbances

9.3.1 Co-seismic Disturbances

The ionospheric responses and disturbances following the 2005 Northern California offshore earthquake are obtained and analyzed using GPS TEC time series (Jin 2018). Figure 9.2 shows the filtered TEC maps from 03:00 UT to 03:09 UT with a time interval of 3 min. The black pentagram represents the location of the earthquake, the colorful solid-dot displays the position of the Sub-Ionospheric Point (SIP) at a certain point and the color values indicate the amplitude of the TEC disturbance or filtered TEC. The filtered TEC is obtained by using the Butterworth filter of the fourth-order zero-phase pulse with a 3–8 mHz window, which is to remove the ionospheric background variations and the trend of SIP's motion. The remaining TEC disturbances are mainly related to the earthquake (Tsugawa et al. 2011). After about 10 min of the main shock, the ionospheric anomalies are detected around the epicenter and 3 min later at 03:03 UT the anomalies became distinct. The TEC disturbances are

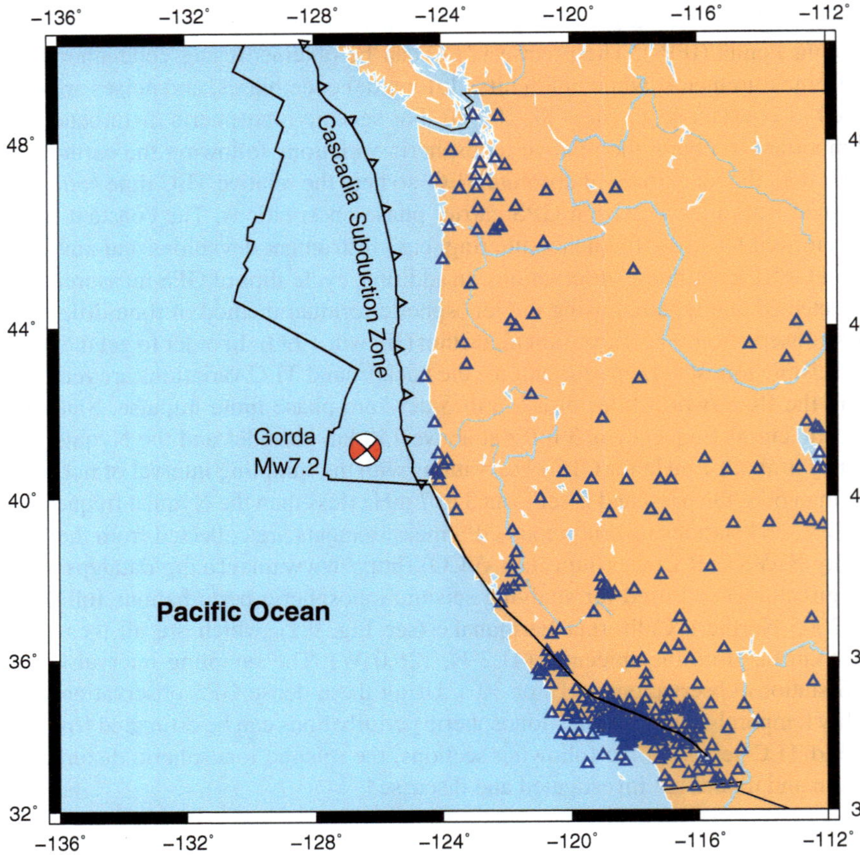

Fig. 9.1 The distribution of GPS stations with the epicenter location in pentagram

mainly located in the south of the epicenter and most of the disturbances appeared in a form of positive anomalies with positive amplitudes. Then the anomalies become stronger at 03:06 UT and the TEC disturbance reaches its maximum amplitude around this time point. An obvious feature is that positive anomalies occur in the vicinity of the epicenter, while negative anomalies are far from the epicenter. The TEC amplitude decreases at 03:09 UT and the TEC anomalies become less distinct. In order to better distinguish and show seismic ionospheric perturbations, the filtered TECs of higher than 0.02 TECU or less than −0.02 TECU are displayed by the last or first value color in the color bar. It can be seen that the amplitude near the epicenter is more than 0.1 TECU. The seismic ionospheric perturbation appears near the epicenter first and then quickly spreads out to 500 km in 15 min far from the epicenter. Also, most significant ionospheric disturbances are found in the south side, while no significant ionospheric disturbances are found in the north side, which may be due to the weak ionospheric disturbances or lower solid Earth and ionospheric coupling in the north side. This

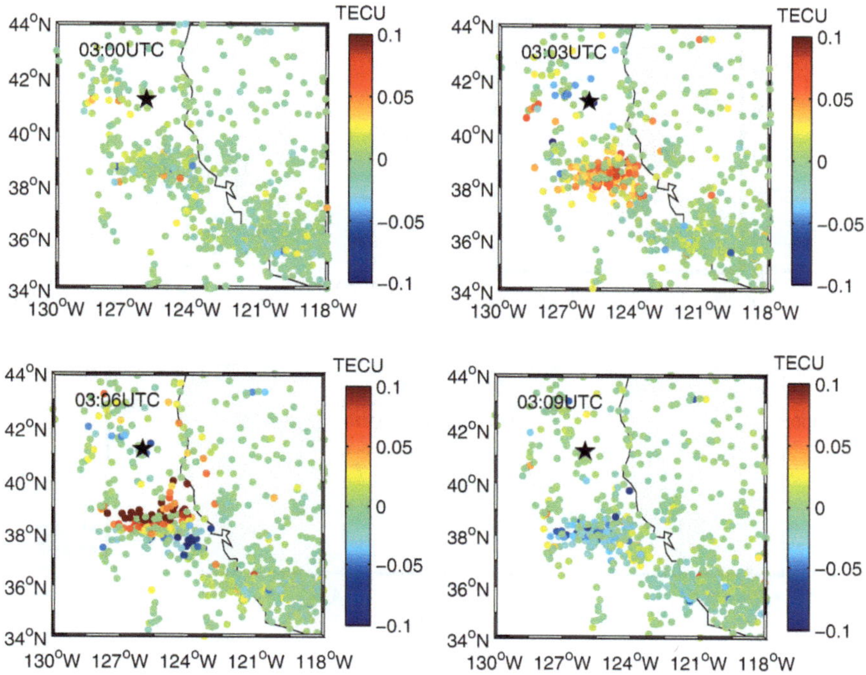

Fig. 9.2 Two-dimensional filtered TEC maps during 03:00–03:09 UT (Universal Time). The pentagram shows the location of the epicenter and the colorful solid-dot presents the position of the SIP. The filtered TEC amplitudes are colored using the color maps presented at the right

is also contributing to the northern attenuation observed in the TEC pattern from satellite 24 (Heki and Ping 2005). In addition, the local geomagnetic field in the north side may cause insignificant ionospheric disturbances. We estimate the ionospheric coupling factor using a radial wave vector with 10° zenith angle (Rolland et al. 2011). The ionospheric coupling factor shows that the ionospheric coupled wave is more attenuated in the north than in the south, indicating the southeastward directivity and the lack of signal in the north at near field (Afraimovich et al. 2010). However, it is still complex for the north–south anisotropy, which is mostly related to the combination of three factors: the seismic source, the geomagnetic field, and the observation geometry. In the future, it needs to further study with more real observations, particularly effect of nonuniform observation geometry of GPS stations (Occhipinti et al. 2008, 2013).

Figure 9.3 shows detailed features of TEC anomalies detected from PRN10. The left of the upper panel displays the SIP traces of each GPS station from 02:30 to 04:00 (UT), the blue dot represents the location of SIPs when the earthquake occurred, the dashed red curve shows the trace of SIPs, and the black pentagram is the epicenter. The distributions of SIP tracks are mainly located in the south of the epicenter with traveling from the north to the south during the selected period. The right of the upper panel displays the average filtered TEC in TECU with the distance, and the

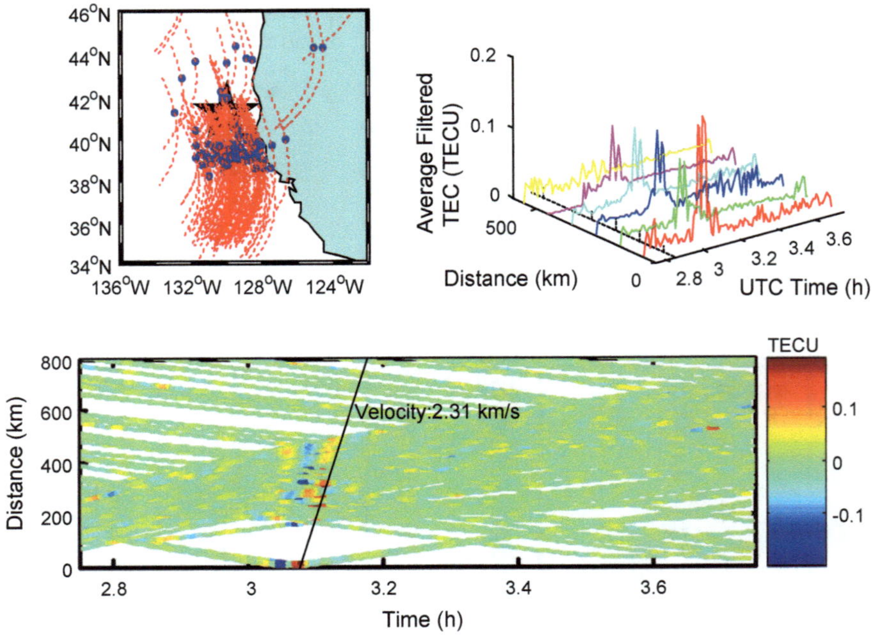

Fig. 9.3 Travel time diagram of seismic ionospheric disturbance following the Mw = 7.2 Northern California earthquake for GPS PRN10 satellite (Bottom panel). The left of the upper panel shows the SIP traces of each GPS station from 02:30 to 04:00 (UT), the blue dot represents the location of SIPs when the earthquake occurred, and the dashed red curve shows the trace of SIPs. The right of the upper panel displays the average filtered TEC with the distance

average amplitude is calculated from all the cases except that the amplitude is less than 0.04 TECU. The maximum of the average filtered TEC for each range appeared near 03:00 and the amplitude of average filtered TEC anomalies decreased with the distance away from the epicenter. The lower panel displays the travel time diagram for PRN08 and PRN10. The fitting speed from this diagram is 2.31 km/s.

Figure 9.4 shows IPP tracks at station P267 with PRN08 (a) and HOTK with PRN10 (b) from 02:30 to 04:00, filtered TEC time series, the elevation angle, and distance changes. The upper panel is the SIP trace from UT 02:30 to 04:00, the black pentagram represents the epicenter, and the red triangle displays the location of station P267 and HOTK. The elevation angle (in blue line) and the distance (in yellow line) between the epicenter and SIPs are also presented in the bottom panel. The TEC anomalies are clearly observed after about 12 min of the main shock in the form of typical N-shape. The smaller amplitude with about 0.04 TECU is found for station P267 and satellite 08 in the left panel with the elevation angle range of 40–80°, while the larger amplitude around 0.10 TECU is found for station HOTK and satellite 10 in the right panel with the elevation angle range of 15–25°. The GPS TEC observation with lower elevation angle is sensitive to co-seismic ionospheric disturbances induced by Rayleigh waves (e.g., Jin et al. 2017a).

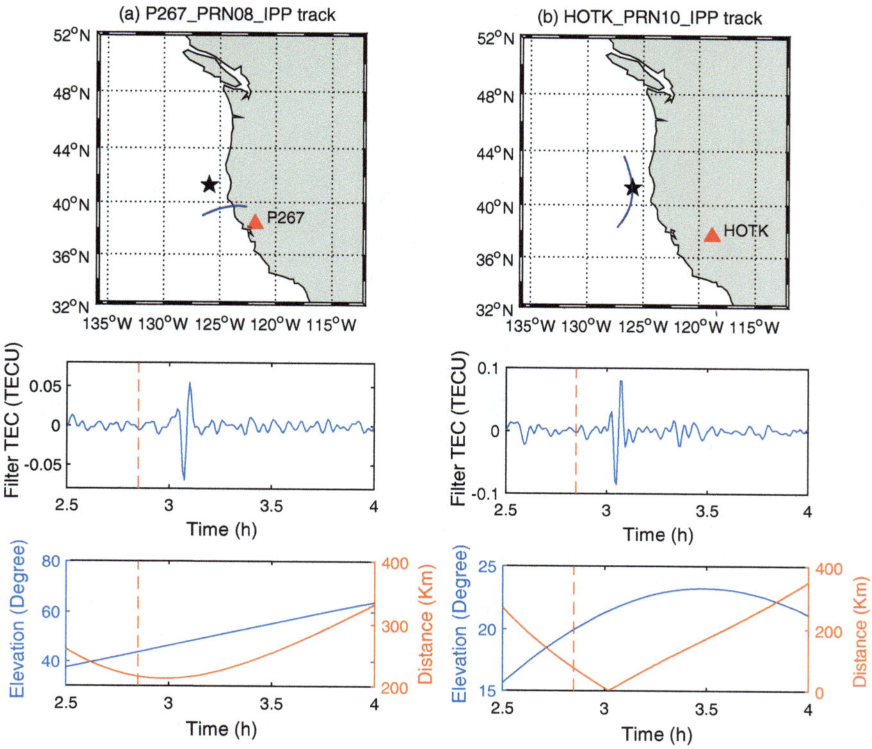

Fig. 9.4 IPP tracks at station P267 with PRN08 (**a**) and HOTK with PRN10 (**b**) from 02:30 to 04:00, filtered TEC time series, the elevation angle, and distance changes

9.3.2 Two-Mode Disturbances

The seismic ionospheric disturbances with the travel time and the epicentral distance have clear linear relationship, which may indicate different SID characteristics and sources. In order to know seismic ionospheric perturbations patterns and modes, the SID velocities following the 2005 Northern California offshore earthquake are estimated and investigated from the filtered TEC perturbations time series at different Ionospheric Pierce Points (IPPs). Here, the propagation speeds of seismic ionospheric perturbations are obtained using the linear fitting with the epicentral distance of the ionospheric TEC disturbance at maximum amplitudes following the time delay. Figure 9.5 shows the traveling graphs of filtered TEC time series following this earthquake for GPS PRN10 and PRN08 satellites. It is clearly seen that there are two linear relationships between the SID travel time and epicentral distance (Jin 2018). Two significantly distinguished ionospheric perturbation modes are found following this earthquake. One is the slow seismic ionospheric perturbation that is spreading at 1.51 km/s and the other is the fast seismic ionospheric perturbation at 2.31 km/s. Such ionospheric TEC perturbations are probably generated by the acoustic waves

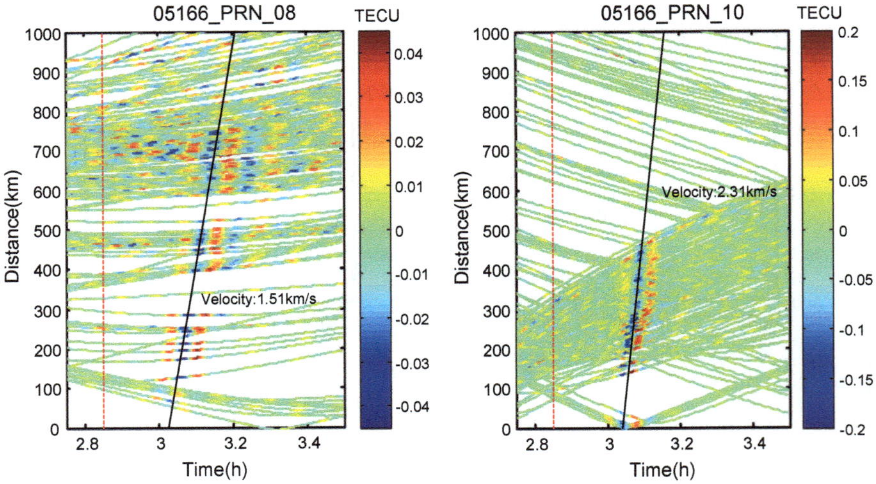

Fig. 9.5 Travel time diagrams of seismic ionospheric disturbance following the Mw = 7.2 Northern California earthquake for GPS PRN08 and 10 satellites. The horizontal axis represents the time from the beginning of the earthquake, i.e., 02:50:54 UT. The vertical axis represents the epicentral distance. Colors show the disturbance amplitudes extracted from GPS TEC series with Butterworth bandpass filter. The blue line is the first-order fitting line for the first peak disturbance

with velocity 500–1500 m/s and Rayleigh waves with velocity 2000–4000 m/s due to dynamic coupling.

The slow mode SID speed with 1.51 km/s is nearly similar to the acoustic velocity of about 1 km/s at 300–500 km height. The significant slow mode propagating perturbations are observed from nearby epicenter to 800 km away and reaches the first peak after 9.5 min of the earthquake, while the 9.5 min is close to the time of a pulse disturbance that propagates to the ionospheric level. Therefore, the slow-propagating mode of seismic ionospheric disturbances is the acoustic wave spreading at the height of the ionosphere, which is caused by the fault dislocation near the earthquake-rupturing region.

The fast mode of seismic ionospheric perturbations is significantly found in 10–20 min after this earthquake from the epicenter to 500 km away, whose velocity is 2.31 km/s. The fast-propagating mode signal has normally superimposed on the slow-propagating mode, which will separate with the increase of the epicentral distance (Astafyeva and Heki 2009). The fast and slow modes have the similar time delay after the earthquake, and therefore the upward propagation of the two-mode ionospheric disturbances should be the same. However, the fast mode propagating speed is close to that of the seismic Rayleigh wave propagation, but faster than the acoustic velocity at the ionospheric height. Therefore, the fast-propagating mode of ionospheric disturbances is more probably the up propagating secondary acoustic wave, which is induced by the seismic Rayleigh wave. Also, the fast-propagating Rayleigh wave disturbance and the slow-propagating acoustic wave disturbance are very clear within 500 km far from the epicenter. After about 500 km, the amplitudes of seismic ionospheric disturbances due to seismic Rayleigh wave and the acoustic wave decrease gradually with the decay of earthquake energy release.

In addition, the significant fast and slow modes SIDs are mainly located in the southeast of the epicenter (Figs. 9.2 and 9.5), particularly at the direction of about epicentral azimuth 140–220° (Jin 2018). Figure 9.6 shows the slip map of the 2005 Mw = 7.2 NC earthquake from California Integrated Seismic Network (http://www.cisn.org/special/evt.05.06.15). The slip solution matches the northeast–southwest trending fault plane with strike of 228°, which is consistent with the historic earthquake events and the distribution of the larger aftershocks in the Gorda plate. The slip appears to be unilateral with larger rupture in the SW. The seismo-ionospheric disturbances are almost parallel to the fault plane with large anomalies in the south, which is consistent with large ruptures of that region. Furthermore, the fast mode is observed at the elevation angles 15–25° of corresponding Line Of Sight (LOS), while slow mode is observed at the elevation angles 45–50° of corresponding Line Of Sight (LOS) (Fig. 9.7). As the GPS TEC is affected by lots of factors, GPS TEC time series are most sensitive to the perpendicular disturbances along the LOS. The fast mode of seismic ionospheric perturbations is the up propagating secondary acoustic wave triggered by the Rayleigh wave, so the low elevation angle LOS TEC is more sensitive to vertical ionospheric disturbance. Since the slow mode ionospheric disturbance is propagating horizontally as the acoustic wave, the high elevation angle LOS TEC is sensitive to horizontal ionospheric disturbance. Therefore, the fast mode is observed at the lower elevation angles of corresponding Line Of Sight (LOS), while the slow mode is observed at the higher elevation angles of corresponding Line Of Sight (LOS). The distribution of the GPS PRN10 and PRN08 elevation angles provide good observation geometry to distinguish the horizontal and vertical acoustic wave propagations. Therefore, the two modes of seismic ionospheric perturbations are clearly observed and distinguished in near-field GPS observations following the 2005 Northern California offshore earthquake. Unlike the 2004 Mw = 9.0 Sumatra–Andaman earthquake and the 2011 Mw = 9.1 Tohoku–Oki earthquake, a smaller magnitude and ground uplift occurred for this strike-slip earthquake. The seismic ionospheric effect is more similar to a point-source disturbance, which makes the two modes highly distinguishable in the near field. The distance distribution also has much difference between the fast mode and the slow mode. Figure 9.8 shows the distance distribution of the maximum filtered TEC of each time series for fast and slow modes, respectively. The red triangles represent the fast mode and the blue circles represent the slow mode. The maximum amplitude of the fast mode filtered TEC time series is much bigger than that of the slow mode. And most cases of the fast mode have a maximum amplitude larger than 0.04 TECU while most amplitudes of the slow mode are less than 0.04 TECU. In addition, the significant fast mode seismo-ionospheric disturbances are less in the range of 500–600 km away from the epicenter, while the slow mode ionospheric disturbances are still significant with more than 600 km far from the epicenter. The fast mode is the up propagating secondary acoustic wave triggered by the Rayleigh wave, so the seismo-ionospheric

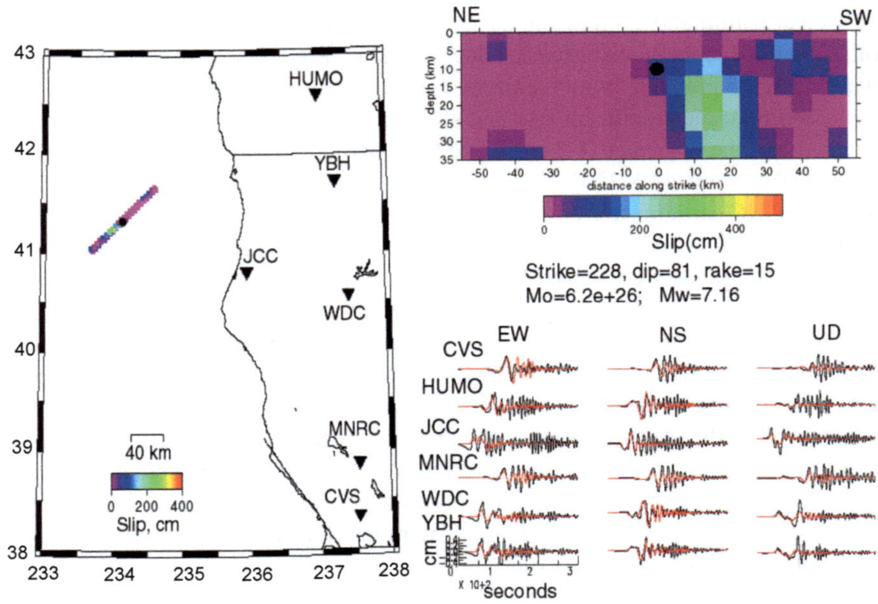

Fig. 9.6 Slip map for the 2005 Mw = 7.2 earthquake from California Integrated Seismic Network

disturbance ranges are smaller. The slow mode is propagating horizontally as the acoustic wave, which has larger range effects.

9.3.3 *Waveform and Spectral Analysis*

The distinct TEC disturbances are detected after the earthquake with PRN08 and PRN10. Figure 9.9 presents ten cases of each satellite to show the feature of the TEC anomalies. The left panel is for PRN08 and the right panel is for PRN10. The red dotted line shows the location of the main shock and the station name of each selected cases is located in the right of each panel. The significant TEC disturbances appear after about 10 min of the main shock and for most selected cases the anomalies last less than 15 min. The ionospheric anomalies appear in the form of an inverted N-shape instead of a regular N-shape, which indicated that the first TEC disturbances are negative anomalies instead of positive anomalies (Jin 2018). The polarity of the ionospheric anomalies is closely related to the mechanism and earthquake process.

Based on previous findings (Afraimovich et al. 2010), the typical polarity distribution of anomalies is often found following the reverse motion earthquakes or normal motion earthquakes. These two types of earthquakes generate vertical co-seismic crustal displacement. For most cases, the normal motion mainly causes subsidence of the surface, while the reverse motion causes uplift of the surface. For example,

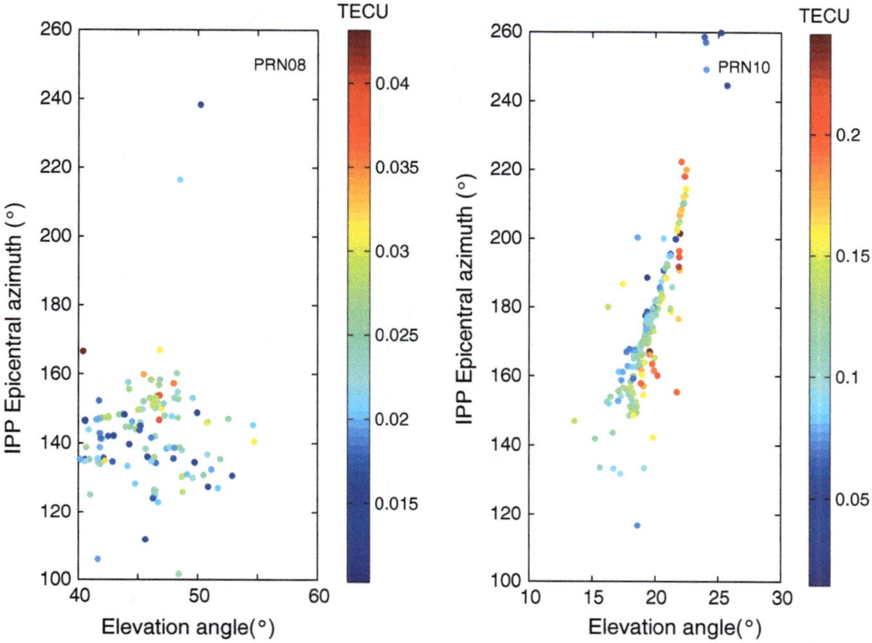

Fig. 9.7 IPP epicentral azimuth and corresponding LOS satellite elevation angles distribution for maximum filtered TEC in each GPS observation arc. Here, the undisturbed arcs (maximum filtered TEC < 0.01 TECU) are not considered

Fig. 9.8 The distance distribution of the maximum filtered TEC of each time series for fast and slow modes

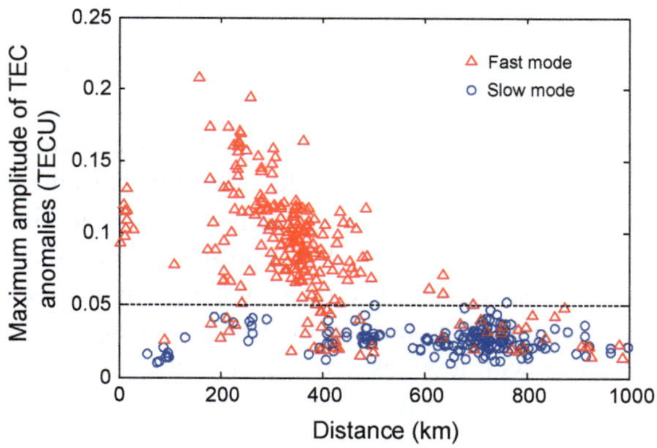

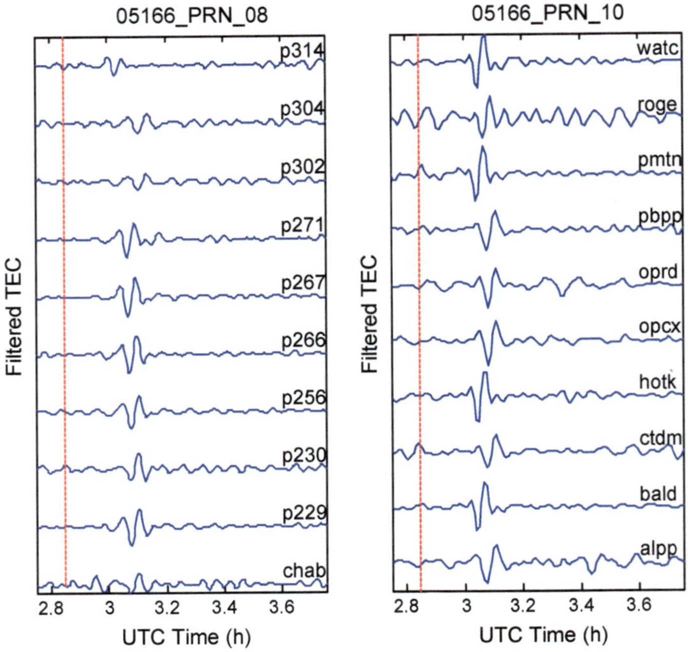

Fig. 9.9 Flitered TEC time series after the earthquake

Astafyeva and Heki (2009) found that there were both regular N-waves and inverted N-waves after the October 4, 1994 earthquake. The N-waves (positive changes) appeared in the southeast of the epicenter above the uplifted area, while the inverted N-waves (negative changes) appeared northwest of the epicenter above the area of subsidence. The type of the 2005 Northern California offshore earthquake was strike-slip motion earthquake, and the crustal displacement was mainly in the horizontal direction. Figure 9.10 shows the polarity distribution of the perturbations, and the red triangles and the blue dots represent the negative TEC change and positive TEC change, respectively. The major ionospheric anomalies are located in the south of the epicenter in the form of the negative changes. The polarities of ionospheric disturbances with less positive and more negative spreading are probably due to high and low acoustic wave velocities of the compression and rarefactions, respectively. The polarities of ionospheric disturbances may reflect co-seismic vertical crustal movements (i.e., uplift or subsidence). The negative changes are located in the subsidence areas, indicating that the ground subsidence induces co-seismic ionosphere disturbances starting with negative changes (Astafyeva and Heki 2009). Therefore, the co-seismic ionospheric disturbances may provide a potential of information on focal processes of earthquakes. However, there remain several physical problems in the formation and propagation of the inverted N-type waves, e.g., other geomagnetic field effects (Sunil et al. 2017).

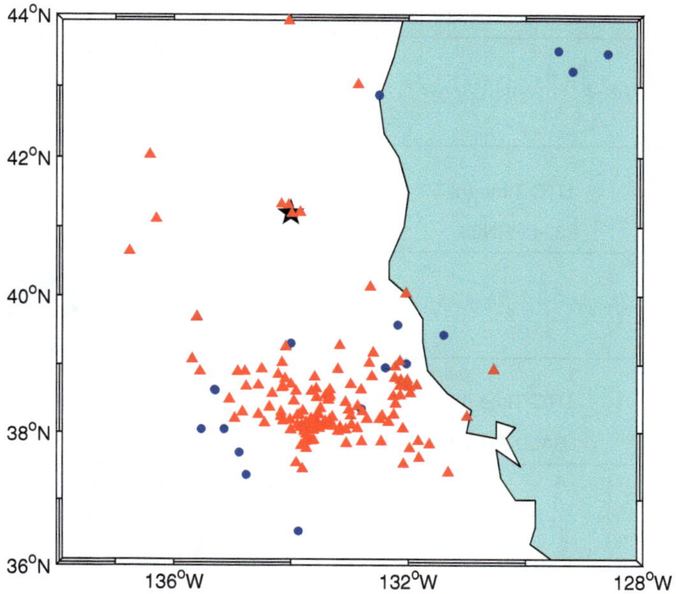

Fig. 9.10 The polarity distribution of seismo-ionospheric perturbations. The red triangles represent the negative anomalies and the blue dots represent the positive anomalies

Figure 9.11 shows the spectrograms of filtered TEC time series for selected stations and satellites. From up to bottom are PRN08 for station CHAB and station P304, PRN10 for station BALD and station ROGE. The left panel shows the filtered TEC time series in blue and distance time series in green, respectively, and the right panel displays the spectrograms of the corresponding cases. The frequency is centered at about 3.7 mHz for CHAB and P304, which is case of the slow speed mode, while the frequency centered at about 5.6 mHz for BALD and ROGE is case of the fast mode. The frequencies are all in the range of the infrasonic wave.

9.4 Discussion

The GPS TEC is used to detect the ionospheric anomalies following the 2005 Northern California offshore earthquake. The significant TEC disturbances are detected after about 10 min of the main shock. The maximum amplitude of the filtered TEC time series is around 0.2 TECU and generally the amplitude decreases with the distance between the epicenter and the SIPs. The ionospheric anomalies are mainly located in the south of the epicenter and last for less than half an hour for most cases. The distinct TEC disturbances are detected from satellite PRN08 and satellite PRN10 and two seismo-ionospheric disturbance modes are found. Furthermore, two modes

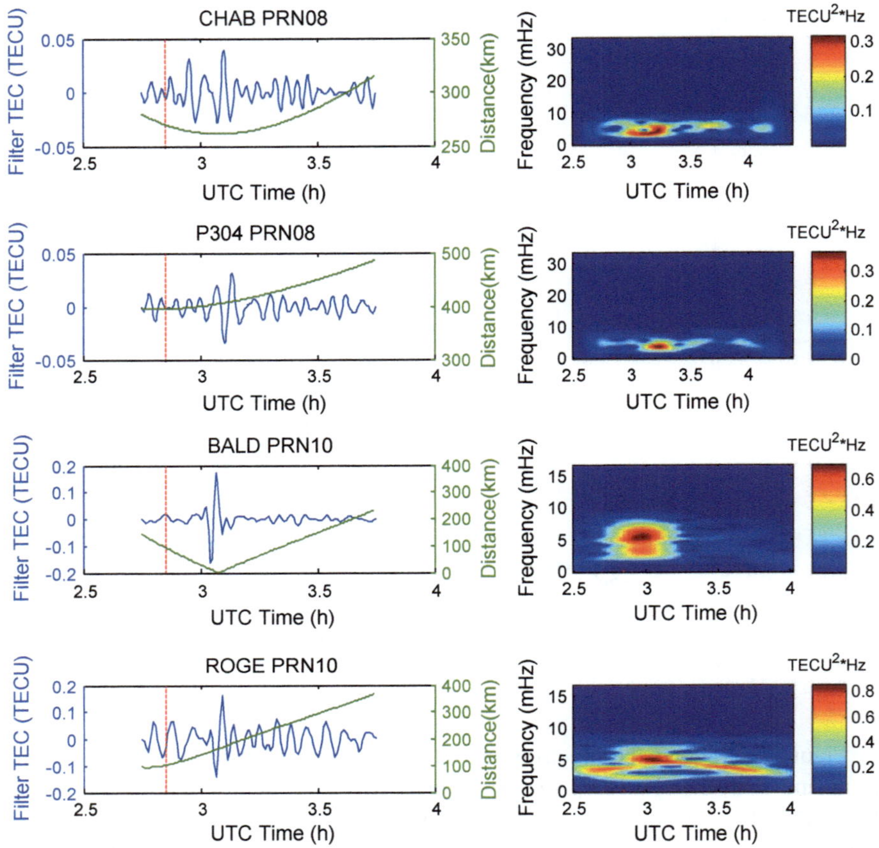

Fig. 9.11 Spectrograms for station CHAB of satellite PRN08, station P304 of satellite PRN08, station BALD of satellite PRN10, and station ROGE of satellite PRN10. The left panel displays the filtered TEC time series in blue and the distances between the earthquake and the IPP of corresponding stations and satellites in green

are compared in the velocity, amplitude, elevation angle or azimuth, and frequency. The TEC anomalies detected from satellite PRN08 travel at a propagating speed of 1.51 km/s, while the TEC anomalies from PRN10 are spreading at a velocity of around 2.31 km/s. The slow mode has an amplitude of 0.02–0.04 TECU, while the fast mode has a larger amplitude of 0.1–0.2 TECU. The most dramatic ionospheric disturbances are located at about 140° of epicentral azimuth, while the elevation angle is different for corresponding Line Of Sight (LOS). The elevation angles of LOS are 45–50° for the observed slow mode and about 15–25° for the fast mode, respectively.

The 2005 Northern California offshore earthquake was a result of strike-slip motion. According to our findings, the strike-slip earthquake can also trigger the ionospheric anomalies detected by GPS TEC. The detected fast mode and slow mode indicated that strike-slip earthquakes can cause different types of ionospheric disturbances. Although unlike reverse motion earthquakes or normal motion earthquakes, there was less vertical crustal displacement with about 2 m, the typical polarity distribution of the ionospheric perturbations is found (e.g., Astafyeva and Heki 2009).

As GPS satellites are moving and ground GPS stations are limited, the observation geometry of GPS plays a vital part in detecting seismic ionospheric anomalies. The seismic ionospheric disturbance modes can be observed from continuous GPS TEC time series. However, the observation geometry of GPS is not uniform, which may affect SID estimations. To better understand seismo-ionospheric perturbation characteristics and evolution, a more accurate morphology of ionospheric disturbances following the earthquakes should be further investigated with more GPS observations. With the rapid development of more and more GNSS constellations, e.g., China's BeiDou and updating Russia's GLONASS as well as European Union's Galileo, it will provide more chances to study and understand seismo-ionospheric perturbations in the future, including effects of the GNSS observation geometry on SID.

9.5 Summary

In this chapter, significant ionospheric disturbances following the 2005 Northern California offshore earthquake are observed about 10 min after the onset by denser GPS measurements with 504 stations. Two clear propagating modes of seismic ionospheric disturbances are found, namely, the fast-propagating mode with a speed of 2.31 km/s and the slow-propagating mode with a speed of 1.51 km/s. The fast mode of ionospheric disturbances is detected in the range of less than 500–600 km away from the epicenter, while the slow mode of ionospheric disturbances is found more than 600 km far from the epicenter during 10–20 min. The maximum amplitude of the fast mode TEC disturbances is much bigger than that of the slow mode. Most cases of the fast mode have a maximum amplitude larger than 0.04 TECU, while most amplitudes of the slow mode are less than 0.04 TECU. Furthermore, the frequency spectrogram of the slow-propagating mode is around 3.7 mHz, while typical frequency of the fast-propagating mode is about 5.3 mHz. In addition, the seismic ionospheric disturbances are much stronger in the southeast when compared to the northwest. The fast-propagating mode of ionospheric perturbations is the up propagating secondary acoustic wave induced by the seismic Rayleigh wave, while the slow-propagating mode of ionospheric disturbances is the horizontal acoustic wave induced by the focal dislocation.

References

Afraimovich, E. L., Ding, F., Kiryushkin, V. V., Astafyeva, E., Jin, S. G., & San'kov, V. (2010). TEC response to the 2008 Wenchuan earthquake in comparison with other strong earthquakes. *International Journal of Remote Sensing, 31*(13), 3601–3613. https://doi.org/10.1080/01431161003727747.

Afraimovich, E. L., Astafyeva, E. I., Demyanov, V. V., Edemskiy, I. K., Gavrilyuk, N. S., Ishin, A. B., et al. (2013). A review of GPS/GLONASS studies of the ionospheric response to natural and anthropogenic processes and phenomena. *Journal of Space Weather & Space Climate, 3,* A27. https://doi.org/10.1051/swsc/2013049.

Aoyama, T., Iyemori, T., Nakanishi, K., Nishioka, M., Rosales, D., Veliz, O., et al. (2016). Localized field-aligned currents and 4-min TEC and ground magnetic oscillations during the 2015 eruption of Chile's Calbuco volcano. *Earth, Planets and Space, 68*(1), 148. https://doi.org/10.1186/s40623-016-0523-0.

Astafyeva, E., & Heki, K. (2009). Dependence of waveform of near-field coseismic ionospheric disturbances on focal mechanisms. *Earth, Planets and Space, 61*(7), 939–943.

Astafyeva, E., Rolland, L. M., & Sladen, A. (2014). Strike-slip earthquakes can also be detected in the ionosphere. *Earth and Planetary Science Letters, 405,* 180–193.

Blewitt, G. (1990). An automatic editing algorithm for GPS data. *Geophysical Research Letters, 17*(3), 199–202.

Brunini, C., & Azpilicueta, F. J. (2009). Accuracy assessment of the GPS-based slant total electron content. *Journal of Geodesy, 83*(8), 773–785.

Calais, E., & Minster, J. B. (1995). GPS detection of ionospheric perturbations following the January 17, 1994, Northridge Earthquake. *Geophysical Research Letters, 22*(9), 1045–1048. https://doi.org/10.1029/95GL00168.

Catherine, J. K., Maheshwari, D. U., Gahalaut, V. K., Roy, P. N. S., Khan, P. K., & Puviarasan, N. (2017). Ionospheric disturbances triggered by the 25 April 2015 M7.8 Gorkha earthquake, Nepal: Constraints from GPS TEC measurements. *Journal of Asian Earth Sciences, 133,* 80–88.

Dautermann, T., Calais, E., & Mattioli, G. S. (2009). Global Positioning System detection and energy estimation of the ionospheric wave caused by the 13 July 2003 explosion of the Soufrière hills volcano. *Montserrat, Journal of Geophysical Research, 114*(B2), 311–320. https://doi.org/10.1029/2008JB005722.

Grawe, M. A., & Makela, J. J. (2016). The ionospheric responses to the 2011 Tohoku, 2012 Haida Gwaii, and 2010 Chile tsunamis: Effects of tsunami orientation and observation geometry. *Earth & Space Science, 2*(11), 472–483.

Heki, K., & Ping, J. (2005). Directivity and apparent velocity of the coseismic ionospheric disturbances observed with a dense GPS array. *Earth and Planetary Science Letters, 236,* 845–855. https://doi.org/10.1016/j.epsl.2005.06.010.

Heki, K., Otsuka, Y., Choosakul, N., Hemmakorn, N., Komolmis, T., & Maruyama, T. (2006). Detection of ruptures of Andaman fault segments in the 2004 great Sumatra earthquake with coseismic ionospheric disturbances. *Journal of Geophysical Research Solid Earth, 111*(B9), 535–540. https://doi.org/10.1029/2005JB004202.

Jin, S. G. (2018). Two-mode ionospheric disturbances following the 2005 Northern California offshore earthquake from GPS measurements. *Journal of Geophysical Research: Space Physics, 123*(10), 8587–8598. https://doi.org/10.1029/2017JA025001.

Jin, S. G., Cho, J., & Park, J. (2007). Ionospheric slab thickness and its seasonal variations observed by GPS. *Journal of Atmospheric and Solar-Terrestrial Physics, 69*(15), 1864–1870. https://doi.org/10.1016/j.jastp.2007.07.008.

Jin, S. G., Han, L., & Cho, J. (2011). Lower atmospheric anomalies following the 2008 Wenchuan Earthquake observed by GPS measurements. *Journal of Atmospheric and Solar-Terrestrial Physics, 73*(7–8), 810–814. https://doi.org/10.1016/j.jastp.2011.01.023.

Jin, S. G., Jin, R., & Li, J. H. (2014). Pattern and evolution of seismo-ionospheric disturbances following the 2011 Tohoku earthquakes from GPS observations. *Journal of Geophysical Research: Space Physics, 119*(9), 7914–7927. https://doi.org/10.1002/2014JA019825.

Jin, S. G., Jin, R., & Li, D. (2016). Assessment of BeiDou differential code bias variations from multi-GNSS network observations. *Annales Geophysicae, 34*(2), 259–269. https://doi.org/10.5194/angeo-34-259-2016.

Jin, S. G., Jin, R., & Li, D. (2017a). GPS detection of ionospheric Rayleigh wave and its source following the 2012 Haida Gwaii earthquake. *Journal of Geophysical Research: Space Physics, 122*(1), 1360–1372. https://doi.org/10.1002/2016JA023727.

Jin, S. G., Jin, R., & Kutoglu, H. (2017b). Positive and negative ionospheric responses to the March 2015 geomagnetic storm from BDS observations. *Journal of Geodesy, 91*(6), 613–626. https://doi.org/10.1007/s00190-016-0988-4.

Jin, S. G., Occhipinti, G., & Jin, R. (2015). GNSS ionospheric seismology: recent observation evidences and characteristics. *Earth-Science Reviews, 147,* 54–64.

Jin, S. G., & Park, J. U. (2007). GPS ionospheric tomography: A comparison with the IRI-2001 model over South Korea. *Earth Planets Space, 59*(4), 287–292. https://doi.org/10.1186/BF03353106.

Jin, S. G., Park, J., Wang, J., Choi, B., & Park, P. (2006). Electron density profiles derived from ground-based GPS observations. *Journal of Navigation, 59*(3), 395–401. https://doi.org/10.1017/S0373463306003821.

Jin, S. G., Wang, J., Zhang, H., & Zhu, W. Y. (2004). Real-time monitoring and prediction of the total ionospheric electron content by means of GPS observations. *Chinese Journal of Astronomy and Astrophysics, 28*(3), 331–337. https://doi.org/10.1016/j.chinastron.2004.07.008.

Liu, J. Y., Tsai, Y. B., Chen, S. W., Lee C. P., Chen, Y. C., Yen, H. Y., et al. (2006). Giant ionospheric disturbances excited by the M9.3 Sumatra earthquake of 26 December 2004 *Geophysical Research Letters, 33*(2), 356–360.

Occhipinti, G., Kherani, E. A., & Lognonné, P. (2008). Geomagnetic dependence of ionospheric disturbances induced by tsunamigenic internal gravity waves. *Geophysical Journal International, 173*(3), 753–765. https://doi.org/10.1111/j.1365-246X.2008.03760.x.

Occhipinti, G., Rolland, L., Lognonné, P., & Watada, S. (2013). From Sumatra 2004 to Tohoku-Oki 2011: The systematic GPS detection of the ionospheric signature induced by tsunamigenic earthquakes. *Journal of Geophysical Research Space Physics, 118*(6), 3626–3636.

Rolland, L. M., Lognonné, P., & Munekane, H. (2011). Detection and modeling of Rayleigh wave induced patterns in the ionosphere. *Journal Geophysical Research, 116,* A05320. https://doi.org/10.1029/2010JA016060.

Rolland, L. M., Vergnolle, M., Nocquet, J. M., Sladen, A., Dessa, J. X., F. Tavakoli, et al. (2013). Discriminating the tectonic and non-tectonic contributions in the ionospheric signature of the 2011, Mw7.1, dip-slip Van earthquake, Eastern Turkey. *Geophysical Research Letters, 40*(11), 2518–2522.

Saito, A., Tsugawa, T., Otsuka, Y., Nishioka, M., Iyemori, T., Matsumura, M., et al. (2011). Acoustic resonance and plasma depletion detected by GPS total electron content observation after the 2011 off the Pacific coast of Tohoku Earthquake. *Earth, Planets and Space, 63*(7), 863–867.

Shinagawa, H., Iyemori, T., Saito, S., & Maruyama, T. (2007). A numerical simulation of ionospheric and atmospheric variations associated with the Sumatra earthquake on December 26, 2004. *Earth, Planets and Space, 59*(9), 1015–1026.

Sunil, A. S., Bagiya, M. S., Catherine, J., et al. (2017). Dependence of near field co-seismic ionospheric perturbations on surface deformations: A case study based on the April, 25 2015 Gorkha Nepal earthquake. *Advances in Space Research, 59,* 1200–1208.

Tsugawa, T., Saito, A., Otsuka, Y., Nishioka, M., Maruyama, T., Shinagawa, H., et al. (2011). Ionospheric disturbances detected by GPS total electron content observation after the 2011 Tohoku earthquake. AGU Fall Meeting.

Vukovic, J., & Kos, T. (2017). Locally adapted NeQuick 2 model performance in European middle latitude ionosphere under different solar, geomagnetic and seasonal conditions. *Advances in Space Research, 60*(8), 1739–1750.

Chapter 10
Seismo-ionospheric Rayleigh Waves

10.1 Introduction

Monitoring and understanding of seismic ionospheric disturbances can provide insights on earthquake rupture, solid Earth and ionospheric coupling, and ionospheric variations' behaviors. The early study on ionospheric responses to the earthquakes began at 1960s (Leonard and Barnes 1965; Davies and Baker 1965). The seismic ionospheric effects were the first time observed from ionosondes near the epicenter of the 1964 Alaskan Earthquake. However, due to the limit continuous observations, it was hard to get a profound and comprehensive understanding of the seismic ionospheric disturbance morphologies and coupling processes between the ground motion and ionospheric dynamics. Until 1990s, GPS has become a powerful tool not only for monitoring Earth's ground deformation (Jin and Park 2006) but also for sounding the Earth's space environment variations (Jin et al. 2008, 2012, 2014; Afraimovich et al. 2013). Since the first time the seismic ionospheric anomalies were observed by GPS Total Electron Content (TEC) after the 1994 Northridge earthquake in California, USA (Calais and Minster 1995), a number of studies have been performed to study pre-/co-/post-seismic ionospheric disturbances related to earthquakes and Tsunamis (Afraimovich et al. 2001; Liu et al. 2010; Astafyeva et al. 2013; Rolland et al. 2013; Occhipinti et al. 2013). It is proved the existence of acoustic traveling ionospheric disturbance originated in the epicenter (Jin et al. 2015). Nowadays, the increasing number of continuously operating GPS stations provides us abundant observations to monitor the detailed pattern and evolution of ionospheric perturbations following the earthquakes (Afraimovich et al. 2010; Ducic et al. 2003; Rolland et al. 2011; Jin et al. 2014). Recently, the detailed seismic ionospheric disturbances for the 2011 Tohoku earthquake (Mw = 9.0) were investigated and revealed by over 1000 GNSS observations from GPS Earth Observation Network (GEONET) in Japan (Rolland et al. 2011; Occhipinti et al. 2013; Jin et al. 2014). The dense TEC disturbances clearly showed acoustic-gravity waves generated at the epicenter, acoustic waves

S. Jin et al., *GNSS Atmospheric Seismology*,
https://doi.org/10.1007/978-981-10-3178-6_10

coupled with Rayleigh wave, as well as gravity waves coupled with the propagation of the following tsunami (Occhipinti et al. 2013).

Ground motion at millimeters per second could induce the neutral atmospheric disturbance at hundreds of meters per second at ionospheric height (Ducic et al. 2003; Jin et al. 2015). On the one hand, the Earth's atmosphere is natural "low pass" filter with a few milli-Hertz (mHz) as the high-frequency attenuation, while the attenuation of low-frequency signal during its upward propagation is rather small (de Groot-Hedlin 2011). The acoustic-gravity waves at several mHz (milli-Hertz) following the earthquake are detectable at ionospheric height due to the atmospheric amplification and low attenuation for low-frequency disturbance. The previous research indicated that the ionospheric perturbations could be detected in three categories: acoustic wave above the focal region, tsunami-generated gravity wave following the oceanic earthquake, and secondary ionospheric perturbation induced by the Rayleigh wave that is a seismic surface wave causing the ground to shake in an elliptical motion (Ducic et al. 2003; Heki et al. 2006; Jin et al. 2014, 2015). With the decrease of the neutral atmospheric mass density from the ground to the ionospheric height, the atmosphere amplifies the disturbance during its upward propagation exponentially (Artru et al. 2004). However, as the divergence of the epicenter surroundings and source types (such as wind, epicenter terrain and focal mechanism), it remains challenging to get a full picture of seismic ionospheric disturbances and to know the ionospheric disturbance mechanism (Astafyeva et al. 2014; Cahyadi and Heki 2015; Reddy and Seemala 2015), although the appearance and pattern of Co-seismic Ionospheric disturbances (CIDs) following the earthquake were detected under certain conditions (Astafyeva et al. 2011 and 2013). For example, Astafyeva (2013) detected the multisegment structure of the seismic fault of the 2011 Tohoku–Oki earthquake from high-rate ionospheric GPS data.

However, the mechanism of solid Earth–atmosphere–ionosphere coupling and disturbance sources are still not clear. Normally, the vertical displacement in the earthquake zone plays a more important role in the formation of seismic ionospheric disturbance (e.g., Jin et al. 2015). The relationship between the earthquake displacement and the ionospheric variation is not understood comprehensively up to now. Furthermore, the relationship between the seismic ionospheric disturbances and ground motion is still debated (e.g., Ducic et al. 2003; Rolland et al. 2011; Maruyama et al. 2012). In this chapter, the Co-seismic Ionospheric Disturbances (CIDs) following the 2018 Mw = 7.9 Alaska earthquake and the 2012 Mw = 7.8 Haida Gwaii earthquake are investigated from the near-field dense GPS observations. The CIDs propagation features and directional divergences are analyzed and discussed as well as the correlation and coupling between the ionospheric disturbance and ground vertical movement.

10.2 Observations and Methods

10.2.1 2018 Alaska Earthquake

The Alaska earthquake with the magnitude of Mw = 7.9 occurred at 9:31 UT on January 23, 2018. The epicenter (56.00°N, 149.17°W) was located at 280 km southeast of Kodiak, Alaska, with a focal depth of 14.1 km and the strike-slip faulting earthquake rupture is about 230 × 30 km (Fig. 10.1). Shortly after the strong earthquake, aftershocks of the magnitude Mw = 4.7 and Mw = 5.0 occurred in the sea near Alaska. The curvature of the arc results in a westward transition of relative plate motion from trench-normal in the east to trench-parallel in the west, accompanied by westward variations in seismic activity. The Pacific Plate subducts into the North America plate at the Alaska-Aleutians trench. The largest earthquakes in this region are generally shallow, predominantly strike-slip events. The location and mechanism source of the Alaska earthquake are consistent with the fault within the Pacific plate.

GPS data are obtained from University NAVstar Consortium (UNAVCO) at ftp://data-out.unavco.org/pub. Figure 10.1 shows the distribution of GPS stations, seismographs, and National Data Buoy Center (NDBC) stations. The group velocity of surface Rayleigh wave propagation in different directions was estimated from 91 Alaska's broadband seismographs (BBS). Here only the z-component data of the seismographs are used to estimate the vertical displacement. The ground vertical motion detected by the bottom pressure record (BPR) on Deep-ocean Assessment and Reporting of Tsunamis (DART) buoy is used to study the speed of Rayleigh wave propagation on the sea surface.

10.2.2 2012 Haida Gwaii Earthquake

The Mw = 7.8 earthquake occurred in Haida Gwaii region, Canada (52.61°N, 132.06°W) at UT 03:04:24 on October 28, 2012 as recorded in the Global Centroid Moment Tensor (CMT) catalog. GPS observations from hundreds of stations are available near the epicenter from University NAVSTAR Consortium (UNAVCO) data center. In this study, we aim to investigate the near-field seismic ionospheric disturbances using data within 2000 km near the epicenter. Figure 10.2 shows the Ionospheric Pierce Points (IPPs) distribution in the left-top panel, epicenter, and GPS stations. A good TEC coverage from GPS dual-frequencies observation is available around the Mw = 7.8 Gwaii earthquake, especially in the southeast of epicenter. In addition, the ground motion data in near field are also used from IRIS (Incorporated Research Institutions for Seismology, http://www.iris.edu) seismic stations and DART (Deep-ocean Assessment and Reporting of Tsunamis) Bottom Pressure Records (BPRs) nearby (Fig. 10.2) (National Oceanic and Atmospheric Administration 2005). These continuous multi-observations of ground motion and ionospheric variations with high precision and resolution (especially for temporal resolution)

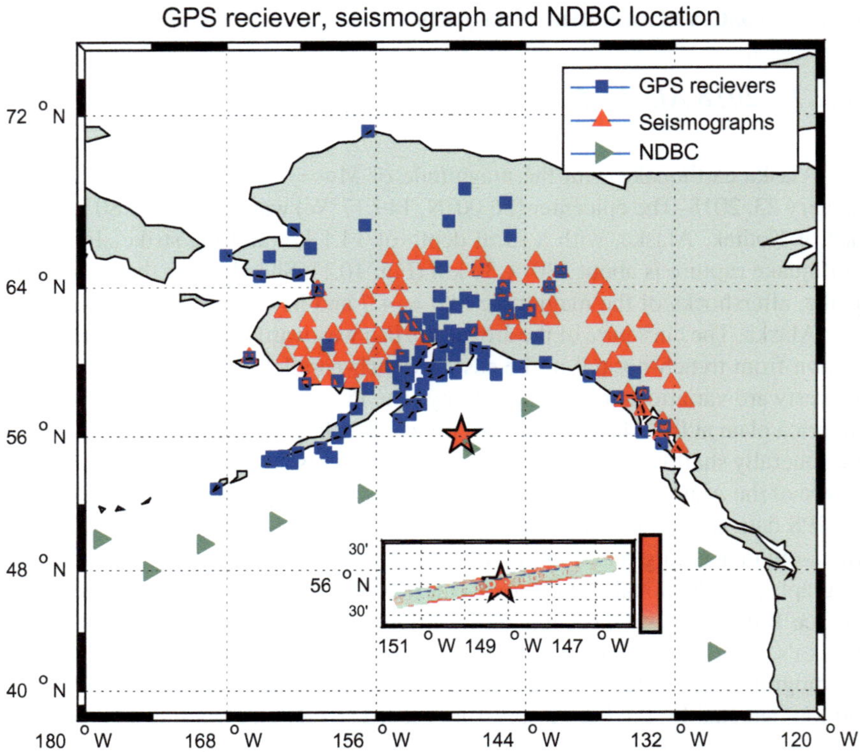

Fig. 10.1 Distribution of GPS stations, seismographs, and NDBC stations around the epicenter. The red pentagram represents the epicenter. The square point stands for GPS stations, the red triangle represents seismographs stations, and the green triangle shows NDBC stations. The sub-picture shows the finite fault of the earthquake

provide us a good opportunity to study the seismic ionospheric response to the 2012 Mw = 7.8 Haida Gwaii earthquake.

10.2.3 Estimation Methods

Ignoring the high-order ionospheric effect in Appleton–Hartree equation, the ionospheric Total Electron Content (TEC) along the signal path could be derived from the dual-frequency GPS carrier phase and pseudorange measurements (Jin et al. 2008, 2016). Here, carrier phase measurements are used to study the seismic ionospheric disturbances with the high precision. A simple cosine mapping function is used to convert slant TEC to vertical TEC for the disturbance propagation analysis. A zero-phase Butterworth filter is used to remove the background variation in the TEC series as its flat frequency response in the passband. Fourth order is selected for the filter that

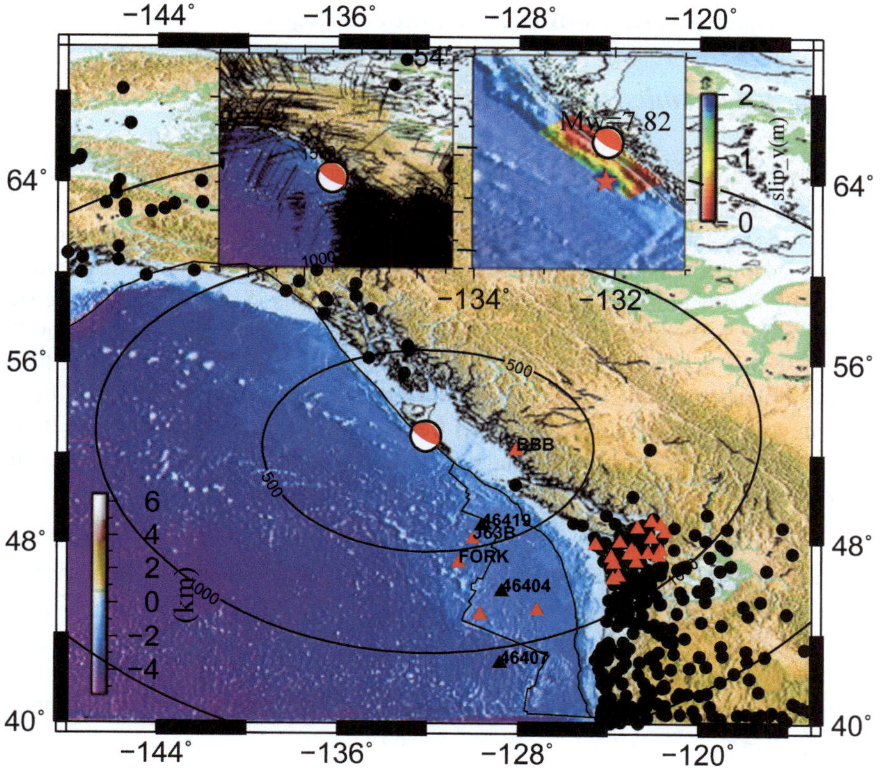

Fig. 10.2 Distribution of GPS stations, seismic stations, and BPRs near the Gwaii earthquake epicenter. The black dots show the locations of GPS stations, the red and black triangles show the locations of seismic stations and BPRs, respectively. The focal mechanism is provided by Global Centroid Moment Tensor (CMT) project (http://www.globalcmt.org). The left-top subfigure presents the IPP distribution during UTC03:00–04:00. The right-top subfigure presents the vertical slips in focal region, which was calculated using the slip model of the 2012 Mw = 7.8 Gwaii Earthquake (Lay et al. 2013). The background topography data used here are the ETOP1 model provided by NOAA

has been proved to be effective for seismic ionospheric disturbance extraction from TEC series (Calais et al. 1998). In this study, we focus on the relationship between Rayleigh wave seismo-ionospheric disturbance and ground motion. So the bandpass is selected as 2 mHz (near the acoustic cutoff frequency around 150 km altitude) to 8 mHz (half of Nyquist frequency for GPS observations with 30-s sampling rate). The selected GPS stations provide observations with 15 or 30 s interval. Here, we choose 8 mHz as the upper cutoff frequency to avoid distortion signals caused by the filtering.

10.3 Rayleigh Waves of the Alaska Earthquake

10.3.1 Co-seismic Ionospheric Disturbances

GPS observations from UNAVCO are used to estimate co-seismic ionospheric disturbance characteristics (Liu and Jin 2019). Figure 10.3 shows the distribution of the Sub-Ionospheric Points (SIPs) track by station AV26 and PRN 05 during the 2018 Alaska earthquake and the detrended TEC with different cutoff frequencies. In Fig. 10.3a, the blue line is the Sub-Ionospheric Points (SIPs) track, and the star and the point, respectively, show the location of the epicenter and AV26 station. In Fig. 10.3b, the blue line shows the elevation angle of the satellite and the red line represents the distance between the SIPs and the epicenter. Figure 10.3c is the filtered TEC time series through different cutoff frequencies. It can be seen that the filtered TEC with cutoff frequencies of 3 and 8 mHz excludes larger interference terms and shows better results.

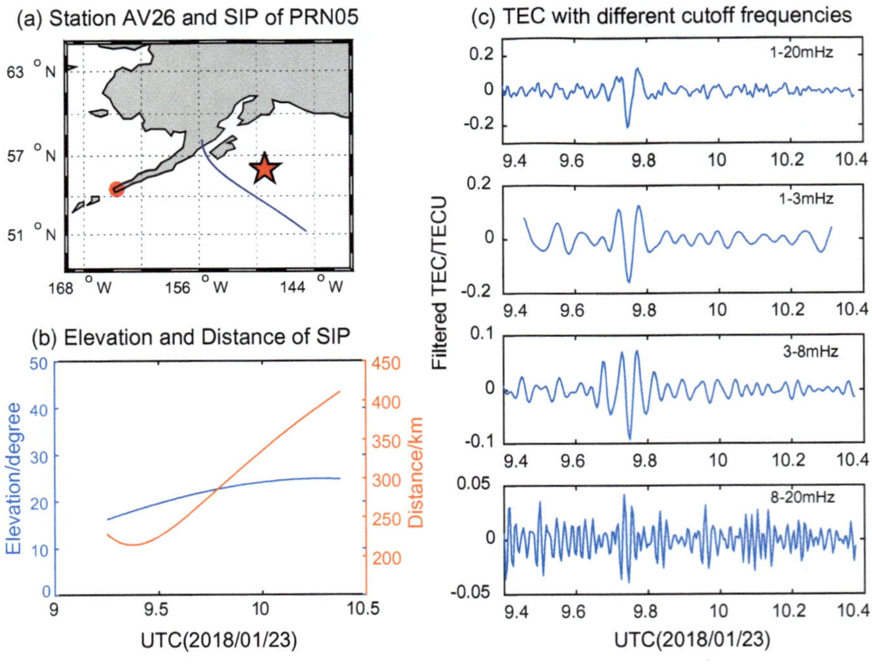

Fig. 10.3 Distribution of the Sub-Ionospheric Points (SIPs) track by station AV26 and PRN 05 during the 2018 Alaska earthquake and the detrended TEC with different cutoff frequencies. **a** The sub-ionospheric points (SIPs) track (blue line), the epicenter (the star), and the location of AV26 station (the point), **b** the elevation angle of the satellite (blue line) and the distance between the SIPs and the epicenter (red line), and **c** the vertical TEC d through different cutoff frequencies

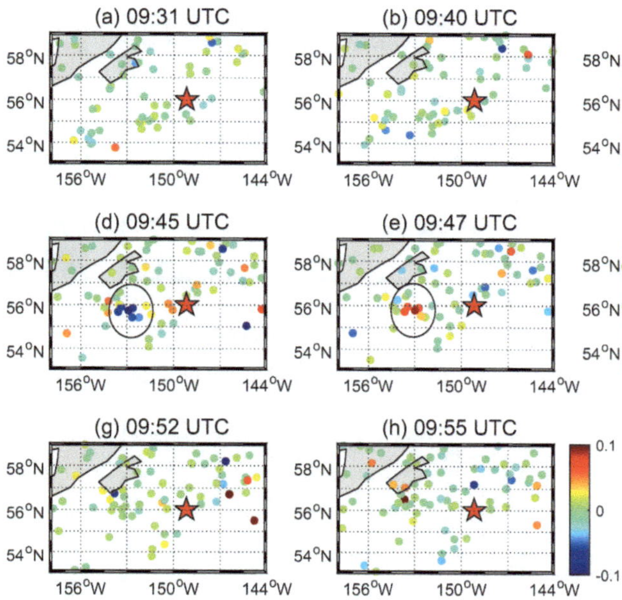

Fig. 10.4 Detrended TEC variations following the 2018 Alaska earthquake. The red pentacle indicates the location of the epicenter of the Alaska earthquake and the dot represents the position of the Sub-Ionospheric Points (SIPs). The color of the dots indicates the amplitude of the filtered TEC disturbances in TECU

Figure 10.4 shows the detrended TEC variations following the 2018 Alaska earthquake from 9:31 to 9:55 UTC. The ionospheric anomalies are detected about 10 min after the main shock. At around 9:43 UTC, the TEC anomalies appeared a few hundred kilometers from the epicenter. The amplitude of seismic ionospheric TEC disturbance can reach the magnitude of 0.1 TECU. The TEC disturbances turn positive to negative and then positive for the same region as shown at 09:45 UT and 09:47 UT. After 09:52 UT, no TEC disturbance can be detected. Both positive and negative disturbances were detected in the southwest of the epicenter, while there is no obvious disturbance in the northeast.

Figure 10.5 shows the co-seismic ionospheric TEC perturbation. The left panel is the SIP tracks between the PRN05 and the station, and the right panel is the travel time diagram of filtered TEC series. The dashed line indicates the time of the earthquake, and the diagonal line is the fitting of the TEC disturbances. From Fig. 10.5a, c, it can be seen that the SIP tracks are nearly covering the epicenter. Figure 10.5b, d shows the travel time diagram of filtered TEC series, which is obtained by taking the time as the x-axis and the distance between the SIP tracks and the epicenter as the y-axis. The filtered TEC series are divided into two parts by the SIPs in the southwest and northeast of the epicenter. About ten minutes after the earthquake, significant seismic ionospheric disturbances with amplitude of about 0.1 TECU are observed from GPS observations in the southwest side of the epicenter. A linear fitting is performed

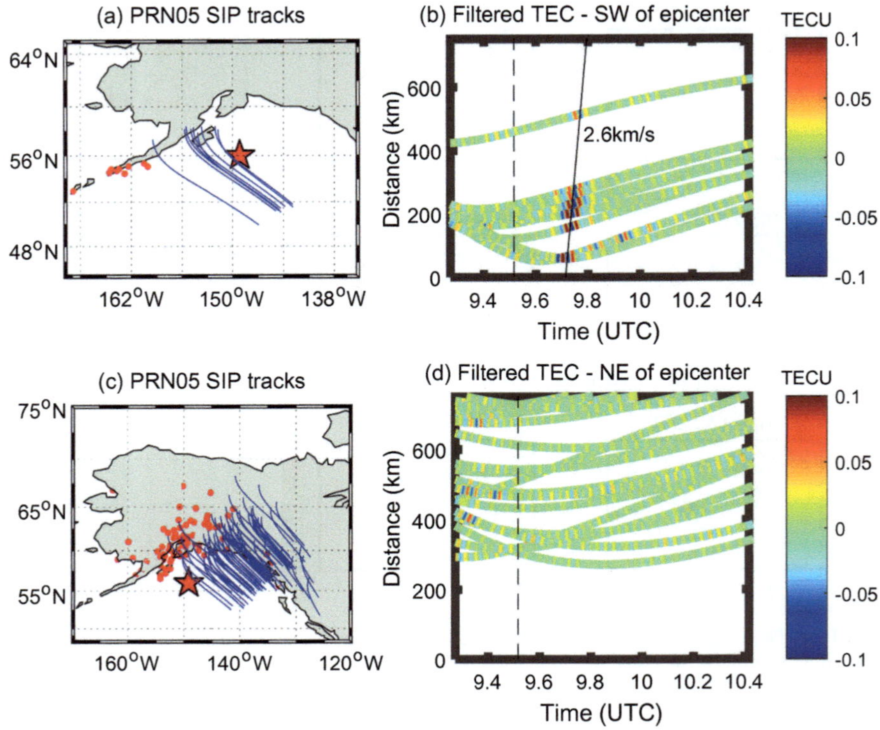

Fig. 10.5 The ionospheric responses to the 2018 Alaska earthquake. The left panel is the SIP tracks between the PRN05 and the station, and the right panel is the travel time diagram of filtered TEC series. The dashed line indicates the time of the earthquake, and the diagonal line is the fitting of the TEC disturbances. Different colors show the magnitude of TEC disturbances

based on the point of the maximal disturbance, and the propagation velocity of the TEC disturbance is obtained with about 2.6 km/s, which is Rayleigh waves, while GPS observations in the northeast side of the epicenter have no obvious seismic ionospheric disturbances.

10.3.2 Ground Vertical Motion

The bottom pressure values from DART BPRs are corrected for temperature effects, which are converted to the height of the ocean surface above the seafloor by using a constant 670 mm/psia. The linear trend and constant terms are removed by least squares, and the remaining is only the height change of the sea level. Figure 10.6 is the detrended sea level change in the vicinity of the epicenter recorded by DART BPRs within 2000 km. The sampling rate of the DART BPR data used in the study was 1 min. Station 46409 is too close to the epicenter, and its seismic impact cannot be

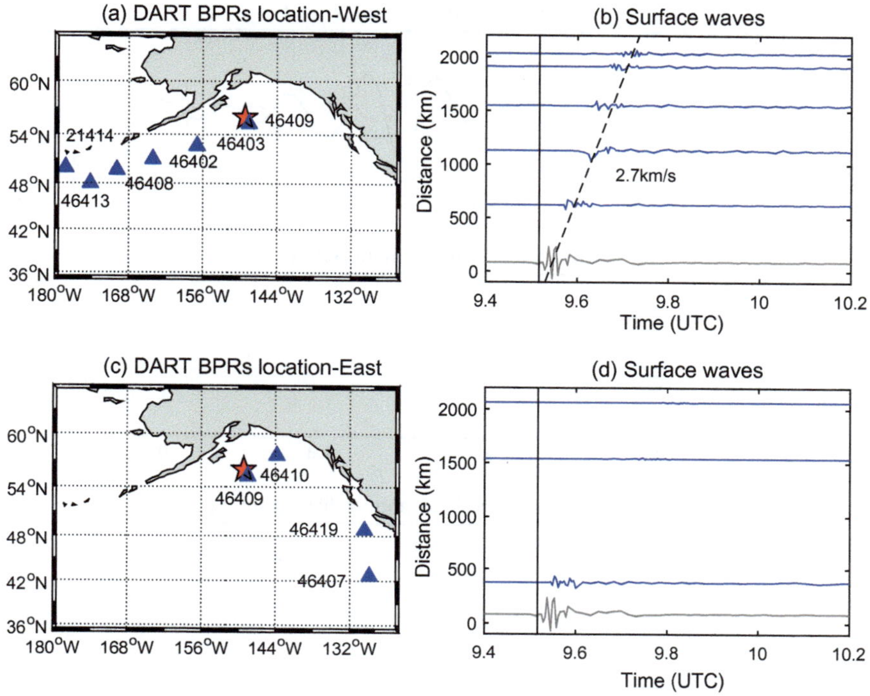

Fig. 10.6 Sea level changes at the west side (**b**) and east side (**d**) near the epicenter recorded by Bottom Pressure Records (BPRs) in the blue triangle (**a**) and (**c**)

regarded as the point source. Therefore, the data of station 46409 are eliminated when fitting. Figure 10.6b shows the vertical motion speed at about 2.7 km/s on the seafloor observed from the DART BPR data in the southwest side of the epicenter, which is consistent with the CIDs propagation velocity detected by GPS data, indicating that the CIDs are caused by the Rayleigh wave on the seafloor vertical motion (Liu and Jin 2019). As the distance increases, the fluctuation of the vertical ground motion is significantly reduced. In the east of the epicenter, station 46403 has observed the disturbance, while the data of station 46409 cannot be fitted and its disturbance characteristics cannot be estimated. The disturbance only exists in the near field, and there is no observation in the far field (Fig. 10.6d).

10.3.3 Correlation Between Vertical Motion and CIDs

Figure 10.7 shows the correlation between the filtered TEC CIDs and DART BPRs vertical motion. Although Rayleigh waves will change obviously with the distance of propagation, the waveforms of the Rayleigh wave propagation are similar within a

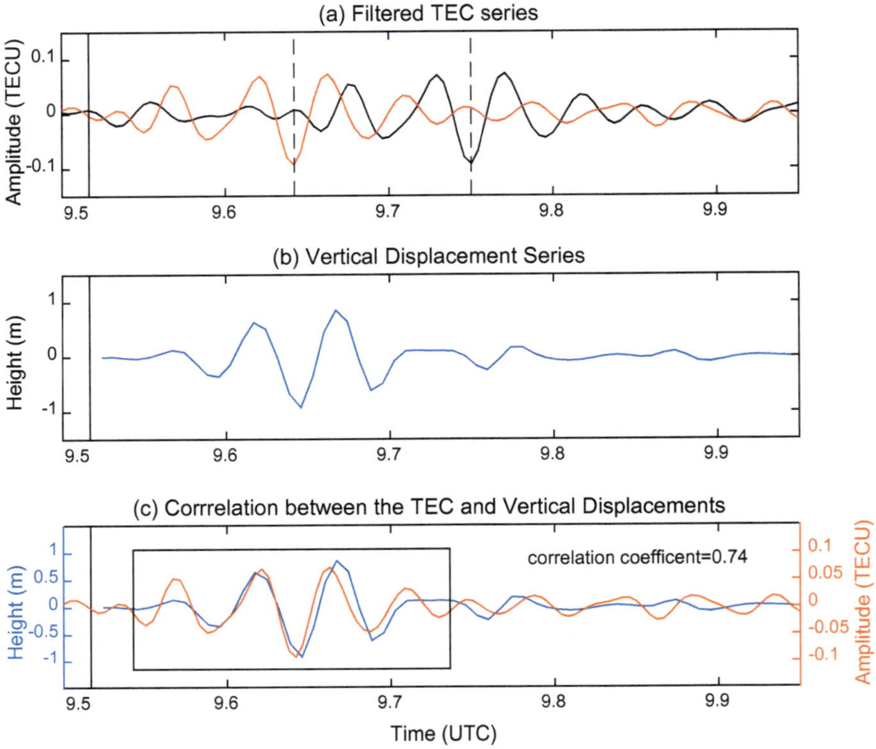

Fig. 10.7 Correlation between filtered TEC CIDs and DART buoy vertical motion. **a** shows the filtered TEC series at station AV26 (black line) and the filtered TEC series in 6.7 min advance (red line), **b** shows the vertical displacement series of station 46403, and **c** shows the correlation between filtered TEC series in 6.7 min advance and vertical displacement series

distance. The DART station 46403, which is just located below the satellite puncture point, is used for the analyses. The correlation coefficient between TEC disturbance series in 6.7 min advance and vertical ground displacement is 0.74, showing that the vertical displacement of the ground is very similar to the waveform of the TEC disturbance series (Liu and Jin 2019). It is expected that the distortion of observed waveforms could be induced by the satellite movement. Fortunately, the radial movement is small when compared to the CID scale in 10 min. For the quasi-radial propagation disturbance, the satellite motion effect was ignored (Jin et al. 2017). The ionospheric disturbance is the same as the Rayleigh wave, but it takes a certain time for the excited acoustic wave to propagate from the surface to the ionosphere. Therefore, the co-seismic ionospheric Rayleigh waves are mainly caused by the vertical ground displacement.

The TEC perturbation time series obtained from GPS dual-frequency observations can estimate the amplitude, duration, and propagation velocity of the disturbance. However, the time-domain analysis can only reflect the change of the amplitude of the

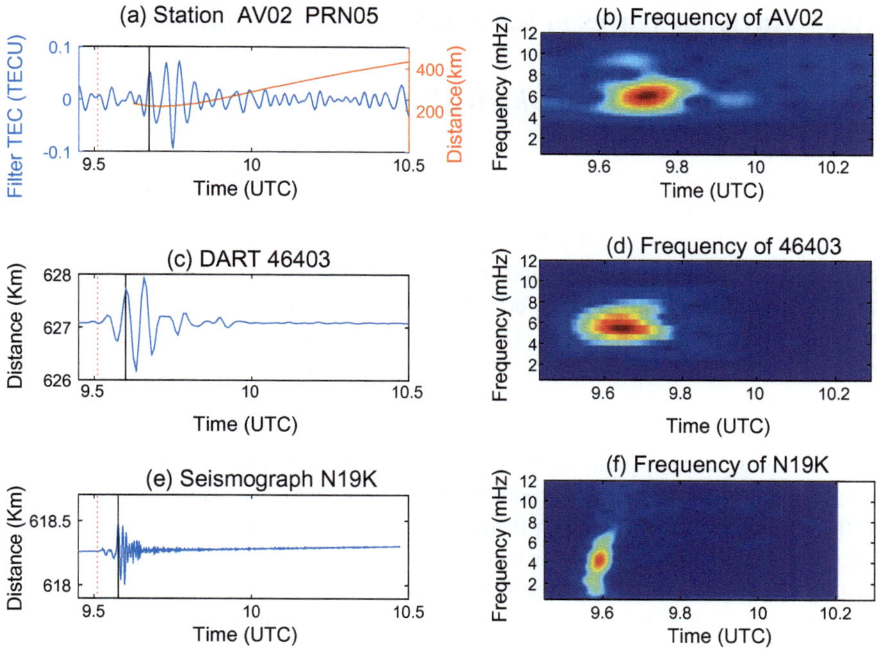

Fig. 10.8 Disturbance time series extracted from the observations and the spectrum distribution from Fourier transform. **a**, **c**, and **e** show the data from the GPS, DART, and seismograph, respectively, and **b**, **d**, and **f** show the spectrum distribution of corresponding observations

signal, so the frequency-domain analysis of the disturbance is performed by using the Fast Fourier Transform (FFT). Figure 10.8 shows the disturbance series extracted from the observations in the left panel, and the spectrum distribution obtained by Fourier transform in the right panel. The GPS TEC series at station AV26 are analyzed and the corresponding frequency of the power extremum is 6.1 mHz, while the frequency of vertical motions from DART BRP has a corresponding frequency of the power extremum with 5.8 mHz. Therefore, the perturbation frequency of the CIDs observed by the GPS satellite is similar to the BPR result, indicating a strong correlation between the two disturbances. The frequency of vertical ground motions from seismometers with a center of 4.0 mHz, which has a little difference since seismometers are located in land, while GPS CIDs observations and DART BRP are covering the sea.

10.4 Rayleigh Waves of the Haida Gwaii Earthquake

10.4.1 Co-seismic Ionospheric Disturbances

Significant ionospheric disturbances in TEC are found from dense GPS observations following the earthquake. For example, Fig. 10.9 shows the filtered TEC series derived from station NANO's PRN01 observations in onset day, previous, and next days. TEC series in the previous and next days are delayed and advanced four minutes for the same IPPs (ionospheric pierce points) location in the ECEF (Earth-Centered Earth-Fixed) coordinate system. The TEC temporal variations in four minutes are relatively small as compared to the TEC horizontal gradient caused by IPPs motion at the low satellite elevation angle. Pronounced ionospheric disturbances with an amplitude up to 0.15 TECU appear following the onset, while the filtered TEC variations in previous and next days are within 0.02 TECU dominated by the GPS measurement noise. After reaching the peak around UT03:15, the disturbance amplitude decreases with the time and the distance far from the epicenter. No geomagnetic or solar events are recorded during this period as well as other sources that could trigger ionospheric disturbances in this region. Therefore, the large ionospheric disturbances above the earthquake's epicenter region are most related to the earthquake. As shown by the filtered TEC from 3 consecutive days in similar positions, the zero-phase fourth-order Butterworth filter with 2–8 mHz catches well seismo-ionospheric disturbances by removing the local TEC background variation.

Snaps of TEC disturbances within 1500 km far from the epicenter during UT03:09–03:21 are presented in Fig. 10.10. The filtered TEC amplitudes are shown from −0.05 to 0.05 TECU. TEC disturbances with over the amplitude range are displayed with the end color of the color bar for a better view of the wave-like perturbation. Actually, the amplitude could be higher than 0.1 TECU in the focal regions as shown in Fig. 10.2. Pronounced waves like disturbances appear in the southeast in the snaps of UT03:14, UT03:16, and UT03:19. The disturbances are the first observed near the epicenter and spread out to several hundred kilometers away. TEC disturbance turns positive peak to negative (or negative to positive) in three minutes for the same region as shown in the snaps UT03:16 and UT03:19. The disturbance period agrees with the acoustic wave mode in the ionospheric height. Till to UT03:21, almost no TEC disturbance is detected. Comparing to the southeast (SE), almost no disturbances are observed in the northwest (NW). Due to the limitation of IPP distribution, the TEC variation in the northeast and southwest is not available.

The most pronounced ionospheric disturbances are detected by GPS observations from the PRN01 since the corresponding IPP locations in 1 h following the onset are the closest to the epicenter. The PRN01 traveling time diagram of CIDs propagating in the SE following the earthquake is presented in Fig. 10.11. The CID propagation speed and appearance epoch are determined by the linear fitting for the locations and epochs at the maximum amplitudes. The results show that the seismic ionospheric disturbance first appears 10.77 min after the onset and propagates at a speed of 2.20 km/s. Although the earthquake-induced tsunamis with maximum 7.6 m of height

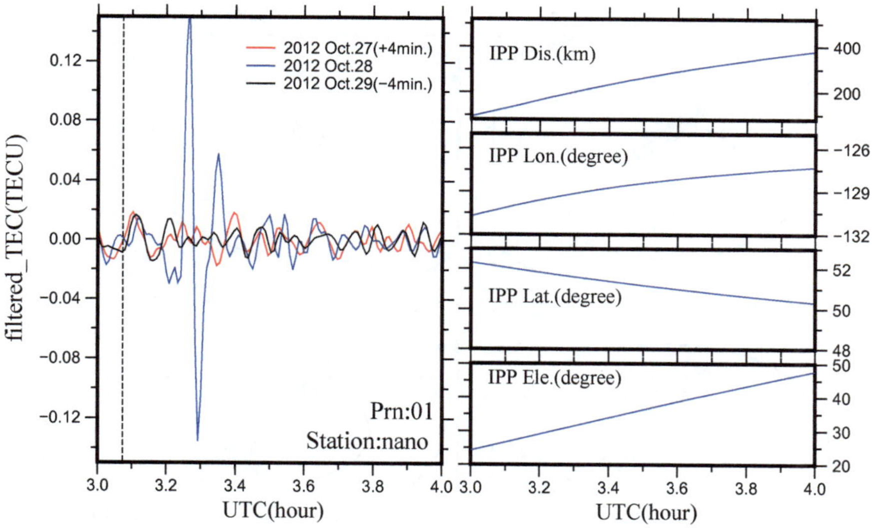

Fig. 10.9 FilteredTEC variations during UT 03:00–04:00 on October 27–29 detected by GPS observations at station NANO from satellite G01 (i.e., PRN01). The Mw = 7.8 Gwaii earthquake onset is marked by the vertical dash line. The right panel shows the epicentral distances, longitudes, latitudes, and satellite elevations for corresponding ionospheric pierce points, respectively

sheltered from storm waves and 13 m at inlet were reported (Leonard and Bednarski 2014), it could not propagate to the ionospheric height from the sea surface in 10 min. While the propagation speed of tsunami-generated gravity wave is much smaller than the one presented here for regions with less than 2000 m of water depth ($v_{tsnuami} = \sqrt{gh}$, g is the acceleration of gravity and h is the water depth) (Occhipinti et al. 2008, 2013). Furthermore, the CIDs should be dominated by the acoustic mode as a result of 2–8 mHz filtering, while the tsunami-related ionospheric disturbance is with a central frequency of 1–2 mHz at gravity mode (Jin et al. 2014; Occhipinti et al. 2013). Grawe and Makela (2015) reported the tsunami-generated CIDs following the Mw = 7.8 earthquake, while it appeared in the far field with more than three thousand kilometers away during UT 08:30–09:00. The co-seismic horizontal acoustic wave above the focal region is another possible source of seismo-ionospheric disturbance. It is induced by the refraction of upward atmospheric pulse due to the vertical neutral density gradient. However, the sound speed at the ionospheric height is around 1 km/s. Therefore, the detected CIDs are mostly related to the Rayleigh wave that spreads out with a much higher speed. Note the locations of CID, seismometers, and DART BPRs are shown in Figs. 10.2 and 10.10, and the 2.20 km/s propagating seismic ionospheric disturbance is just above the region where the Rayleigh wave group velocity is about 2.22 km/s. However, other superposition of acoustic and Rayleigh waves maybe affect CIDs and make the apparent velocity between the acoustic and the Rayleigh waves (e.g., Astafyeva et al. 2013), which needs to be further investigated with available data in the future.

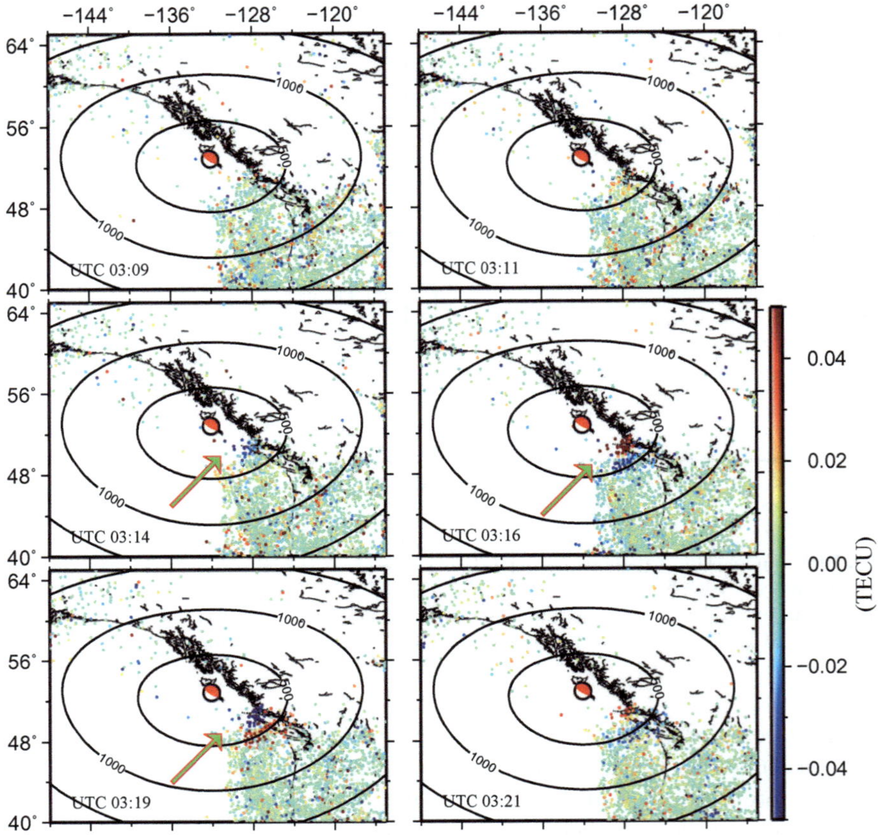

Fig. 10.10 TEC perturbations within 1500 km far from the epicenter following the Gwaii earthquake detected by GPS data during UT03:09–03:21. The green arrows show pronounced waves like disturbances in the southeast in the snaps of UT03:14, UT03:16, and UT03:19

10.4.2 Observation Geometry and CID Amplitudes

It is well known that the intersection between GPS signal paths and Earth's ionospheric electron concentrated region varies with the satellite motion. The GPS observation geometry has an effect on amplitudes of the observed CIDs (Jin et al. 2016). As the omnidirectional GPS observations are not available around the epicenter, we mainly discussed the effect of satellite elevation angle on the observed CID amplitudes. Figure 10.12 presents the ionospheric disturbance peaks of TEC series obtained from PRN01 and PRN11 observations during UT03:08–03:20 when the pronounced seismic ionospheric disturbance appears. Here, we only present the result of PRN01 and PRN11 since the corresponding IPPs are mainly distributed in the region with the most pronounced CIDs. Obvious amplitude attenuation with the epicentral distance is shown in both the observations of PRN01 and PRN11. Owing to the station

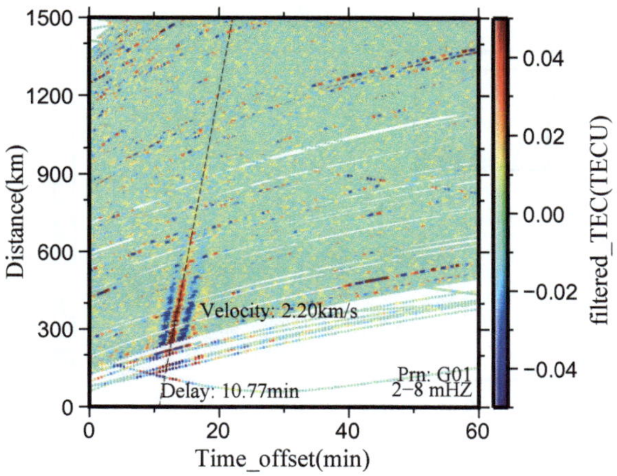

Fig. 10.11 Travel time diagram of CIDs detected by GPS observations of PRN01. The zero in horizontal and vertical axes is corresponding to the onset and the epicenter location of the Mw = 7.8 Gwaii earthquake in 2012. The slope line is determined by the corresponding locations and epochs of separated disturbance peaks. Here, the "delay" and "velocity" are corresponding to the first appearance epoch and group propagation velocity of seismic disturbances following the Gwaii earthquake at 2–8 mHz

distribution and satellite position, satellite elevation angles are similar to each other for separating satellite observations. It is a good opportunity to see the CID distance delay without the satellite elevation angle effect. The third-order polynomial fitting is applied to fit the relationship between epicentral distances and disturbance amplitudes as shown in the left-top panel of Fig. 10.12. The result of higher order polynomial fitting is similar to the one from the third-order fitting, implying that the CID amplitudes attenuate cubically with the epicentral distance approximately in the near field. As the epicentral distance increase, the ionospheric temporal–spatial dynamic background becomes much more divergence that could lead to different attenuate patterns. It is expected that the quasi-cubical attenuate is not the case in the far field. Comparing with PRN01, the disturbance observed by PRN11 observation is much weaker, especially in 350–550 km as marked by the dashed rectangle in Fig. 10.12. Although the IPPs have similar locations, the observed disturbance amplitudes are higher for observations with lower satellite elevation angles. It is consistent with the statistical features of the Rayleigh wave-related seismo-ionospheric disturbance amplitudes from six GPS TEC series following different earthquakes proposed by Rolland et al. (2011). However, the cosine mapping function from slant TEC to TEC may have some effects. GPS TEC observations are not sensible to the acoustic perturbation propagating in parallel with the signal path direction, as GPS ionospheric measurements are integration along the line of sight between GPS satellites and receivers. The GPS TEC disturbance effect from the local ionospheric electron density increase or decrease induced by the wave-like disturbance would be attenuated when it propagates along the signal path. For regions within 200 km, the

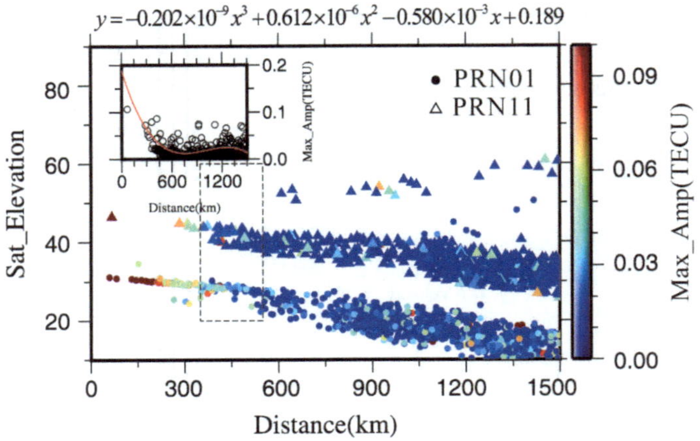

Fig. 10.12 Maximum ionospheric disturbance amplitudes of filtered TEC series obtained from PRN01 and PRN 11 observations during the UT03:08–03:20. The colored circle is the peak amplitudes for TEC series from PRN01 observations and the triangle is the peak amplitudes for TEC series from PRN 11. The third-order polynomial fitting between epicentral distances and disturbance amplitudes is shown in the left-top panel

elevation angle effect is not obvious. The maximum disturbance amplitude is similar for 30–50° elevation angles. It indicates that ionospheric disturbances induced by the oblique acoustic wave existed in the focal region. The Rayleigh wave-induced ionospheric disturbance is not the unique dominant seismo-ionospheric effect. In addition, interaction between the movements of charged particles in Rayleigh waves and magnetic fields might selectively attenuate the seismic ionospheric disturbances (Afraimovich et al. 2001; Heki and Ping 2005; Rolland et al. 2013).

10.4.3 Main Sources of the Rayleigh Wave

10.4.3.1 Ground Vertical Motion

In order to confirm the CID's Rayleigh wave dependence, post-seismic vertical ground motion in near field is investigated. Using the same bandpass filtering, the normalized vertical ground motions from the observations of 23 broadband high-gain seismometers and 3 DART BPRs within 1200 epicentral distance are presented in Fig. 10.13. The stations' locations are marked in Fig. 10.2. Not like the high sampling rate seismic data, the BPRs series released by DART are normally 15-min interval for the tidal report at quiet state, while the sampling rate is up to 1 min or 15 s when a sea bottom pressure disturbance is detected (Meinig et al. 2005). Here, the sampling rate of the released DART BPRs during UT03:00–04:00 is 1 min, so we interpolated the BPRs at quiet time to 1 min interval with the Spline interpolation

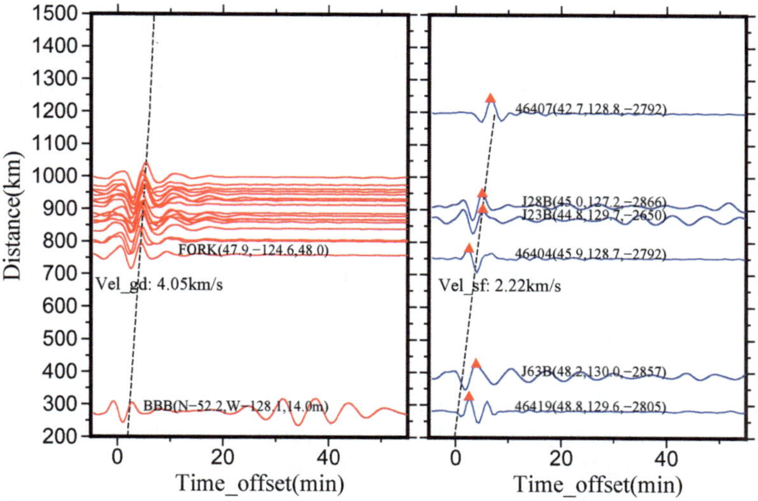

Fig. 10.13 Vertical ground motion in 2–8 mHz detected by the vertical broadband high-gain seismometers and DART BPRs (46419, 46404, and 46407) near the epicenter. The locations of selected stations are shown in Fig. 10.2. The vertical displacements at solid ground surface and seafloor are presented separately on the left and right panel. The series are normalized to the corresponding maximum disturbance amplitudes. The zero line for each series is adjusted to the station epicentral distance (in vertical axis). The locations and elevations of DART BPRs, and selected seismometers are marked. The unmarked stations are mainly distributed near the Seattle coast with 0–300 m elevation

unifying the data interval and avoiding the filtering alias. Considering the Rayleigh wave propagation anisotropy related to the crust structure diversity, we present the vertical displacement disturbance of solid ground surface and seafloor separately in the left and right panels of Fig. 10.13. The Rayleigh waves with similar waveform in 2–8 mHz are detected by both seismic and DART BPR data. The southwestern Rayleigh wave group velocity is determined with the wave crest appearance epochs and locations. The ground surface Rayleigh waves show a high propagation velocity up to about 4.05 km/s, while the one on the seafloor is about 2.22 km/s that agree well with the CIDs' propagation speed detected by GPS data. It indicates that the observed CIDs are induced by the Rayleigh wave generated upward acoustic wave. Note that the island station FORK is also located in the CID regions, while the corresponding Rayleigh signature in the vertical displacement series appears earlier in a faster mode. The observed CID in the southeast prefers to be related to the Rayleigh wave propagation in the seafloor. It should be pointed out that the Rayleigh wave appearance epoch is almost at the onset according to the linear fitting estimation from the vertical ground displacement records on the seafloor. The results are based on the point source and constant propagation disturbance. While the real fault dip slip is far away from the point source as shown in Fig. 10.2, the constant propagation should not be the real case within the scale of the focal region. Here, we focus on the most pronounced CIDs in 300–700 km. Although not enough GPS observations are available to see the ionospheric disturbances above Seattle region, we could not validate the existence of the ionospheric Rayleigh signature with a speed of 4.05 km/s for corresponding region following this earthquake. The Rayleigh wave-induced CID at such a speed is possible (Maruyama et al. 2012). It is believed that the anisotropy of the Rayleigh wave could be an important source of CIDs propagation divergence in different directions.

10.4.3.2 Correlation Between Ground Motion and CIDs

Although directional divergence of the Rayleigh wave has a significant effect on its propagation velocity, the waveform is similar within 1000 km far from the epicenter as discussed above. The vertical ground displacement series should agree with each other after time delay adjustment. The stacked series are used to reduce the subtle disturbance resulted from other factors, such as measurement noise, local environment effect, etc. The DART station 46419, which is just located below the CID, is set as the reference series for the determination of time delay between different series using the cross-correlation technique. The same processing is used to get the stacked TEC series. The TEC with the highest sum of the correlation coefficients is selected as the reference series (Kiryushkin et al. 2007). It is expected that the distortion of observed waveforms could be induced by the satellite movement (Garrison et al. 2007). Fortunately, the radial movement is small when compared to the CIDs scale as shown in Figs. 10.10 and 10.11 in ten minutes. For the quasi-radial propagation disturbance, we ignored the satellite motion effect. Both stacked displacement and TEC series are normalized to corresponding maximum values as shown in Fig. 10.14. The 2–8 mHz ionospheric CIDs observed by GPS data show high consistency with the vertical ground motion. The correlation coefficient is up to the maximum, i.e., 0.83 with 8.8 min delay, and the small difference with the linear fitting results could be induced by the noise in the TEC series.

The wave front and trajectories of upward propagation disturbances are modeled from a simplified acoustic ray tracing for a horizontally stratified atmosphere (Garcés et al. 1998). In the vertically layered atmosphere without considering propagation loss, acoustic wave propagation equation can be simplified as

$$\begin{aligned} &\tfrac{dx}{dt} = k_x c + u \quad \tfrac{dy}{dt} = k_y c + v \quad \tfrac{dz}{dt} = k_z c \\ &\tfrac{dk_x}{dt} = n k_x k_z \quad \tfrac{dk_y}{dt} = n k_x k_z \quad \tfrac{dk_z}{dt} = n(k_z^2 - 1) \\ &n = \frac{dc}{dz} + k_x \frac{du}{dz} + k_y \frac{dv}{dz} \end{aligned} \tag{10.1}$$

where $\hat{k}(k_x, k_y, k_z)$ is the unit wave vector, $\vec{r}(x, y, z)$ is the wave front position, (u, v) is the horizontal component of the background wind field, and c is the speed of sound. In the ideal atmosphere, the speed of sound c can be expressed as

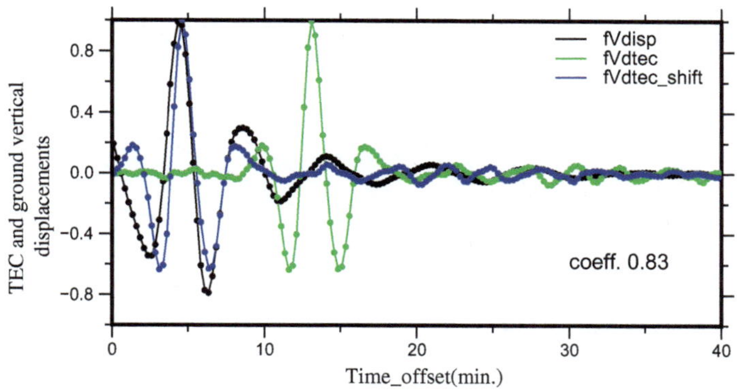

Fig. 10.14 Correlation between the stack TEC and ground vertical displacements in the southwest of the epicenter. The black line is the normalized stacked vertical displacements series, the green line is the normalized stacked filtered TEC series, and the blue line is a copy of the green one but 8.8 min in advance. Here, the vertical displacement series is opposite to the records, namely, negative value meaning the upward motion, for intuitive illustration of the correlation

$$c = \sqrt{\frac{\gamma RT}{M}} \tag{10.2}$$

where γ is the ratio of constant pressure specific heat capacity to constant volume specific heat capacity (about 1.4), R is the universal gas constant (8.31×10^3 J/(mol K), M is the average molecular molar amount (g mol^{-1}) that can be got from NRLMSISE-00 model (Picone et al. 2002), and T is the atmospheric temperature. Figure 10.15 shows the acoustic ray tracing for zenith 60–90° from a point perturbation source. Here, the sound and wind velocity profiles for the focal region at 03:15 UT, October 28, 2012 are derived from NRLMSISE-00 atmospheric model and HWM07 (Drob et al. 2008). The red line is the equal-time contours with 2-min interval. As shown in Fig. 10.15, the upward acoustic perturbation propagates to around 150 km altitudes after 8.8 min, where the electron density is high enough to induce a detectable TEC disturbance for GPS observations. It is up to the ionospheric F2 peak height after about 14 min that is consistent with the observed TEC peak epoch as shown in Fig. 10.14. Note the propagation direction of oblique acoustic wave is deflected dramatically as it goes through atmospheric layers with large mass density gradient, and trapped around the 200 km altitude. Then, it could trigger the traveling ionospheric disturbance at the sound speed. So, it should not be the dominant source of seismo-ionospheric effect detected by GPS data in southeast of epicenter. As we are known that the vertical coupling at interface between the atmosphere and Earth surface is much more efficient than the horizontal one. Usually, the initial disturbance at the bottom of the atmosphere from ground motion is an upward pulse except for epicenter region where oblique acoustic disturbance is possible as the large amplitude slips induced by oblique fault rupture (Astafyeva et al. 2011). The high correlation coefficient between the 2–8 mHz ground vertical displacements and

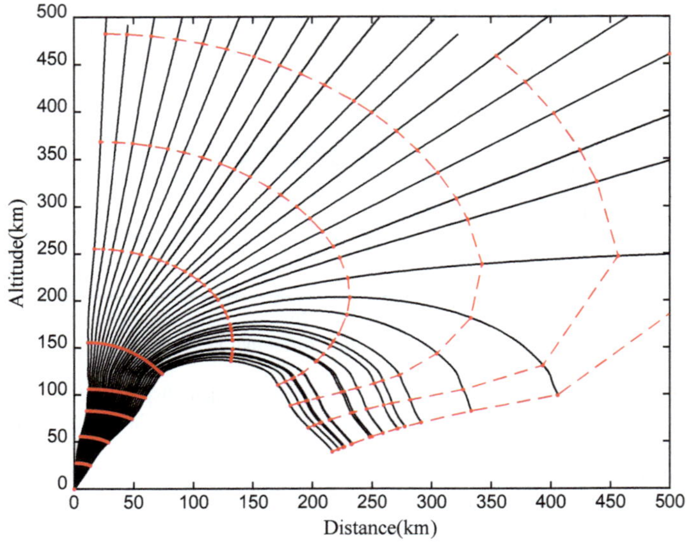

Fig. 10.15 Acoustic ray tracing for zenith 60–90° from a point perturbation source. Here, the sound and wind velocity profiles for the focal region at October 28 UT03:15 are derived from NRLMSISE-00 atmospheric model and HWM07 (Drob et al. 2008). The red line is the equal-time contours with 2-min interval

seismo-ionospheric disturbance indicates that the Rayleigh ionospheric disturbance is the result of vertical ground motion at Earth–atmosphere coupling frequencies in corresponding regions. Therefore, it has the potential of ionospheric disturbance detection using ground motion observations.

10.5 Modeling of the Rayleigh Wave–Acoustic Wave

10.5.1 Disturbance Source

The neutral atmospheric disturbance triggered by the ground displacement can be simulated with the state equation, momentum, and mass conservation (de Groot-Hedlin 2007; Rolland et al. 2011). Utilizing the electron continuity and momentum equation (Shinagawa et al. 2007), the electron disturbance could be simulated. In our simulation, the wind effect, viscosity, and heat conduction are not considered, which do not make great sense for the 2–8 mHz acoustic wave propagation from the ground to the ionospheric height. A Finite-Difference Time-Domain (FDTD) method is applied to get the neutral perturbation induced by the Rayleigh wave in the two-dimensional stratified atmosphere centered at epicenter (de Groot-Hedlin et al. 2011). NRLMSISE-00 model is used to get the vertical profiles of neutral particle density and sound velocity (Picone et al. 2002). The wind velocity profiles are derived from the HWM07 (Drob et al. 2008). The perturbation source is simulated using the product of a cosine and sins function as follows:

$$v_{disp,z=0}(x,t) = A\,\mathrm{sinc}\left[2\pi f_b\left((t-t_0)-\frac{|x|}{v_{Reigh}}\right)\right]\cos\left[2\pi f_c\left((t-t_0)-\frac{|x|}{v_{Reigh}}\right)\right] \quad (10.3)$$

where x is the epicentral distance, t_0 is the earthquake onset time, f_c is the central frequency (5 mHz), f_b is the central frequency (2 mHz), and v_{Reigh} is the Rayleigh wave propagation speed (2.2 km/s). The background ionospheric electron density was from IRI-2012 (Bilitza et al. 2014). Here, we take the 2012 Mw = 7.8 Gwaii earthquake as example, t_0 is 03:04:24 UT, October 28, 2012. Figure 10.16 is the simulated frequency spectrum (upper) and normalized displacements (bottom) of the 2012 Mw = 7.8 Gwaii earthquake, which shows that 4.2 mHz of the power spectrum peak is corresponding to the frequency.

The seismo-ionospheric disturbances on the southeast side after the Mw = 7.8 Gwaii earthquake on October 28, 2012 mainly occur over the offshore area with a water depth of approximately 2,800 m. The seawater can be approximated as an incompressible fluid, so the vertical movement of the seafloor causes a vertical motion similar to that of the sea surface in the corresponding area (i.e., the interface of the

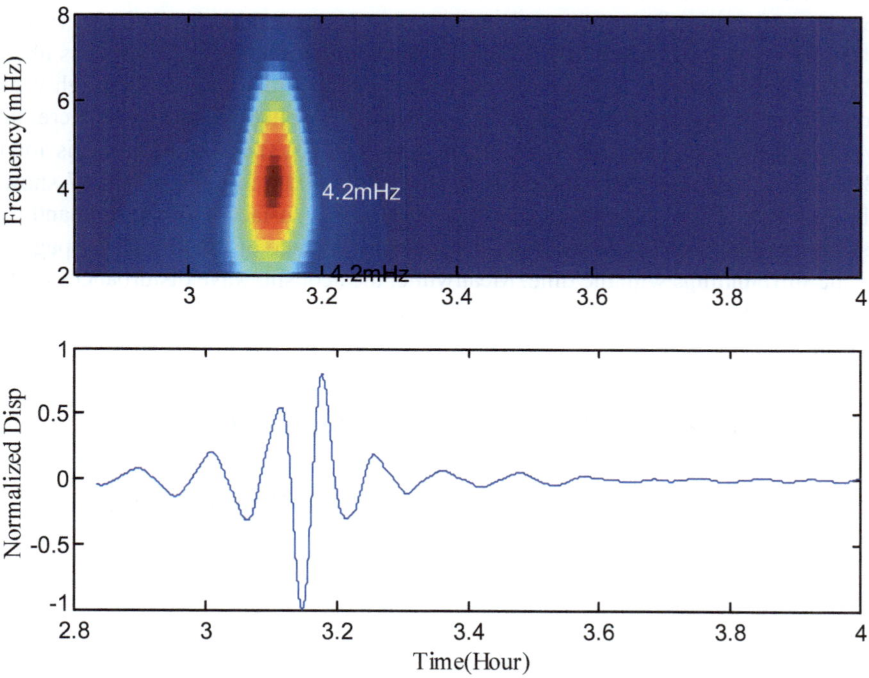

Fig. 10.16 Simulated frequency spectrum (upper) and normalized displacements (bottom) of the 2012 Mw = 7.8 Gwaii earthquake

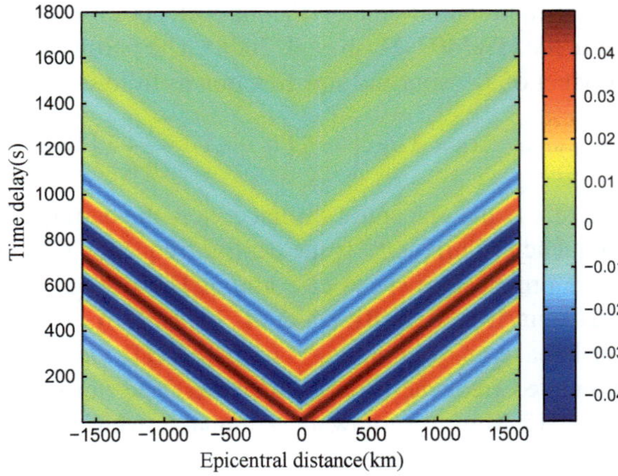

Fig. 10.17 The amplitude variation of the disturbance propagating from the epicenter at the central frequency of 4.2 mHz with the speed of 2.2 km/s. The horizontal axis represents the epicentral distance, and the vertical axis represents the time difference between the current epoch and the main shock moment

Earth's surface and the bottom atmosphere). Since the sound velocity in seawater is about 1.5 km/s, and the sea surface vertical motion is delayed by a time scale of several seconds, the influence of the water layer can be neglected, namely, the vertical deformation of the solid Earth detected by the seabed is regarded as the vertical deformation at the interface of the sea surface and the bottom of the atmosphere. Based on the perturbation propagation characteristics obtained from the GPS observation data above, v_{Reigh} is taken at 2.2 km/s on the southeast side (water depth is about 2800 m offshore area). To simplify the perturbation simulation, the Rayleigh wave attenuation is neglected in the near-field seismic ionospheric disturbance. Here, the vertical slip of the earthquake is about 2 m (Lay et al. 2013), and the duration is about 37 s (www.globalcmt.org), then A is approximately 0.05 m/s. Figure 10.17 shows the variation of the disturbance source on the Earth's surface with the time and the epicentral distance. The vertical deformation disturbance of the ground propagates to the surroundings with the time. Meanwhile, the acoustic wave disturbances of the upward propagation are sequentially generated.

10.5.2 Coupling of Rayleigh Waves and Ionosphere

Acoustic wave in the atmospheric can be described as (Yang and Chen 2007)

$$\begin{aligned} \frac{Dp}{Dt} &= -\rho c^2 \nabla \cdot \vec{V} \\ \rho \frac{D\vec{V}}{Dt} &= -\nabla p + \eta \nabla^2 \vec{V} + \left(\zeta + \frac{\eta}{3}\right) \nabla (\nabla \cdot \vec{V}) + \rho \vec{g} \\ \frac{D\rho}{Dt} &= -\rho \nabla \cdot \vec{V} \end{aligned}$$

$$\left(\frac{D}{Dt} = \frac{\partial}{\partial t} + (\vec{V} \cdot \nabla)\right) \tag{10.4}$$

where p is the pressure,ρ is the atmospheric density, c is the velocity of sound, $\overrightarrow{V}$ is the speed of acoustic wave, η is the shear viscosity (about 1.8×10^{-5} kg m^{-1} s^{-1}), ζ is the volumetric viscosity (normally can be ignored), and $\overrightarrow{g}$ is the gravity acceleration. The pressure, atmospheric density, and acoustic speed can be described as the background field and disturbance field (subscript 0 is the background field and subscript s is the disturbance field), namely,

$$p = p_0 + p_s \quad \rho = \rho_0 + \rho_s \quad \overrightarrow{v} = \overrightarrow{v_0} + \overrightarrow{v_s} \tag{10.5}$$

For a small disturbance, the high-order term can be ignored as well as the background wind field, so Eq. (10.2) can be written as

$$\begin{aligned}
\frac{\partial p_s}{\partial t} &= -\rho_0 c^2\left(\frac{\partial v_x}{\partial x} + \frac{\partial v_y}{\partial y} + \frac{\partial v_z}{\partial z}\right) + \rho_0 g v_z \\
\frac{\partial v_x}{\partial t} &= \rho_0^{-1}\left[-\frac{\partial p}{\partial x} + \eta\left(\frac{\partial}{\partial x^2} + \frac{\partial}{\partial y^2} + \frac{\partial}{\partial z^2}\right)v_x + \frac{1}{3}\eta\left(\frac{\partial^2 v_x}{\partial x^2} + \frac{\partial^2 v_y}{\partial x \partial y} + \frac{\partial^2 v_z}{\partial x \partial z}\right)\right] \\
\frac{\partial v_y}{\partial t} &= \rho_0^{-1}\left[-\frac{\partial p}{\partial y} + \eta\left(\frac{\partial}{\partial x^2} + \frac{\partial}{\partial y^2} + \frac{\partial}{\partial z^2}\right)v_y + \frac{1}{3}\eta\left(\frac{\partial^2 v_x}{\partial y \partial x} + \frac{\partial^2 v_y}{\partial y^2} + \frac{\partial^2 v_z}{\partial y \partial z}\right)\right] \\
\frac{\partial v_z}{\partial t} &= \rho_0^{-1}\left[-\frac{\partial p}{\partial z} + \eta\left(\frac{\partial}{\partial x^2} + \frac{\partial}{\partial y^2} + \frac{\partial}{\partial z^2}\right)v_z + \frac{1}{3}\eta\left(\frac{\partial^2 v_x}{\partial z \partial x} + \frac{\partial^2 v_y}{\partial z \partial y} + \frac{\partial^2 v_z}{\partial z^2}\right) - \rho_s g\right]
\end{aligned} \tag{10.6}$$

where the neutral atmospheric background field can be got from NRLMSISE-00 model. A Finite-Difference Time-Domain (FDTD) method is applied to get the neutral perturbation induced by some corresponding sources (de Groot-Hedlin et al. 2011). Using the simplified single-fluid ionospheric model, the electron continuity and momentum equation can be obtained as (Shinagawa et al. 2007)

$$\begin{aligned}
\frac{\partial n_e}{\partial t} &= -\nabla \cdot (n_e v_{e||})_i \\
v_{e||} &= -\frac{1}{m_e n_e v_{en}}\nabla_{||} p + \frac{g_{||}}{v_{en}} + v_{n||}
\end{aligned} \tag{10.7}$$

where n_e is the electron density, v_e is the speed of electron disturbance, m_e is the electron mass, v_{en} is the electronic collision frequency, and || shows the parallel direction of magnetic field. Here, background magnetic field can be got from IGRF12 (http://www.ngdc.noaa.gov/geomag/models.shtml) and background electron density can be obtained from IRI-2012 (Bilitza et al. 2014), and the neutral electron density collision frequency is computed based on the empirical formula 6.8 (Kelley 2009). So the speed of electron disturbance v_e can be obtained as

$$v_e = 5.4 \times 10^{-10} n_n T_e^{0.5} + \left[34 + 4.18 \ln\left(\frac{T_e^3}{n_e}\right)\right] n_e T_e^{-1.5} \tag{10.8}$$

where v_e is the speed of electron disturbance, n_e, n_n, and T_e are the electron density, neutral atmospheric molecular density, and electron temperature, respectively. The perturbation of the electron density can be obtained by integrating the continuity equation for the charge density from the explosion time t_0 to the observation time t (Dautermann et al. 2009):

$$\delta N(\vec{r}, t) = -\frac{\partial N}{\partial z} \int_{t_0}^{t} v_{e,z}(\vec{r}, t)dt - N(z) \int_{t_0}^{t} \nabla \cdot v_{e,z}(\vec{r}, t)dt \qquad (10.9)$$

where δN is the disturbance of the electron number density (N).

Figure 10.18 shows the simulated ionospheric electron density disturbances at 9–20 min after the onset in the NE–SW direction (epicentral azimuth = 135°). Consistent with the time delay from observations and acoustic ray tracing mentioned above, the pronounced ionospheric disturbance appears about 9 min at around 150 km altitudes. The perturbation enhances with its upward propagation induced by the decreasing neutral particle density and increasing electron density. As an example, the signal path between station NANO and satellite PRN01 is marked by the blue lines in Fig. 10.18 (the satellite elevation angle varies 25–34°). It is obvious to see that the ground motion-induced ionospheric waves like disturbances can be observed the signal path across the positive and negative disturbance regions in sequence. With the ideal GPS observation geometry (i.e., the single path is quasi-parallel to the disturbance wave front), the observed wave disturbance should be near the real case. Then the high correlation between the vertical ground motion and the disturbance at ionospheric height is expected. The geomagnetic field effect on the seismic ionospheric disturbance, i.e., the higher disturbance amplitude in southwest direction, is shown especially after the disturbance arrived at the F2 layer height. While the effect is weak that probably cannot explain the absence of seismic ionospheric effect in the northeast. Lack of efficient GPS observations for seismic ionospheric effect detection is the major reason. Apparently, the disturbance signal would be attenuated if the GPS signal path is near the perpendicular direction of the wave front as shown in Fig. 10.18.

10.6 Conclusions

In this Chapter, the seismo-ionospheric disturbances following the 2018 Mw = 7.9 Alaska earthquake and the 2012 Mw = 7.8 earthquake are investigated and studied using GPS data near the epicenter. Results show that the seismo-ionospheric disturbances are mainly related to the Rayleigh wave signature in the ionospheric height. The seismic ionospheric effect is a result of the upward propagation acoustic wave induced by the Rayleigh wave rather than the direct effect of the focal rupture. With the "ideal" GPS observation geometry, higher observed CID amplitudes are found for the observations with lower satellite elevation angles, i.e., the signal path near the

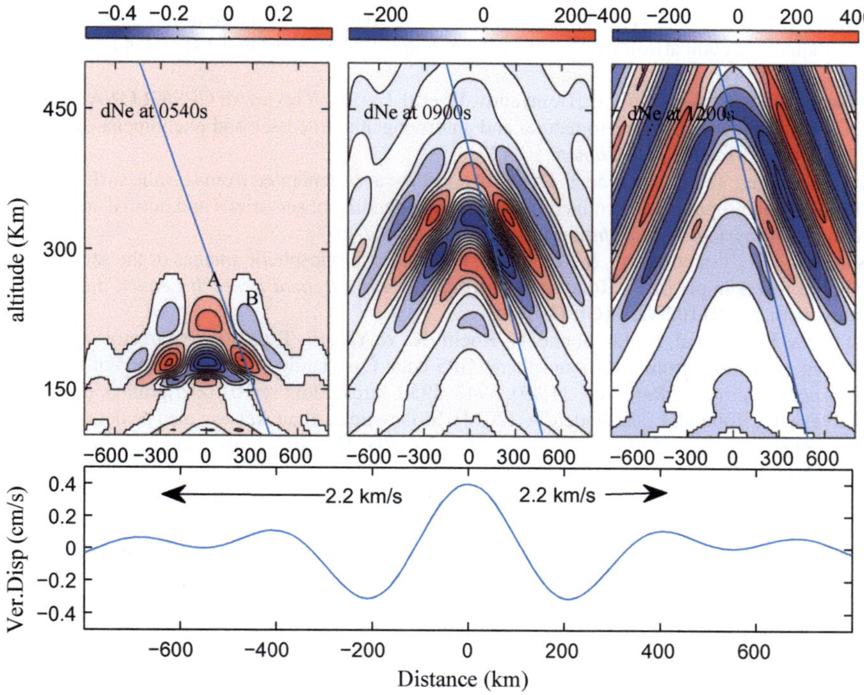

Fig. 10.18 The electron density disturbances (10^{10} electron/m^3) following the moving vertical ground motion source as shown in the bottom. The blue lines in the contour plots are the signal path for station NANO PRN01

direction of the disturbance wave front. This is also evidence that the observed CIDs are the result of the upward acoustic wave rather than the horizontal acoustic wave that could induce ionospheric disturbance traveling with the sound speed. CIDs propagation agrees well with the group velocity of the Rayleigh wave in corresponding regions. Rayleigh wave isotropy should be an important source for the divergence of CID's propagation. After removing the noise by stacking, the strong correlation between the vertical ground motion and CIDs variation is found. The vertical ground motion should be the dominant source of the CIDs and affects the CID waveform directly. Our simulation results confirm the features of electron density perturbation following this earthquake. High correlation could be expected in the GPS TEC series if the signal path is nearly parallel to the disturbance wave front.

References

Afraimovich, E. L., Perevalova, N. P., Plotnikov, A. V., & Uralov, A. M. (2001). The shock-acoustic waves generated by the earthquakes. *Annales Geophysicae, 19*(4), 395–409.

Afraimovich, E. L., Ding, F., Kiryushkin, V., et al. (2010). TEC response to the 2008 Wenchuan Earthquake in comparison with other strong earthquakes. *International Journal of Remote Sensing, 31*(13), 3601–3613.

Afraimovich, E. L., Astafyeva, E., Demyanov, V., et al. (2013). A review of GPS/GLONASS studies of the ionospheric response to natural and anthropogenic processes and phenomena. *Journal of Space Weather and Space Climate, 3,* A27.

Artru, J., Farges, T., & Lognonne, P. (2004). Acoustic waves generated from seismic surface waves: Propagation properties determined from Doppler sounding observations and normal-mode modeling. *Geophysical Journal International, 158*(3), 1067–1077.

Astafyeva, E. I., Lognonne, P., & Rolland, L. (2011). First ionospheric images of the seismic fault slip on the example of the Tohoku-oki earthquake. *Geophysical Reseach Letters, 38,* L22104. https://doi.org/10.1029/2011GL049623.

Astafyeva, E., Rolland, L., Lognonne, P., Khelfi, K., & Yahagi, T. (2013). Parameters of seismic source as deduced from 1 Hz ionospheric GPS data: Case-study of the 2011 Tohoku-oki event. *Journal Geophysical Research, 118*(9), 5942–5950. https://doi.org/10.1002/jgra50556.

Astafyeva, E., Rolland, L., & Sladen, A. (2014). Strike-slip earthquakes can also be detected in the ionosphere. *Earth and Planetary Science Letters, 405,* 180–193. https://doi.org/10.1016/j.epsl.2014.08.024.

Bilitza, D., Altadil, D., Zhang, Y., et al. (2014). The International Reference Ionosphere 2012—A model of international collaboration. *Journal of Space Weather and Space Climate, 4,* A07.

Calais, E., & Minster, J. (1995). GPS detection of ionospheric perturbations following the January 17, 1994, Northridge earthquake. *Geophysical Research Letters, 22*(9), 1045–1048.

Calais, E., Minster, J., Hofton, M., & Hedlin, M. (1998). Ionospheric signature of surface mine blasts from Global Positioning System measurements. *Geophysical Journal International, 132*(1), 191–202.

Cahyadi, M. N., & Heki, K. (2015). Coseismic ionospheric disturbance of the large strike-slip earthquakes in North Sumatra in 2012: Mw dependence of the disturbance amplitudes. *Geophysical Journal International, 200,* 116–129. https://doi.org/10.1093/gji/ggu343.

Dautermann T, Calais, E., Lognonne, P., et al. (2009). Lithosphere–atmosphere–ionosphere coupling after the 2003 explosive eruption of the Soufriere Hills Volcano, Montserrat. *Geophysical Journal International, 179*(3), 1537–1546.

Davies, K., & Baker, D. M. (1965). Ionospheric effects observed around the time of the Alaskan earthquake of March 28, 1964. *Journal of Geophysical Research Atmospheres, 70*(9), 2251–2253.

de Groot-Hedlin, C. (2007). Finite difference modeling of infrasound propagation to local and regional distances. In *Proceedings of the 29th Monitoring Research Review*, Denver, CO (pp. 836–844).

de Groot-Hedlin, C., Hedlin, M., & Walker, K. (2011). Finite difference synthesis of infrasound propagation through a windy, viscous atmosphere: Application to a bolide explosion detected by seismic networks. *Geophysical Journal International, 185*(1), 305–320.

Drob, D. P., Emmert, J., Crowley, G., et al. (2008). An empirical model of the Earth's horizontal wind fields: HWM07. *Journal of Geophysical Research Atmospheres, 113*(A12304), A12304.

Ducic, V., Artru, J., & Lognonne, P. (2003). Ionospheric remote sensing of the Denali Earthquake Rayleigh surface waves. *Geophysical Research Letters, 30*(18), 223–250.

Garcés, M. A., Hansen, R., & Lindquist, K. (1998). Travel times for infrasonic waves propagating in a stratified atmosphere. *Geophysical Journal International, 135*(1), 255–263.

Garrison, J. L., Lee, S., Haase, J., & Calais, E. (2007). A method for detecting ionospheric disturbances and estimating their propagation speed and direction using a large GPS network. *Radio Science, 42*(6), 553–566.

Grawe, M. A., & Makela, J. J. (2015). The ionospheric responses to the 2011 Tohoku, 2012 Haida Gwaii, and 2010 Chile tsunamis: Effects of tsunami orientation and observation geometry. *Earth and Space Science, 2,* 472–483. https://doi.org/10.1002/2015EA000132.

Heki, K., Otsuka, Y., Choosakul, N., Hemmakorn, N., Komolmis, T., & Maruyama, T. (2006). Detection of ruptures of Andaman fault segments in the 2004 great Sumatra earthquake with coseismic ionospheric disturbances. *Journal of Geophysical Research Solid Earth, 111*(B9), 535–540.

Heki, K., & Ping, J. (2005). Directivity and apparent velocity of the co-seismic ionospheric disturbances observed with a dense GPS array. *Earth and Planetary Science Letters, 236,* 845–855.

Jin, S. G., & Park, P. (2006). Strain accumulation in South Korea inferred from GPS measurements. *Earth, Planets and Space, 58*(5), 529–534.

Jin, S. G., Luo, O., & Park, P. (2008). GPS observations of the ionospheric F2-layer behavior during the 20th November 2003 geomagnetic storm over South Korea. *Journal of Geodesy, 82*(12), 883–892.

Jin, R., Jin, S. G., & Feng, G. (2012). M_DCB: Matlab code for estimating GNSS satellite and receiver differential code biases. *GPS Solutions, 16*(4), 541–548.

Jin, S., Jin, R., & Li, J. H. (2014). Pattern and evolution of seismo-ionospheric disturbances following the 2011 Tohoku earthquakes from GPS observations. *Journal of Geophysical Research Space Physics, 119*(9), 7914–7927.

Jin, S. G., Occhipinti, G., & Jin, R. (2015). GNSS ionospheric seismology: Recent observation evidences and characteristics. *Earth-Science Reviews, 147,* 54–64.

Jin, S. G., Jin, R., & Li, D. (2016). Assessment of BeiDou differential code bias variations from multi-GNSS network observations. *Annales Geophysicae, 34*(2), 259–269. https://doi.org/10.5194/angeo-34-259-2016.

Jin, S. G., Jin, R., & Li, D. (2017). GPS detection of ionospheric Rayleigh wave and its source following the 2012 Haida Gwaii earthquake. *Journal of Geophysical Research: Space Physics, 122*(1), 1360–1372. https://doi.org/10.1002/2016JA023727.

Kelley, M. C. (2009). The Earth's ionosphere: Plasma physics and electrodynamics (2nd ed.). San Diego, CA USA: Academic Press (Elsevier). ISBN 978-0-12-088425-4.

Kiryushkin, V. V., & Afraimovich, E. L. (2007). Determining the parameters of ionospheric perturbation caused by earthquakes using the quasi-optimum algorithm of spatiotemporal processing of TEC measurements. *Earth, Planets and Space, 59*(4), 267–278.

Lay, T., Ye, L., Kanamori, H., et al. (2013). The October 28, 2012 Mw7.8 Haida Gwaii under thrusting earthquake and tsunami: Slip partitioning along the queen charlotte fault transpressional plate boundary. *Earth and Planetary Science Letters, 375*(8), 57–70.

Leonard, R. S., & Barnes, R. A. (1965). Observation of ionospheric disturbances following the Alaska earthquake. *Journal of Geophysical Research, 70*(5), 1250–1253.

Leonard, L. J., & Bednarski, J. M. (2014). Field survey following the 28 October 2012 Haida Gwaii tsunami. *Pure and Applied Geophysics, 171*(12), 3467–3482.

Liu, J. Y., Tsai, H., Lin, C., et al. (2010). Co-seismic ionospheric disturbances triggered by the Chi-Chi earthquake. *Journal Geophysical Research, 115,* A08303.

Liu, Y. H., & Jin, S. G. (2019). Ionospheric Rayleigh wave disturbances following the 2018 Alaska earthquake from GPS observations. *Remote Sensing, 11*(8), 901. https://doi.org/10.3390/rs11080901.

Maruyama, T., Tsugawa, T., Kato, H., Ishii, M., & Nishioka, M. (2012). Rayleigh wave signature in ionograms induced by strong earthquakes. *Journal of Geophysical Research, 117*(117). https://doi.org/10.1029/2012ja017952.

Meinig, C., Stalin, S. E., Nakamura, A. I., & Milburn, H. B. (2005). Real-time deep-ocean tsunami measuring, monitoring, and reporting system: The NOAA DART II description and disclosure rep., NOAA, Pacific Marine Environmental Laboratory.

National Oceanic and Atmospheric Administration. (2005). Deep-Ocean Assessment and Reporting of Tsunamis (DART(R)), National Geophysical Data Center, NOAA. https://doi.org/10.7289/v5f18wns.

Occhipinti, G., Kherani, E., & Lognonne, P. (2008). Geomagnetic dependence of ionospheric disturbances induced by tsunamigenic internal gravity waves. *Geophysical Journal International, 173*(3), 753–765.

Occhipinti, G., Rolland, L., Lognonne, P., et al. (2013). From Sumatra 2004 to Tohoku-Oki 2011: The systematic GPS detection of the ionospheric signature induced by tsunamigenic earthquakes. *Journal of Geophysical Research: Space Physics, 118*(6), 3626–3636.

Picone, J. M., Hedin, A. E., Drob, D. P., & Aikin, A. C. (2002). NRLMSISE-00 empirical model of the atmosphere: Statistical comparisons and scientific issues. *Journal Geophysical Research, 107*(A12), 1468. https://doi.org/10.1029/2002JA009430.

Reddy, C. D., & Seemala, G. K. (2015). Two-mode ionospheric response and Rayleigh wave group velocity distribution reckoned from GPS measurement following Mw7.8 Nepal earthquake on 25 April 2015. *Journal of Geophysical Research: Space Physics, 120,* 7049–7059. https://doi.org/10.1002/2015JA021502.

Rolland, L. M., Vergnolle, M., Nocquet, J.-M., Sladen, A., Dessa, J.-X., Tavakoli, F., et al. (2013). Discriminating the tectonic and non-tectonic contributions in the ionospheric signature of the 2011, Mw7.1, dip-slip Van earthquake, Eastern Turkey. *Geophysical Reseach Letters, 40,* 2518–2522. https://doi.org/10.1002/grl.50544.

Rolland, L. M., Lognonne, P., & Munekane, H. (2011). Detection and modeling of Rayleigh wave induced patterns in the ionosphere. *Journal Geophysical Research, 116,* A05320. https://doi.org/10.1029/2010JA016060.

Shinagawa, H., Iyemori, T., Saito, S., & Maruyama, T. (2007). A numerical simulation of ionospheric and atmospheric variations associated with the sumatra earthquake on December 26, 2004. *Earth, Planets and Space, 59*(9), 1015–1026.

Yang, X. R., & Chen, Y. (2007). *Atmospheric acoustics*. Science Press.

Chapter 11
Epicenter from Ionospheric Disturbances

11.1 Earthquake Epicenter

Epicenter is one of the core parameters in seismology. Accurate and rapid determination of the epicenter has important guidance significance for evaluation and rescue of the earthquake destruction. Meanwhile, the epicenter parameter is also the fundamental for the study of earthquake precursory, evolution process, and long-term material migration. In traditional study of seismology, the epicenter is determined by the seismic wave detected by several seismographs, but still has large uncertainty and long-time delay. Recently, with the accuracy and near-real-time improvement of GNSS observations as well as the spatial and temporal resolution, GNSS became a new technology for seismology research. On one side, high-rate GNSS ground deformation data observation results, namely, GNSS seismograph, have been widely used in seismic wave detection, revealing the rupture process, etc (Larson et al. 2003; Bilich et al. 2008). On the other side, the propagation characteristics of obvious co-seismic ionospheric disturbances have provided new opportunity for the determination of the epicenter. Here, we use the co-seismic ionosphere disturbance for the inversion of the epicenter and its validation.

The ionospheric $TEC_i(t)$ time series can be described as the sum of the background change and the seismic ionospheric disturbance from the dual-frequency GNSS observations:

$$TEC_i(t) = TEC_{T,i}(t) + dTEC_{S,i}(t, v, \lambda, \varphi, h) + dTEC_{n,i(t)} \quad (11.1)$$

where the first item denotes the ionospheric background change and the second item denotes the seismic ionospheric disturbance signal related to the time, disturbance propagation velocity, the longitude, latitude, and height of the disturbance source. In Eq. 11.1, the background change can be eliminated by filtering, deduction of the trend, and polynomial fitting. Thereafter, the seismic ionospheric disturbance from GNSS can be described as (Jin et al. 2015, 2017; Jin 2018):

S. Jin et al., *GNSS Atmospheric Seismology*,
https://doi.org/10.1007/978-981-10-3178-6_11

$$dTEC_i(t) = dTEC_{S,i}(t, v, \lambda, \varphi, h) + dTEC_{n,i(t)} \tag{11.2}$$

Therefore, dense dual-frequency GPS network can determine the seismic ionospheric disturbances. If such disturbances are generated by the earthquake and propagating from the epicenter, it may determine the epicenter.

11.2 Hypothesis of Epicenter Estimation

The spatial and temporal variation of the Earth's ionospheric TEC is complex and the seismic ionospheric disturbances and coupled process are complicated. Hence, a series of rational simplification are needed in estimating the epicenter from the seismic ionosphere disturbances. First, the single-layer hypothesis of the ionospheric TEC estimation is used, namely, that the electrons are accumulated in the thin layer in 300–500 km height (Jin et al. 2016). This hypothesis formulates the variation of the electrons in the ionosphere in two dimensions with the temporal change (Jin and Park 2007). Although this hypothesis omits the vertical change of the ionosphere, the research showed that the two-dimensional ionospheric TEC could describe accurately the horizontal and temporal characteristics of the ionospheric disturbance (Jin et al. 2008, 2012). In the ionosphere single-layer hypothesis, the ionospheric spatial and temporal change can be formulated as

$$TEC(\lambda_{IPP}, \varphi_{IPP}, t) = TEC_0(\lambda_{IPP}, \varphi_{IPP}, t) + dTEC(\lambda_{IPP}, \varphi_{IPP}, t) + \varepsilon_{TEC(\lambda_{IPP}, \varphi_{IPP}, t)} \tag{11.3}$$

where λ_{IPP} and φ_{IPP} denote the position of the ionospheric puncture point, t is the epoch of the observation, $\varepsilon_{TEC(\lambda_{IPP}, \varphi_{IPP}, t)}$ is the noise of observations, and TEC_0 and $dTEC$ denote the change of the background and the ionospheric disturbance change, respectively. For TEC_0, the spatial-temporal regular changes mainly include the solar activity, the long-term existing and slow-changing magnetic field factor, and daily variations. However, for $dTEC$, the disturbance driving forces are unpredictable and subject to various instantaneous sudden incident. Commonly the regular change of the background TEC can be removed by the envelope method, polynomial fitting method, filtering, moving-average method, etc., and then the ionospheric disturbance change can be extracted from the GNSS observation sequence.

The second hypothesis is that the earthquake epicenter is the ionospheric disturbance source. This hypothesis is the foundation for the epicenter inversion using co-seismic ionospheric disturbances. On one side, the past and previous studies from the co-seismic ionospheric disturbance characteristics showed that one of the most obvious characteristics of co-seismic ionospheric disturbances are propagating with circling the epicenter. It is commonly accepted that the co-seismic ionospheric disturbances are caused by the coupling of the earthquake and the ionosphere. The spatial scale of the earthquake rupture fault is normally within 1500 km, much smaller with respect to the scale of the disturbance of the ionosphere. Within the scale of several hundred to thousands of kilometers, co-seismic ionospheric disturbance source is the point-like disturbance located in the epicenter, which is reasonable. On the other side, other kinds of ionospheric disturbances with different frequencies are removed by the filtering. The hypothesis is one of the key characteristics.

Study from the GNSS seismic ionospheric disturbances showed that the velocity of propagation of co-seismic ionospheric disturbance ranges from several hundred to thousand meters per second (Jin et al. 2014). For a specific earthquake in specific location and time period, the ionospheric disturbance velocity is normally stable. Hence, it is acceptable to assume that the seismic ionospheric disturbance velocity is a constant value for specific earthquake epicenter location estimation. However, the seismic ionospheric disturbance velocity is commonly unknown, which needs to be estimated by the epicenter at the same time. Under the hypothesis of epicenter disturbance source and the constant propagation velocity hypothesis, the inversion parameters are the location of the point source of disturbance (namely, the latitude and longitude of the epicenter) and the propagation velocity of the co-seismic ionospheric disturbance.

11.3 Method of Epicenter Estimation

Here, the Quasi-Optimum Algorithm (QOA) is used for epicenter estimation (Kiryushkin and Afraimovich 2007). The seismic ionospheric disturbance is supposed to be spherical fluctuation with uniform velocity in different orientations, and the ionospheric disturbance source is the secondary source of the epicenter ground deformation above the epicenter with specific height, then the ionospheric disturbance source location could be inversed by QOA method from the sequence data of disturbance from GNSS observation stations. Kiryushkin and Afraimovich (2007) validated the effectiveness of the QOA method by the Hokkaido earthquake on September 25, 2003 and the Honshu earthquake in September 5, 2004. The validation results showed the ionospheric disturbance source location from the seismic ionospheric disturbance fits well with the epicenter longitude and latitude, and also located in 300–400 km height, with disturbance velocity at 400–900 m/s. However, the two earthquakes are both acoustic disturbance location in earthquake sky with oblique propagation mode (the red disturbance mode in Fig. 11.1). The acoustic wave triggered by the earthquake fault dislocation in the process of propagation is on account of the change of atmospheric density and the direction of propagation deflects. Its propagation direction is nearly horizontal when arrived at the height of the ionospheric layer. Considering that the acoustic wave with being much bigger zenith angle cannot reach the ionospheric height, the horizontal wave detected by the ionosphere is mainly caused by the acoustic wave with much smaller acoustic wave zenith angle (nearly vertical). Hence, the horizontal element of the disturbance is subtle before obvious deflection of the disturbance, and the range of disturbance is around the sky above the epicenter. The secondary source above the epicenter

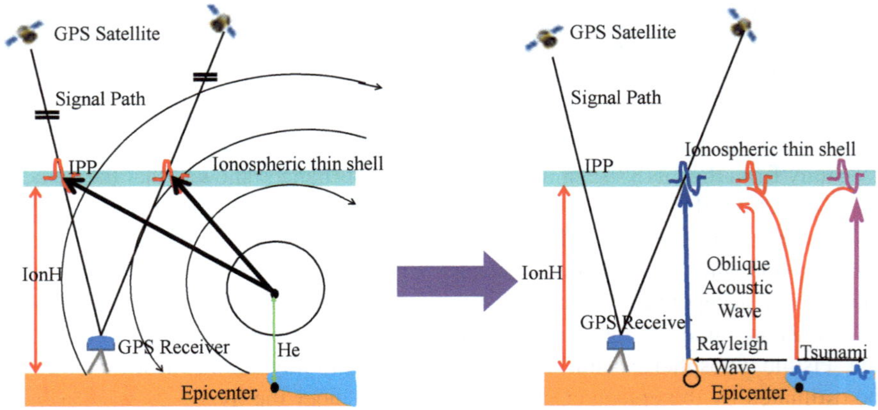

Fig. 11.1 Co-seismic ionospheric disturbances for epicenter inversion

hypothesis is reasonable. However, the seismic ionospheric disturbance consists of not only horizontal acoustic wave disturbance but also Raleigh wave disturbance (the disturbance in blue color in Fig. 11.1) and the tsunami gravity wave disturbance (the disturbance in pink color in Fig. 11.1) (Heki 2006; Jin et al. 2014).

Since the co-seismic Rayleigh wave or the tsunami gravity wave detected by GNSS was not essentially from the horizontal propagated disturbance at the height of ionosphere layer, but from ground/sea surface Rayleigh wave or gravity wave, the secondary source hypothesis is not suitable for the case. In a small region (range less than 1000 km from the epicenter), the Earth background atmosphere layer has strong spatial correlation with the ionosphere layer, and the vertical propagation of the disturbance triggered by the earthquake is almost the same. The time delay of the disturbance from the ground to the ionosphere is nearly identical. The ionospheric disturbance detected by GNSS is from the ground or the sea surface disturbance. Using the QOA method for inversion of the co-seismic ionospheric Rayleigh wave or tsunami gravity wave, the source location will distract from the real location, namely, the epicenter.

The horizontal propagation velocity maintains the same vertical propagation velocity of the disturbance from the ground to the ionosphere, which showed that the considerable variation is due to the change of the parameters such as the atmospheric density. Under this case, the uniform propagation velocity model cannot fit with the real case in ionospheric disturbance. Hence, during the process of epicenter location inversion, we omit the hypothesis of ionospheric secondary disturbance and consider that the source of the seismic ionospheric disturbance is, namely, the epicenter. Although different modes of seismic ionospheric disturbances have different formation mechanisms, the core cause is the earthquake epicenter fault rupture. From the aspect of the physical formation mechanism of disturbance, the hypothesis seems to be reasonable. Besides, considering the fact that the velocity of disturbance changes during its vertical propagation process, normally supposed that the horizontal propagation velocity of disturbance is fixed to substitute the uniform spherical propagation velocity hypothesis in QOA algorithm.

Based on the epicenter point source and the uniform horizontal propagation velocity hypothesis, according to the seismic ionospheric disturbance extraction method, the disturbance sequence from the GNSS detected ionospheric sequence deducted from the background trend can be formulated as

$$dTEC_i(\lambda_{IPP}, \varphi_{IPP}, t) = TEC_{s,i}(\vec{r}, v_h, t) + TEC_{n,i}(\lambda_{IPP}, \varphi_{IPP}, t) \tag{11.4}$$

where $TEC_{s,i}(\vec{r}, v_h, t)$ is the seismic ionospheric disturbance at the time of t generated by the epicenter rupture and propagated at the horizontal speed of v_h to the puncture point, $\vec{r}$ is the horizontal distance between the epicenter and the puncture point, and i denotes the number of sequence data of the GNSS ionospheric observation. From the previous discussion, the frequency of the ionospheric disturbance is at the level of several milli-Hertz (mHz), lasting only for several minutes. The drift of the puncture point caused by the satellite movement is much smaller. Hence, the horizontal gradient element in the ionospheric observation sequence is negligible in the following discussion. $dTEC_i(\lambda_{IPP}, \varphi_{IPP}, t)$ and $TEC_{n,i}(\lambda_{IPP}, \varphi_{IPP}, t)$ are simplified as $dTEC_i(t)$ and $TEC_{n,i}(t)$.

Supposed that the high-frequency noise in the observation sequence is Gaussian noise, the convolution of the observation sequence and the initial disturbance signals is expressed as (Afraimovich et al. 2001)

$$\begin{aligned} q &= \frac{1}{N_n}\int_{t_1}^{t_2}\sum_{i=1}^{M}[dTEC_i(t)B_i(\tau_i)]R_0(t)dt \\ &= \sum_{i=1}^{M}\frac{1}{N_n}\int_{t_1}^{t_2} dTEC_{\Sigma,i}(t, \tau_i)R_0(t)dt \\ &= \sum_{i=1}^{M} q_i \end{aligned} \tag{11.5}$$

where N_n denotes the noise frequency spectrum and $B_i(\tau_i)$ denotes the phase gain of the TEC sequence. For the non-harmonic signal, $B_i(\tau_i)$ denotes the ith time delay of the disturbance sequence $dTEC$, $R_0(t)$ denotes the initial disturbance signal, namely, the ionospheric disturbance source, $dTEC_{\Sigma,i}$ denotes the additional sequence after the time delay correction of the ith observation, and M denotes the number of observations. Normally, the initial disturbance is unknown and $dTEC_{\Sigma,i}$ substitutes it in real processing, then the signals can expressed as

$$q_i = \frac{1}{N_n}\int_{t_1}^{t_2}[dTEC_{\Sigma,i}(t,\tau_i)]^2 dt$$
$$= \frac{1}{N_n}\left[\int_{t_1}^{t_2} dTEC_{\Sigma,i-1}(t)^2 dt + 2\int_{t_1}^{t_2} dTEC_{\Sigma,i-1}(t) dTEC_i(t) dt + \int_{t_1}^{t_2} dTEC_i(t)^2 dt\right] \quad (11.6)$$

where q_i stands for the power of the observation noise of the ith observation sequence and the time delay τ_i can be determined by $dq_i/d\tau_{i_{\hat{\tau}_i=\tau_i}} = 0$.

The first item in the right end of Eq. (11.6) is the energy accumulation of the $i-1$ sequence, and the third item is the energy of the ith TEC sequence. Both of them are irrelevant form the time delay. The second item is twice of the correlation of the $i-1$th additional item and the ith TEC item. Supposed that the correlation is r_i, q_i reached the maximum value when r_i reached the biggest, where $dr_i/d\tau_{i,\widehat{\tau_i}=\tau_i} = 0$. The time delay is computed from the observation sequence data, namely, stacking operation after the time delay correction:

$$dTEC_{\Sigma,i}(t) = dTEC_{\Sigma,i-1}(t) + dTEC_i(t - \hat{\tau}_i) \quad (11.7)$$

For the ith TEC sequence, the stacking can be formulated as

$$dTEC_{\Sigma,i}(t) = dTEC_0(t) + \sum_{j=1}^{i} dTEC_j(t - \hat{\tau}_j) \quad (11.8)$$

where $dTEC_0(t)$ is the central reference sequence. In real observation, not all observation sequences could detect the corresponding disturbance. For stacking operation on each sequence, the central parametric sequence should reflect the corresponding seismic ionospheric disturbances. Then, we can determine if it was disturbance sequence through the correlation with the following sequence and make efficient stacking operations. Normally, the sequence with the largest correlation coefficient was selected as the central sequence. We can get the maximum-likelihood value of q through the TEC stacking operation. In the stacking operation, the repeated signals will gain through stacking, while the Gaussian noise will be weakened. The stacked disturbance signal is the average of the ionospheric disturbance from GPS station observations. Here, the characteristic time was selected as the time t_0 for the maximum disturbance signal and the corresponding puncture point $(\lambda_{IPP}, \varphi_{IPP})$ of the maximum disturbance is the location of disturbance.

The epicenter is estimated from co-seismic ionospheric disturbances, namely, solving the epicenter latitude and longitude as well as the horizontal propagation velocity of disturbance v_h. For every combination of parameters, we can calculate the arriving $t_{i,k}$ of the puncture point for each item i of the sequence. Here, $t_{i,k}$ meets the following relation:

$$(t_{i,k} - t_0) - \frac{\Delta d_{i,k}}{v} = \min \quad (11.9)$$

where $\Delta d_{i,k} = d_{i,k} - d_0$ is the distance difference between the supposed earthquake epicenter location to the puncture point for the central reference sequence and the puncture point for the ith sequence at the k epoch, and v is the disturbance propagation velocity. Considering the fact that within several minutes, the vertical propagation velocity can be considered as a constant, the ratio of the vertical velocity and the horizontal velocity is also constant and can be expressed as

$$\frac{\Delta d_{i,k}}{v} = \int_s \frac{\sqrt{\Delta d_{h,i,k}^2(s) + \Delta d_{v,i,k}(s)^2}}{\sqrt{v_v(s)^2 + v_h(s)^2}} ds = \int_s \frac{\Delta d_{h,i,k}(s)}{v_h(s)} ds = \frac{\Delta d_{h,i,k}}{v_h} \tag{11.10}$$

where h and v denote the horizontal and vertical directions, respectively. For horizontal distance $\Delta d_{h,i,k}$, it can be determined by the projection of the puncture point and the epicenter. When Eq. 11.9 gets the minimum, let $\tau_i^m = t_{i,\hat{k}} - t_0$, it is the disturbance time delay solved from the given epicenter and the propagation velocity. Then, the total energy for each TEC sequence given by the supposed epicenter and the disturbance velocity is

$$q^m(\lambda_{ep}, \varphi_{ep}, v_h) = \frac{1}{N_n} \sum_{i=1}^{M} \int_{t_1}^{t_2} [dTEC(t, \tau_i^m(\lambda_{ep}, \varphi_{ep}, v_h)]^2 dt \tag{11.11}$$

If the energy solved by the time delay determined by the supposed epicenter and the disturbance velocity coincide with that solved by the time delay from the cross-correlation, it means that the corresponding parameter was close to the real situation and the observation problem. Supposed that the ratio of the parametric disturbance and the real disturbance energy is

$$C(\lambda_{ep}, \varphi_{ep}, v_h) = \frac{q^m(\lambda_{ep}, \varphi_{ep}, v_h)}{q\mathrm{max}} = \frac{\sum_{i=1}^{M} \int_{t_1}^{t_2} [dTEC_{\Sigma,i}^m(t, \tau_i^m(v, \lambda, \phi, h))]^2 dt}{\sum_{i=1}^{M} \int_{t_1}^{t_2} [dTEC_{\Sigma,i}(t, \hat{\tau}_i)]^2 dt} \tag{11.12}$$

We can get a solution for $C(\lambda_{ep}, \varphi_{ep}, v_h)$ from different supposed epicenter locations and horizontal propagation velocities, when C reaches its extreme value. Then, we can search for the optimal epicenter location and disturbance horizontal velocity. When the disturbance initial location and the propagation velocity are determined, the disturbance starting time could be retrieved from

$$t_e = t_0 - \frac{\rho_{0e}}{v} \tag{11.13}$$

where $\rho_{0,e}$ is the distance between the central reference signal location and the disturbance source.

Since the seismic ionospheric disturbance has different modes, the propagation velocity differs from each other. If it does not differ from the ionospheric disturbance detected by GNSS, the hypothesis that the horizontal disturbance velocity is constant is obviously invalid and will cause the deriving result of epicenter, and the disturbance velocity detracts from the real situation. Therefore, the modes should be distinguished before the inversion of the epicenter. From the discussion of the previous chapters, it is essential to distinguish the disturbance from two aspects. On one side, the disturbance frequency needs to be distinguished. The seismic ionospheric layer disturbance needs to be separated as the acoustic wave and the gravity wave. From the previous studies, the disturbance frequency is 3–5 mHz and later the frequency is less than 3 mHz. Thus, the acoustic wave and the gravity wave could be separated by choosing appropriate filtering band. For the other aspect, one could separate the disturbance wave front. The seismic ionospheric disturbance is not only caused by the acoustic wave disturbance from the seismic Rayleigh wave but also from the oblique propagated ionospheric disturbance existing above the earthquake sky due to the deflection of the acoustic wave. The previous studies showed that the propagation velocity is the Rayleigh wave velocity in the local region, which is much bigger than the acoustic velocity below the atmosphere ionosphere. Hence, the propagated wave front is nearly vertical. The GPS observation from low elevation angle is much more sensitive to the ionospheric disturbance caused by the Rayleigh wave. However, the wave front of the tilt acoustic wave is nearly horizontal, and observation from the higher elevation angle is more sensitive to this kind of disturbance.

11.4 Epicenter Estimation and Validation

Here, we estimate the epicenter for the 2005 Mw = 7.2 Northern California earthquake with the point-source disturbance hypothesis in Fig. 11.1, which is located near the PBO observation network with relatively rich observation data and convenient for the validation. This earthquake happened at 02:50:54 UT, June 15, 2005 with an epicenter location at (N41.8°, W126.0°) inversed by the seismograph observation. Since it has no record for large tsunami after the main shock moments, here we only consider the acoustic disturbance of the earthquake and the zero-phase fourth-order Butterworth filtering is used for the co-seismic ionospheric disturbance extraction. Here, the data for the co-seismic ionospheric disturbance are from 504 stations observations within 1500 km of the epicenter collected by UNAVCO.

11.4.1 Disturbance Modes at Different Elevation Angles

Figure 11.2 shows the time offset of the co-seismic ionospheric disturbance at different elevation angles. Here, the epicenter is provided by the CMT catalogue at Harvard University. The observations showed that, after the earthquake, obviously two types of co-seismic ionospheric disturbances existed near the epicenter, namely, the disturbance with propagation velocity detected at low elevation angles (10–30°) with a speed of 2.2 km/s and disturbance detected at high elevation angles (40–50°) with a speed of 1.4 km/s. The rapidly propagated ionospheric disturbance existed at a distance about 500–600 km from the epicenter and the slowly propagated ionospheric disturbance diminished at a distance less than 300 km from the epicenter. However, observations from the other elevation angles did not show obvious ionospheric disturbances within the same range from the epicenter. Figure 11.3 is the seismic ionospheric disturbance distribution graph 10–20 min after the earthquake. Obvious disturbances happened mainly on the southeast of the epicenter, so the cause for the difference of the co-seismic ionospheric disturbances with different propagation modes at different elevation angles is not the orientation difference. This not only demonstrated that the angle between the signal path and the disturbance wave front directly affects the GNSS seismic ionospheric disturbance amplitude (the smaller the angle is, the weaker the disturbance is, vice versa) but also showed the validity of distinguishing the modes of disturbance using elevation angles.

As showed in Fig. 11.3, co-seismic ionospheric disturbances detected by the GNSS observation with elevation angles at 10°–20° range, 20°–30° range, and 40°–50° range showed obvious epicenter propagation characteristics, with different propagation velocities. Considering the propagation velocity and the wave front of the two disturbance modes, we think the disturbances are Rayleigh wave disturbance and the tilt acoustic wave above the epicenter. The disturbances propagated at a speed of 1.4 km/s are mainly detected by observations with elevation angles at 40°–50° and no such kind of disturbances are detected from observations at higher elevation angle. This showed that the horizontal element of the acoustic wave at the epicenter is very small, mainly tilt acoustic wave instead of nearly horizontal when reaching the ionosphere layer, and thus the horizontally propagated wave is small and the amplitude of the ionospheric disturbance will diminish quickly in the process of reaching the above, which is also one of the reasons that the range of disturbances in this kind is very small. Note that the disturbance in this mode here is in fact bigger than the acoustic wave velocity at the height of the ionosphere (about 1 km/s), possibly related to the background wind field, etc., which needs more analysis and investigations in further step.

11.4.2 Epicenter Estimation Results

From previous chapters, the velocity of the horizontal ionospheric disturbance triggered by the Rayleigh wave has high coincidence with the Rayleigh wave of the ground. Meanwhile, due to the crustal structure of the Earth, the propagation velocity of the Rayleigh wave in different directions of the epicenter has some variations, and thus the ionospheric disturbances caused by the Rayleigh wave are also different. For large area all direction Rayleigh waves disturbance, the hypothesis that the horizontal disturbance velocity is constant can no longer meet well with the real situation.

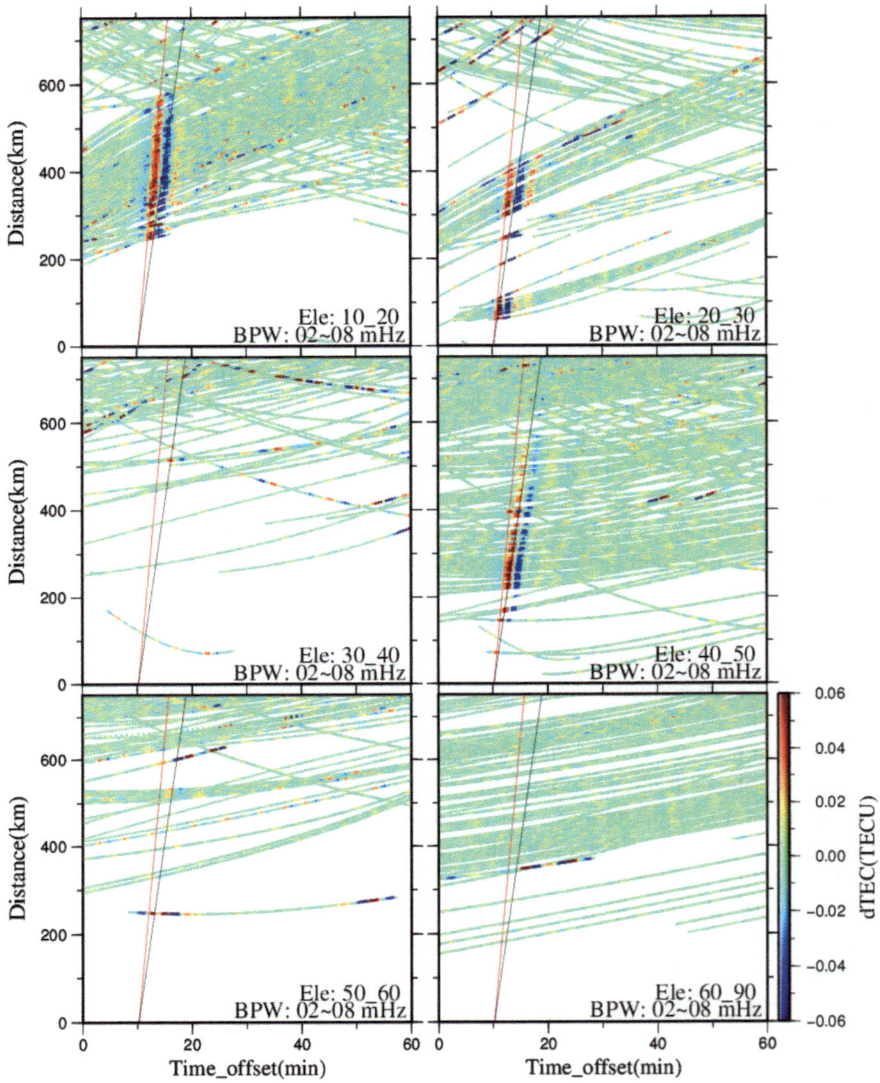

Fig. 11.2 Time offset for the co-seismic ionospheric disturbance from different elevation angles of the Northern California Mw = 7.2 earthquake on June 15, 2005. The horizontal axis stands for the time delay of the observation epoch and the main earthquake time (UTC02:50:54), and the vertical axis is the distance between the puncture point and the epicenter (N41.8°, W126.0°). The red line and the black line are the reference line for the propagation velocity of 2.2 and 1.4 km/s

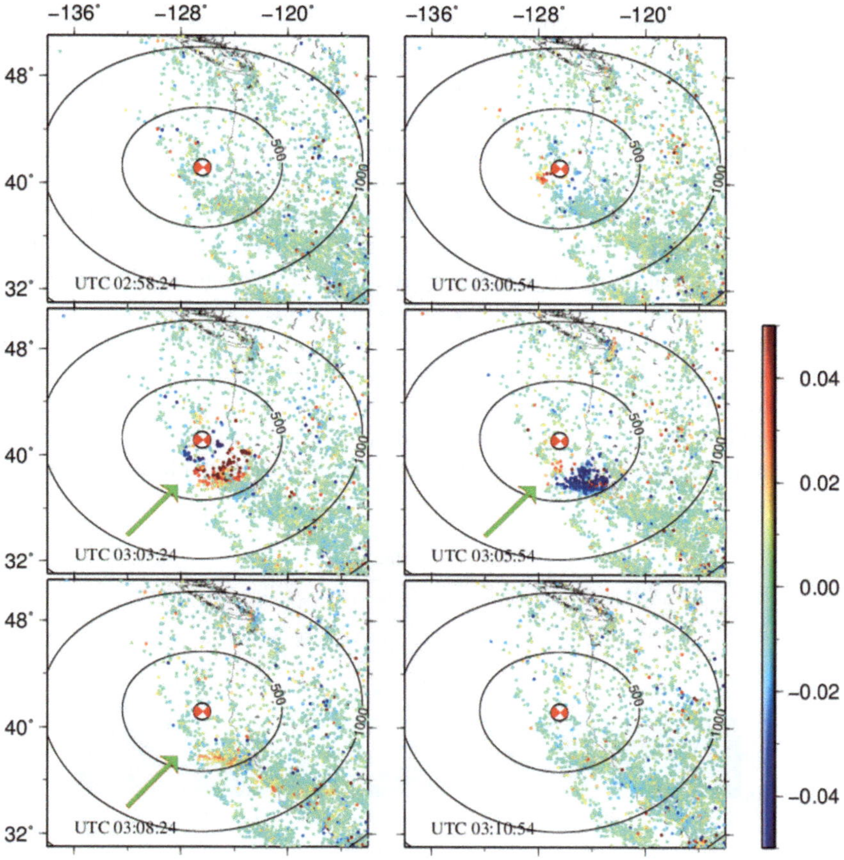

Fig. 11.3 The two-dimensional TEC disturbances around the earthquake epicenter in North California at 02:58:24–03:10:54 UT, June 15, 2005. The bigger circle denotes the equal-spaced lines for 500 and 1000 km

However, in this case the observation from GNSS in lower elevation angles has uniform velocity mode, showing that the difference in this region has minute difference in Rayleigh wave, and therefore the constant hypothesis is reasonable. Furthermore, the velocity of the acoustic wave is not only related to the atmospheric density but also temperature profile. The background field near the epicenter is similar, and thus the nearly constant velocity is reasonable.

Since no strong tsunami record existed after the earthquake and the frequency of the ionospheric disturbance proposed here is in the range of 2–8 mHz. The elevation angles here corresponding with the GNSS observed disturbance with extreme value can be used directly to distinguish different modes of disturbances, which are used for epicenter and disturbance horizontal velocity estimation. For improvement of the searching efficiency, firstly one needs to search parameters with a larger change step, and then a second search is conducted with a smaller step after a coarse location estimation so as to determine the epicenter rapidly.

Figure 11.4 is the epicenter inversion result using the 10–30 elevation angle detected by seismic ionospheric disturbances. The epicenter is estimated at (N42°, W126°), and the ionospheric horizontal disturbance velocity is 2.49 km/s. Here, the horizontal disturbance velocity is the peak velocity corresponding with the assumed velocity and the fitting of *C* value. Since the estimated epicenter has minute difference with the result from the CMT catalogue, for a more precise estimation of the epicenter, the searching range of the epicenter should be narrowed to N38–44°, W129–123° with a step of 0.1°, which is shown in Fig. 11.5. The extreme value of *C* is 0.992, the corresponding epicenter is (N42.4, W126.2), and the horizontal velocity of the disturbance is 2.48 km/s. The result showed that the difference between the epicenter for a twice search and the result from the seismograph is within 1°, and the horizontal disturbance velocity estimated coincides with the time-offset analysis. For similar twice search for the ionospheric disturbance with elevation angles at

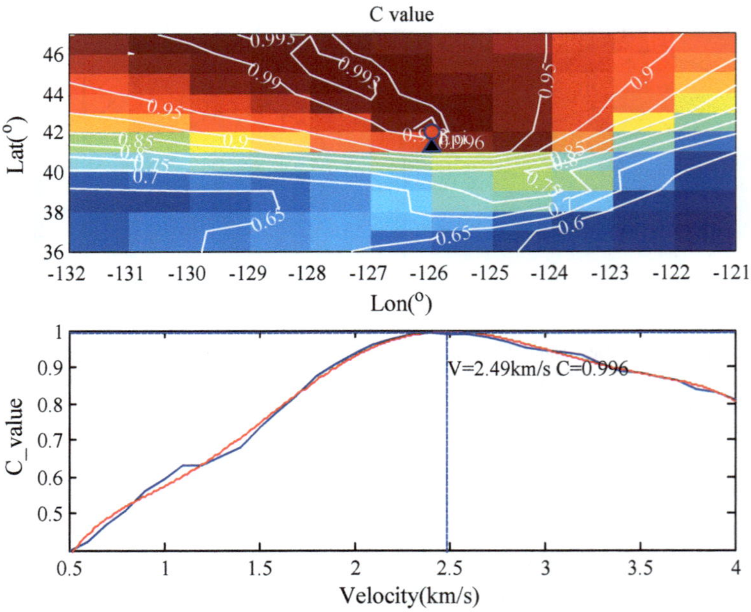

Fig. 11.4 The energy ratio *C* with different assumed epicenters and horizontal disturbance velocities. The upper panel showed the distribution of *C* with different epicenter locations (N36°–52°, W137°–121°) and the disturbance velocities when *C* reaches the extreme value. The bottom panel showed the variation of *C* with different disturbance velocities and the epicenters when *C* reaches the extreme value. The data used here are from observations with elevation angles at 10°–30°, the searching step is 1° in latitude and longitude, respectively, and the searching step for the velocity is 0.05 km/s. The red line below is the polynomial fitting result of *C*. The triangle is the epicenter location from the CMT catalogue. The dot is the epicenter corresponding with the *C* at extreme value

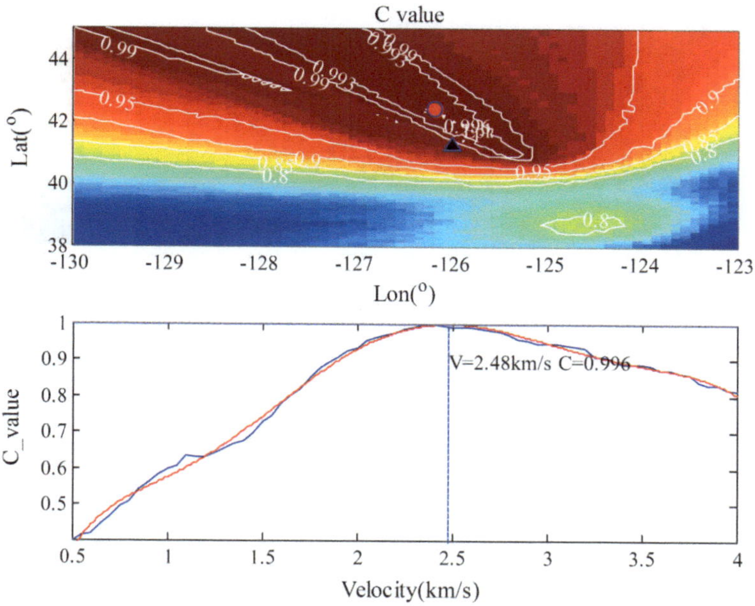

Fig. 11.5 Estimation of the epicenter and the horizontal disturbance velocity from seismic ionospheric disturbances with elevation angles between 40° and 50°. The searching step for the latitude and longitude is 0.1° and the step for the velocity search is 0.05 km/s. The legend is the same as Fig. 11.3

40–50°, which is shown in Fig. 11.6, the extreme value of C is 0.928, the epicenter is (N40.6°, W125.9°), and the disturbance propagation velocity is 1.43 km/s. The epicenter inversion result from the seismic ionospheric disturbances for both modes showed difference less than 1° in longitude and latitude with the epicenter determined by the seismograph, and it can get an accurate inversion result for the ionospheric disturbance horizontal velocity.

However, it needs to point out that, no matter the inversion of the Rayleigh disturbance and the tilt acoustic epicenter inversion, located on the western-north side of the epicenter, the change of C is very small in the process of the epicenter approaching the center of the earthquake. The C value is extremely high in the several degrees range of the western-north side of the epicenter, which may be due to the fact that the seismic disturbance is mainly distributed along the southern-east side of the epicenter (Fig. 11.3). When the given epicenter goes western-north side of the real epicenter, the propagation velocity is high and the result also coincides with the limited observation. Therefore, the method could get epicenter estimation with higher precision when the same type of ionospheric disturbance is in wider range.

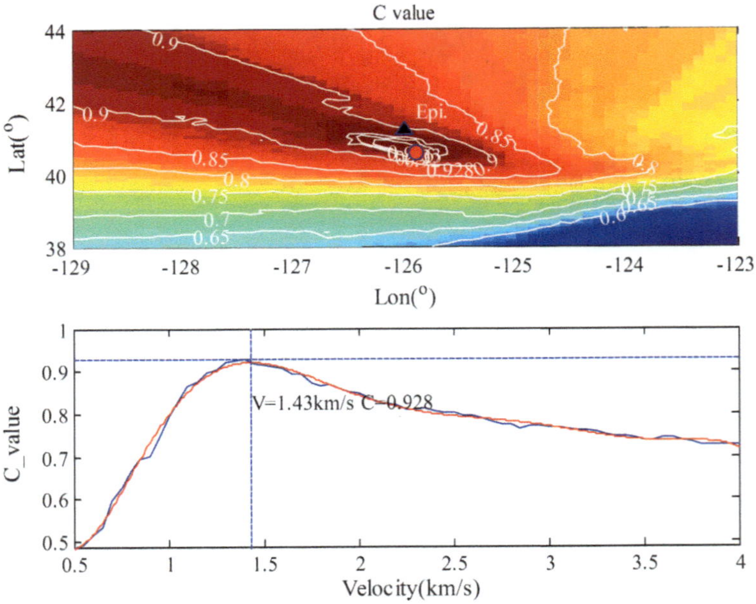

Fig. 11.6 Binary search results for the earthquake epicenter and the horizontal velocity estimation at 40°–50° elevation angle. The legend is the same as Fig. 11.3

11.5 Summary

Based on the QOA algorithm, the epicenter inversion method of the seismic ionospheric disturbances is presented. Based on the principle of model time delay correction for the disturbance superimposed energy and cross-correlation time delay correction for disturbance superimposed energy, the epicenter in two-dimensional location and the disturbance horizontal velocity in three dimensions are estimated for the 2005 Mw = 7.2 North California Earthquake. Besides, we demonstrated that the elevation angle can distinguish the Rayleigh acoustic wave and the tilt acoustic disturbance above the epicenter. The effectiveness of the method has been validated by the seismometer. Both the Rayleigh acoustic wave and the acoustic wave of seismic ionospheric disturbances can inverse the epicenter location and the horizontal disturbance propagation velocity. On the one hand, one can validate the effectiveness of the epicenter point-source hypothesis. On the other hand, it provides a new approach to estimate the epicenter with potential applications.

References

Afraimovich, E. L., Chernukhov, V. V., & Kiryushkin, V. V. (2001). Spatial-temporal characteristics of the ionospheric disturbance induced by shock acoustic waves due to rocket launching. *Journal of Communications Technology & Electronics, 46*(11), 1199–1206.

Bilich, A., Cassidy, J. F., & Larson, K. M. (2008). GPS seismology: Application to the 2002 MW 7.9 Denali fault earthquake. *Bulletin of the Seismological Society of America, 98*(2), 593–606.

Heki, K. (2006). Explosion energy of the 2004 eruption of the Asama Volcano, central Japan, inferred from ionospheric disturbances. *Geophysical Research Letters, 33*(14).

Jin, S. G. (2018). Two-mode ionospheric disturbances following the 2005 Northern California offshore earthquake from GPS measurements. *Journal of Geophysical Research: Space Physics, 123*(10), 8587–8598. https://doi.org/10.1029/2017JA025001.

Jin, R., Jin, S. G., & Feng, G. P. (2012). M_DCB: Matlab code for estimating GNSS satellite and receiver differential code biases. *GPS Solutions, 16*(4), 541–548. https://doi.org/10.1007/s10291-012-0279-3.

Jin, S. G., Jin, R., & Li, J. H. (2014). Pattern and evolution of seismo-ionospheric disturbances following the 2011 Tohoku earthquakes from GPS observations. *Journal of Geophysical Research: Space Physics, 119*(9), 7914–7927. https://doi.org/10.1002/2014JA019825.

Jin, S. G., Jin, R., & Li, D. (2016). Assessment of BeiDou differential code bias variations from multi-GNSS network observations. *Annales Geophysicae, 34*(2), 259–269. https://doi.org/10.5194/angeo-34-259-2016.

Jin, S. G., Jin, R., & Li, D. (2017). GPS detection of ionospheric Rayleigh wave and its source following the 2012 Haida Gwaii earthquake. *Journal of Geophysical Research: Space Physics, 122*(1), 1360–1372. https://doi.org/10.1002/2016JA023727.

Jin, S. G., Luo, O., & Park, P. (2008). GPS observations of the ionospheric F2-layer behavior during the 20th November 2003 geomagnetic storm over South Korea. *Journal of Geodesy, 82*(12), 883–892. https://doi.org/10.1007/s00190-008-0217-x.

Jin, S. G., Occhipinti, G., & Jin, R. (2015). GNSS ionospheric seismology: Recent observation evidences and characteristics. *Earth Science Reviews, 147,* 54–64. https://doi.org/10.1016/j.earscirev.2015.05.003.

Jin, S. G., & Park, J. (2007). GPS ionospheric tomography: A comparison with the IRI-2001 model over South Korea. *Earth Planets Space, 59*(4), 287–292.

Kiryushkin, V. V., & Afraimovich, E. L. (2007). Determining the parameters of ionospheric perturbation caused by earthquakes using the quasi-optimum algorithm of spatiotemporal processing of TEC measurements. *Earth, Planets and Space, 59*(4), 267–278.

Larson, K. M., Bodin, P., & Gomberg, J. (2003). Using 1-Hz GPS data to measure deformations caused by the Denali fault earthquake. *Science, 300*(5624), 1421–1424.

Chapter 12
Tsunami Ionospheric Disturbances

12.1 Introduction

Big earthquakes and tsunamis can excite acoustic resonance between the land or ocean surface and the lower atmosphere. The wave amplitude increases against the decrease of the atmosphere density, making the upper atmosphere a good medium to propagate the gravity waves. Some of the resonance waves leak upward into the ionosphere and trigger ionospheric anomalies, which can be observed as Traveling Ionospheric Disturbances (TIDs) (Crowley et al. 2016; Rolland et al. 2010). These TIDs generated energy by ocean waves were first established by Daniels (1952) and the theory was developed by Hines (1972). There are different methods to detect these TIDs, including radar altimeter, incoherent scatter radar, ground-based GPS network as well as GPS radio occultation performed between a GPS satellite and a Low Earth Orbit (LEO) satellite. For example, Coisson et al. (2015) detected the ionospheric disturbances following the March 11, 2011 earthquake and tsunami off coast of Tohoku using GPS radio occultation. The gravity waves were also detected in neutral atmosphere by the Gravity Field and Steady-State Ocean Circulation Explorer (GOCE) satellite with low orbiting altitude of 250 km (Garcia et al. 2015).

Global Positioning System (GPS) is a powerful tool to observe the atmospheric or ionospheric response to the earthquakes and tsunamis through GPS atmospheric delay signals, especially for regions with dense GPS continuous operating stations. The GPS delays caused by the tropospheric and ionospheric refraction when the GPS signals pass through the Earth's atmosphere include lengthening of the geometric path and variation of signal's group velocity and phase velocity. The GPS dual-frequency (L1 and L2) observations can estimate ionospheric delay or Total Electron Content (TEC). GPS TEC was the first time used as a sensor to estimate seismo-ionospheric disturbances (Calais and Minster 1995). Later GPS TEC was more applied to earthquakes and tsunami study. For instance, the detailed seismo-ionospheric disturbances of the 2008 Wenchuan earthquake and the 2011 Japan earthquake and tsunami were studied by Jin et al. (2015).

S. Jin et al., *GNSS Atmospheric Seismology*,
https://doi.org/10.1007/978-981-10-3178-6_12

The ionospheric disturbance characteristics after the earthquakes or tsunamis were widely investigated based on GPS TEC (e.g., Heki et al. 2006). The TEC disturbances following earthquakes and tsunamis were more complicated with different modes and features. Heki et al. (2006) reported that the seismic ionospheric disturbance following the great Sumatra earthquake in 2004 was mainly related to the direct acoustic wave from the focal area, tsunami-induced gravity wave, and secondary acoustic wave excited by the Rayleigh wave in the far field. The results of the 2011 Tohoku earthquake in Japan also showed three modes with different propagation velocities, which include tsunami-generated gravity waves (0.1–0.3 km/s), acoustic waves (0.3–1.5 km/s), and seismic Rayleigh waves (2–3 km/s) (Jin et al. 2014, 2015).

The features including velocities, propagating direction, and amplitude of TEC anomalies were widely studied in the last decade as well as the coupling mechanism between the earthquake/tsunami and ionosphere (Meng et al. 2015). Three-dimensional model of ocean wave disturbances triggered by earthquakes and tsunamis, the one-dimensional model of atmospheric gravity wave propagation, and the one-dimensional model of ionospheric response to the earthquakes and tsunamis were combined by Occhipinti et al. (2006) to simulate the ionospheric disturbances following the 2004 Sumatra tsunami. The perturbation results were compared with the TEC observations obtained from Jason-1 and Topex/Position satellites and showed a good agreement between the model simulations and satellite observations. However, the influence of atmospheric viscosity on the propagation of gravity waves is not considered in the simulation, which makes the velocity of disturbance not consistent with the fact, and the vertical and horizontal wind fields perturb the velocity to 600 m/s. Occhipinti et al. (2008) simulated the ionospheric response to heavy waves at different latitudes and their results demonstrated that the coupling of neutral and ionospheric ions in low latitudes was significantly stronger than that in mid-latitudes.

The coupling between the tsunamis and the ionosphere is complicated and many factors should not be ignored. Hickey et al. (2009) proposed that the influence of Coriolis force, atmospheric viscosity, and background wind field should be taken into account when simulating the tsunami-induced interaction of gravity waves with the ionosphere. Furthermore, different wave propagation directions and horizontal wind fields related to the depth of the ocean also contribute to the upward propagation of the gravity waves. Therefore, the simulation of the ionospheric disturbances triggered by tsunami-induced atmospheric gravity waves should take full account of atmospheric viscosity and background wind and other natural physical processes, as well as wave propagation direction and geomagnetic properties. Comparison of the simulated data with observed data helps understand the actual propagation of gravity waves and their perturbations in the ionosphere. Galvan et al. (2012) used the coupled model of the atmosphere and ionosphere and the Song model to simulate the ionospheric disturbance caused by the atmospheric gravity wave triggered by the Japan tsunami in 2011 as well as compared the modeling results with the TEC perturbation time series from GPS observation data of Japan's GEONET network. It was found that the maximum disturbance caused by the 2011 Japanese tsunami in the ionosphere is 1.5 TECU in the southeast direction.

Although many progresses in this field were obtained, the core scientific problems regarding the nature of the coupling between the ocean and ionosphere are still not sufficiently understood and the accurate tsunami warning is still difficult. The tsunamis can cause great damage to human beings, e.g., about 20 thousand people killed because of the 2011 tsunami of Japan. Dense scientific GPS observation network is operated in Japan, while the present tsunami forecasting systems are still not effective. Traditional forecasting system based on the long-period waves costs too much time to calculate the accurate information of the rapture area and is hard to predict the near-field tsunami. Dense offshore monitoring network for tsunami warning using GPS buoys and ocean-pressure gages required a large budget and human power. While the TEC disturbances could be detected about 10–20 min after the main shock by GPS network observations in real time, which may have the potential to enhance the current tsunami forecasting system by providing the tsunami speeds and amplitude independently.

12.2 Tsunamis Information

Four tsunamis generated by the 2010 Maule (Chile) earthquake, 2011 Tohuku (Japan) earthquake, 2012 Haida Gwaii (Canada) earthquake, and 2014 Kaikoura (New Zealand) earthquake are studied in this chapter. The basic information including the main shock time, magnitude, depth, location of the epicenter, and the initial tsunami height are shown in Table 12.1. The selected earthquakes are located in the region equipped with dense GPS network. For example, there are more than a thousand continuous GPS stations in Japan and nearly 2 hundred in New Zealand.

12.3 Tsunami Ionospheric Disturbances

12.3.1 2011 Tohoku Tsunami in Japan

TEC time series from GPS observations with more than 1200 continuous GPS stations were filtered with a Butterworth band filter. The ionospheric disturbances are displayed in Fig. 12.1, the black pentagram represents the location of the main shock and the dot with different color denotes the amplitude of the TEC disturbances and the location of Sub-Ionospheric Point (SIP). The SIP is defined as the ground projection of the intersection between the satellite-receiver Line Of Sight and the altitude of the peak electron density. Here, we assume the altitude of the peak electron density is 300 km. The main shock occurred at 05:46:23 UT, as shown in Fig. 12.1a, and there were distinct TEC disturbances with amplitude of about 0.4 TECU near the epicenter. The ionospheric disturbances mainly distributed west–north and west—south to the epicenter. 30 min later as shown in Fig. 12.1b, the disturbances spread

Table 12.1 The information of the earthquakes and tsunamis

Earthquake	Date	Time (UT)	Magnitude (Mw)	Depth (km)	Epicenter		Initial tsunami height (m)
					Latitude (°)	Longitude (°)	
Maule earthquake, Chile	27 Feb. 2010	06:34:14	8.8	35	35.91°S	72.73°W	4.5–5.0
Tohoku earthquake, Japan	11 March 2011	05:46:23	9.0	32	38.32°N	142.37°E	6–15
Haida Gwaii earthquake, Canada	28 Oct. 2012	03:04:08	7.8	14	52.79°N	132.10°W	3.6
Iquique earthquake, Chile	1 Apr. 2014	23:46:45	8.3	20	19.642°S	70.817°W	2.1

north and southwest. Positive anomalies and negative anomalies form a ring structure, indicating that the ionospheric anomalies were caused by the earthquakes and tsunamis propagating from one source to the surroundings. However, due to the different structure and direction of the earthquake fault and the direction of tsunami propagation, the ionospheric disturbances caused by the earthquake and tsunami are different in different directions. As can be seen from Fig. 12.1b, the major fluctuations following the earthquake and tsunami in Japan were mainly in the north and southwest. Figures 12.1c and d show the disturbances at 7:00 UT and 7:30 UT and the ionospheric anomalies became weaker and weaker far from the epicenter.

The ionospheric disturbances following the 2011 Japan earthquake and tsunami could be detected from PRN09, PRN15, PRN18, and PRN27 by filtering the TEC time series. The detailed ionospheric disturbances obtained from these four satellites are displayed in Fig. 12.2. The oblique lines are used for the fitting velocities of the TEC disturbances. The different colors show the amplitude of TEC disturbances. Figure 12.2a displays the results from PRN09, and the ionospheric disturbances could be firstly detected 15 min after the main shock at around 6 o'clock, 200 km away from the epicenter. The amplitude of the TEC disturbances is about 0.4 TECU. It can be clearly seen that there are two modes of ionospheric disturbances, one is within 500 km with the speed of about 1190 m/s and the other appears later around 6:30 o'clock, 600 km away from the epicenter and the amplitude of this disturbance was similar with the first mode, while the travel velocity is about 187 m/s. The first mode results from the acoustic waves generated by the earthquake and the second wave chain is caused by the tsunami-generated gravity waves. Figure 12.2b displays the disturbances detected from PRN15. The ionospheric disturbances appeared earlier and nearer to the epicenter than the results from PRN09. The amplitude of the TEC disturbances is about 0.45 TECU. There are also two types of ionospheric distur-

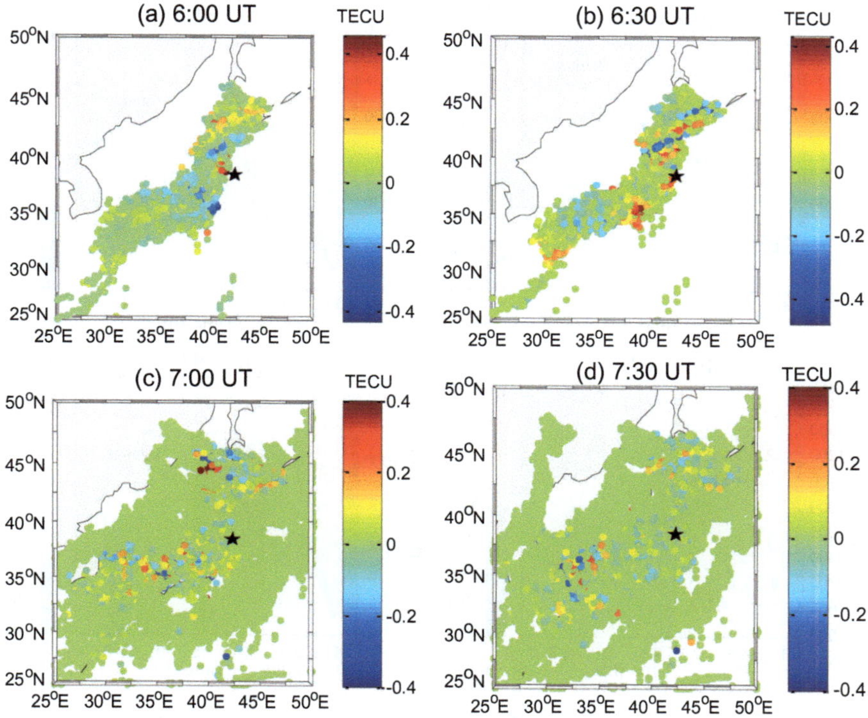

Fig. 12.1 TEC variations following the 2011 Tohoku earthquake and tsunami from GPS observations

bances. One is around 6 o'clock, while the other appears around 6:30 UT. While the velocity of the first mode of TEC disturbance is much bigger than the first mode of PRN09 and the speed is about 2778 m/s. The speed of the second mode is in the range of 176–300 m/s. The first mode was triggered by the Rayleigh waves. The second wave of the PRN15 is the same as PRN09 and generated by the gravity waves.

Figure 12.2c shows the results from PRN18. The ionospheric disturbances could be first detected 15 min after the main shock at around 6 o'clock, 700 km away from the epicenter. The amplitude of the TEC disturbances is about 0.4 TECU. After fitting the typical velocity of the TEC disturbances, three modes of anomalies could be found. The first mode is triggered by the Rayleigh waves with a velocity of 2315 m/s, the second mode is the acoustic waves with a speed of 1110 m/s, and the last mode is generated by gravity waves with a velocity of about 238 m/s. Figure 12.2d shows the results from PRN27. The ionospheric disturbances could be first detected before 6 o'clock in the very near field of the epicenter. The amplitude of the TEC disturbances is about 0.4 TECU. In general, the disturbances of PRN27 is similar to the disturbances obtained from PRN15 and the acoustic triggered with a velocity of 1156 m/s and gravity waves generated with velocity between 165 and 280 m/s could be detected.

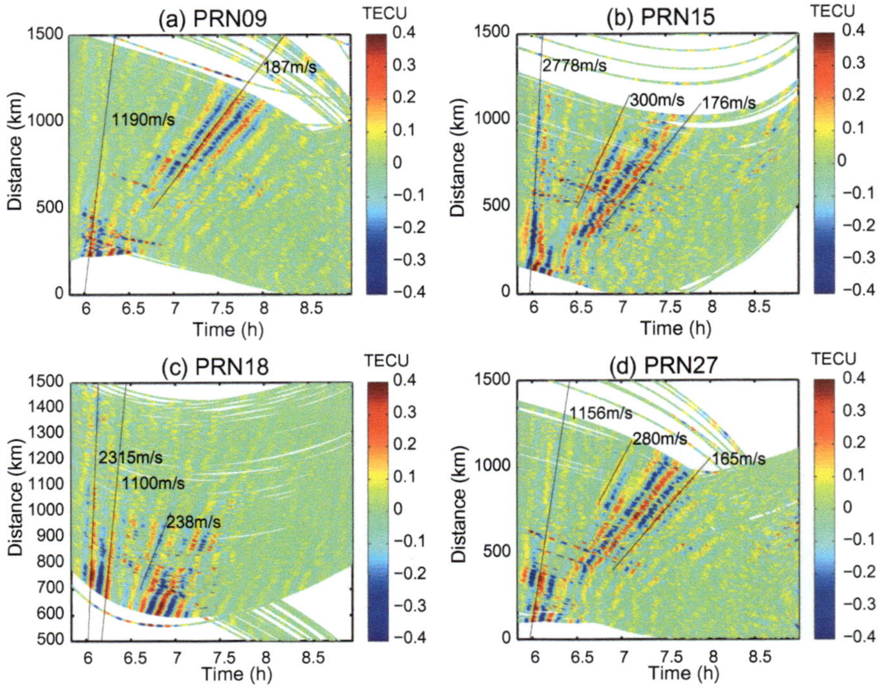

Fig. 12.2 TEC disturbances of PRN09, PRN15, PRN18, and PRN27 following the 2011 earthquake and tsunami

Three types of ionospheric disturbances generated by Rayleigh waves, acoustic waves, and gravity waves, respectively, could be detected from PRN09, PRN15, PRN18, and PRN27. The amplitudes of these anomalies are about the same. Figure 12.3 shows the SIP tracks of PRN09, PRN15, PRN18, and PRN27 from 6 to 9 o'clock. In Fig. 12.3a, we can find out that the satellite PRN09 runs from southwest to northeast. Figure 12.3b shows the travel direction of PRN15 is from northwest to southeast. The moving direction of PRN18 is similar to PRN09. And the satellite PRN27 runs from the northwest to the south as displayed in Fig. 12.3d. These amplitudes and traveling speeds are affected by the geometric features. The modes and velocity of the ionospheric disturbances of PRN09 are similar to PRN27. Both acoustic waves and gravity waves generated disturbances were detected and the fitting traveling speed of the TEC anomalies are about the same. This is consistent with the distribution of the SIP tracks.

In order to investigate the relationship between the amplitude of the disturbances and the distance between the epicenter and the SIPs, the average amplitude of TEC disturbances from 0 to 1600 km with altitude interval of 200 km was calculated. The absolute value of the disturbances was used instead of the true value when obtaining the average value because we concern more on the amplitude value instead of the polarity. The time series of the average amplitude of PRN09, PRN15, PRN18, and

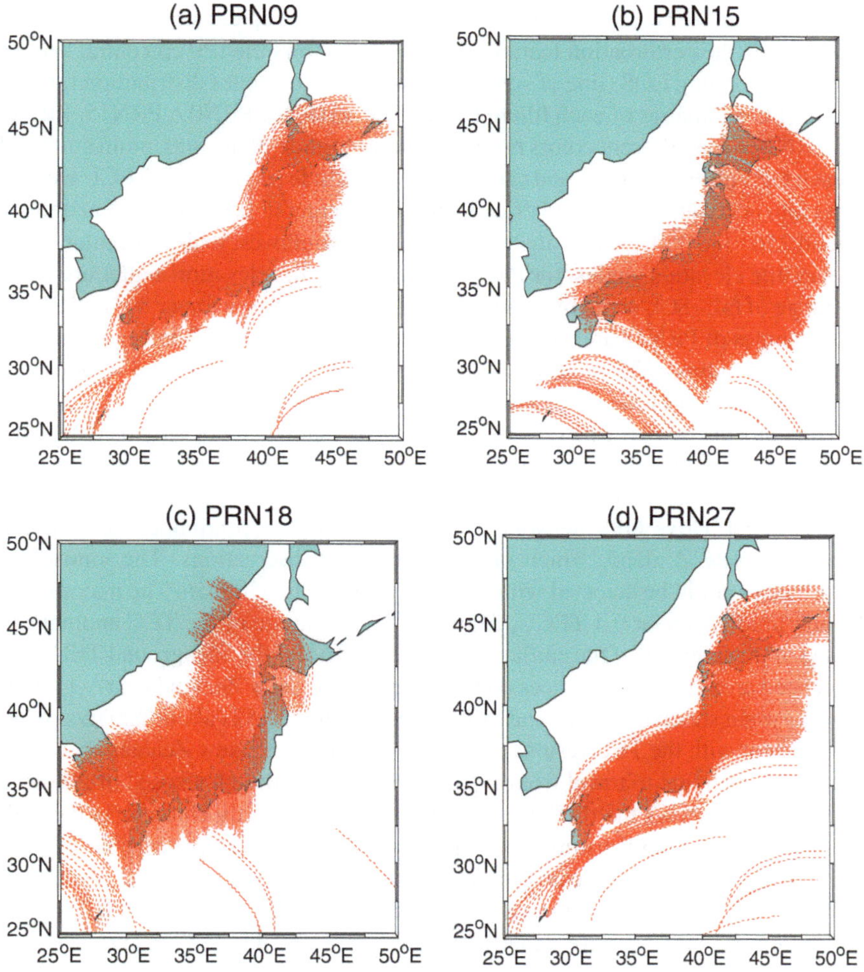

Fig. 12.3 The SIP tracks of the PRN09, PRN15, PRN18, and PRN27

PRN27 are obtained. Both have two general wave chains, the first wave chain is near 6 o'clock and the amplitude of this wave chain is decreasing with the increasing of the altitude. This phenomenon is due to the attenuation of infrasound during propagation. The second wave chain appears at 7 o'clock. The first wave chain is related to the acoustic waves, while the second wave chain results from the gravity waves. Because of the lower speed of the tsunami-generated gravity waves, the second wave chain appears later than the first one. The change trend of the second wave chain is opposite with the first one, which results from the decrease of the atmosphere density and the consequent increase of the wave amplitude.

In addition to the distance, the direction toward the epicenter can also have an influence on the perturbation features. Figure 12.4 shows the IPP epicentral azimuth and corresponding LOS (line of sight) satellite elevation angles distribution for maximum TEC anomalies of each filtered TEC time series for PRN09, PRN15, PRN18, and PRN27. The different colors represent the amplitude of the maximum amplitude of each GPS observation arc and cases whose amplitude less than 0.1 TECU were not taken into account. The ionospheric disturbances center at IPP epicentral azimuth of around 250° and 360°, which are the north and the southwest to the epicenter. As for the satellite elevation angles, for PRN09 and PRN27, the elevation angles are within the range of 60°–80°, while majority of the elevation angles are near 80° and the elevation angles of PRN18 are within the range of 30°–60°. The IPP epicentral azimuth and corresponding LOS satellite elevation angles could be combined to determine the main direction of the propagation of the ionospheric disturbances.

The ionospheric perturbations caused by the tsunami can not only be detected in the near field of the epicenter but also could be detected thousands of kilometers away from the epicenter. Figure 12.5 displays the TEC disturbances obtained from New Zealand native GPS network. There are about 200 continuous GPS stations located in New Zealand, which provide abundant observations. The ionospheric disturbances could be detected within the range of 7800–8500 km. The maximum of the amplitude is about 0.4 TECU. Actually, the amplitude of the TEC anomalies is related to the intensity of the earthquake and tsunami and the background TEC of the influenced area. The disturbances could be detected in New Zealand nearly 11 hours after the main shock in Japan. The fitting speed in Fig. 12.5 is about 200 m/s, which is consistent with the TEC results and appeared at 16:40 in New Zealand, because it lasts about 11 hours to travel from Japan to New Zealand with a speed of 200 m/s.

Figure 12.6 shows the time-domain and frequency-domain characteristics of the near-field and far-field disturbances after the 2011 Japanese earthquake tsunami. The blue curve on the left panel represents the time series of the TEC disturbance term, and the green curve represents the distance between the satellite SIP position and the epicenter. The right panel is a spectrogram of the TEC disturbance with the time. The first row panel is the detection result of the satellite PRN09 at station 0011 in Japan. The distance between the SIP of satellite PRN09 and the epicenter is between 300 and 800 km. The disturbance occurs 15 min after the main shock, about 400 km away from the epicenter. At the same time, the maximum amplitude of the disturbance is about 0.1 TECU. The frequency density distribution of the TEC signal has two extreme points. The first extreme value is about 4.5 mHz, the second extreme value is 4.0 mHz, and both extreme values are greater than 3 mHz. According to the dispersion diagram, the disturbance is not the gravity wave generated by the tsunami wave. From the speed of the perturbation time-space variation, it can be determined that the frequency of the first extreme point corresponds to the disturbance of the first mode that is induced by the Rayleigh wave and the second extreme point is caused by the infrasound wave. The second row panel is the result of station 0683 in Japan with satellite PRN09. From the time series of the disturbance, the disturbance is relatively flat. The maximum amplitude of the disturbance is 0.4 TECU, and the frequency of the signal is concentrated at 1.2 mHz. From the frequency point,

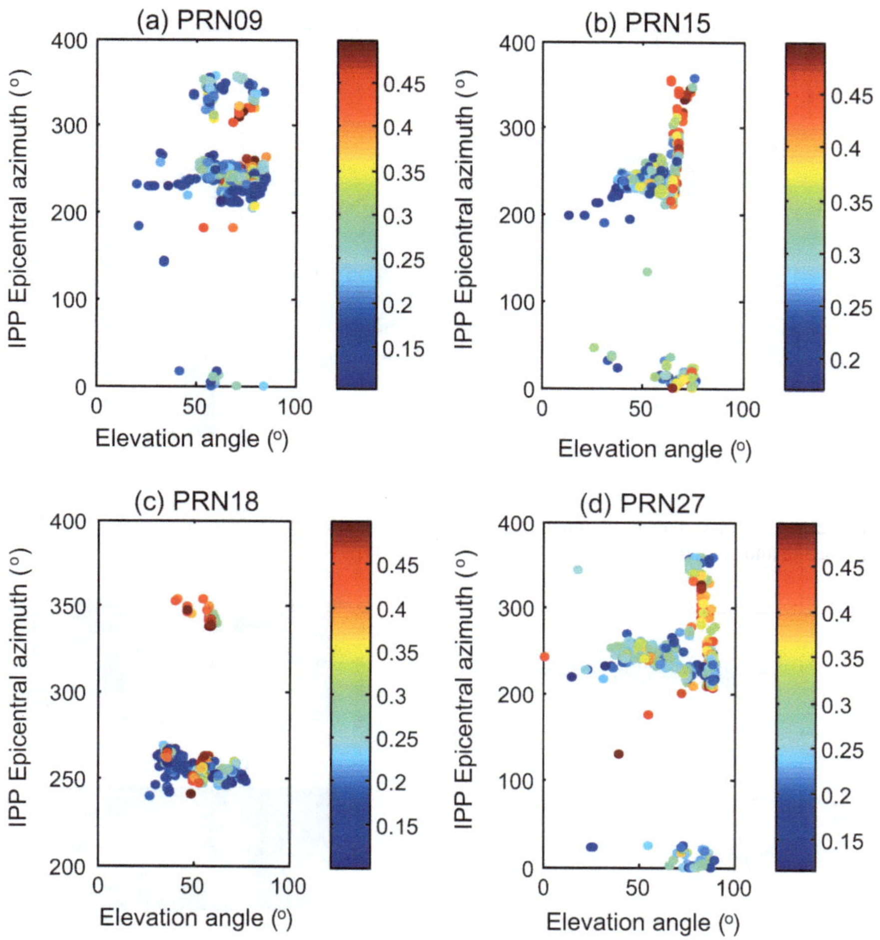

Fig. 12.4 IPP epicentral azimuth and corresponding LOS satellite elevation angles distribution for maximum filtered TEC in each GPS observation arc. Here, the undisturbed arcs (maximum filtered TEC <0.1 TECU) are not considered

the disturbance is excited by the gravity wave. This extreme point appears around 7:30 UT, corresponding to the perturbation of the third mode in the TEC space-time variation, which is excited by gravity waves. The propagation speed of the disturbance is 187 m/s. It can be inferred that the wavelength of the gravity wave is between 100 and 150 km by the relationship between the propagation velocity, the frequency, and the wavelength.

The third row panel results from station JOKA and satellite PRN05 in the far field, Hawaii. The disturbance occurred in the range of 6500–7000 km from the epicenter and the amplitude of the maximum disturbance is 0.35 TECU occurring at 12:30–13:00. The frequency density peak of the disturbance signal is about 1.0 mHz.

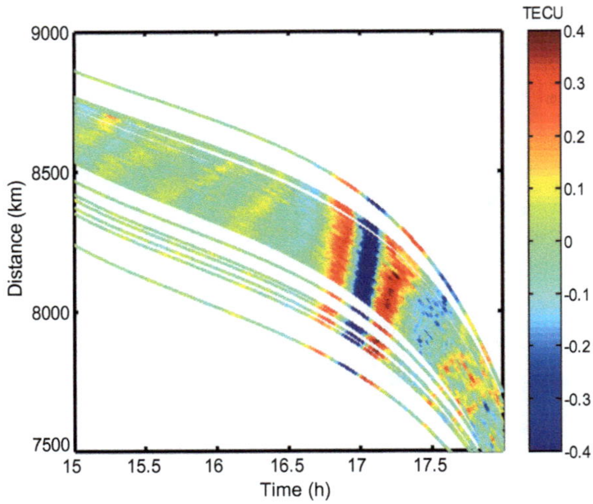

Fig. 12.5 TEC disturbances obtained from GPS observations in New Zealand after the 2011 Japan earthquake and tsunami

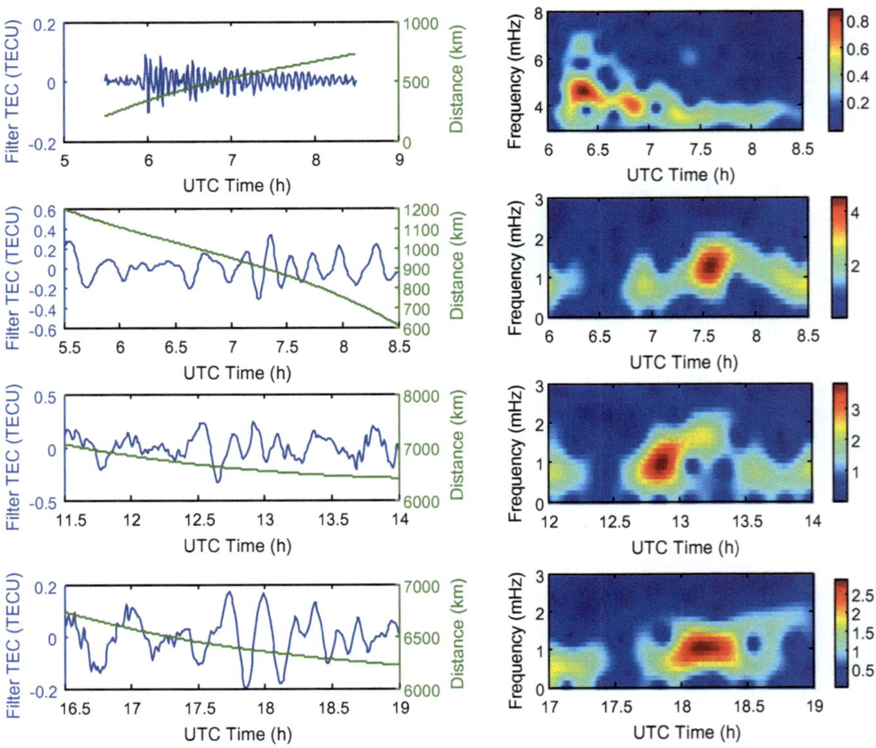

Fig. 12.6 Spectrogram of the filtered TEC from GPS observations in near field (Japan) and far field (Hawaii)

From its propagation speed (220 m/s), it can be determined that the disturbance is excited by gravity waves. The fourth row panel results from satellite PRN14 at station CRIM in Hawaii, which is the second wave of disturbance detected over the Hawaii after the earthquake tsunami in Japan, about 12 h after the main shock. This disturbance is caused by the tsunami due to aftershocks. The amplitude of the disturbance is about 0.2 TECU slightly lower than the amplitude of the first wave disturbance. This difference should be caused by the tsunami intensity. The characteristic frequency of the TEC disturbance is 1.2 mHz, which is within the range of atmospheric gravity waves. In summary, after the earthquake tsunami in Japan, 4.5 Hz Rayleigh wave, 4.0 mHz infrasound wave, and 1.2 mHz atmospheric gravity wave can be detected in the near field. The excitation source of far-field ionospheric disturbance is only atmospheric gravity wave.

Gravity waves can transfer energy to thousands of kilometers. The heating effect on the ionosphere of the energy carried by the gravity waves is much greater than that of the solar ultraviolet. The energy carried by the gravity wave plays an important role in the activity of the ionosphere. In order to study the ionospheric response at thousands of kilometers away from the tsunami eruption, we selected Hawaii as the example. There are about 60 consecutive GPS observation stations on the Hawaiian Islands. Filtering the TEC time series are calculated by GPS dual-frequency carrier phase observations using a Butterworth bandpass filter with cutoff frequencies of 1 and 8 mHz. For example, Fig. 12.7 shows the TEC disturbances from GPS stations in Hawaii after the 2011 Japan earthquake and tsunami. Two modes of ionospheric disturbances can be detected using the GPS data on the island of Hawaii after the earthquake tsunami in Japan.

The first disturbance can be detected around 12:00 UT by PRN21, PRN29, and PRN31. The left side of Fig. 12.8 is the result of the first disturbance, while the right side is satellite's SIP track during the time period of 11:30–14:00 UT. The blue, red, and green dots represent for PRN21, PRN29, and PRN31, respectively. The distance between the Hawaiian Islands and the epicenter of the Japanese earthquake is between 5700 and 6200 km. Through the speed fitting, the propagation speed of the disturbance is about 280 m/s, and the maximum amplitude of the disturbance is about 0.25 TECU. Taking the disturbance at 6100 km from the epicenter as an example, the disturbance is detected at 12 o'clock, and it takes about 6 h to propagate 6100 km at a speed of 280 m/s, meaning that the disturbance is transmitted from the epicenter to the Hawaiian Islands at 6:00 UT. This time is coincided with the time of the Japanese earthquake's main shock, indicating that the first disturbance detected on the Hawaiian Islands was caused by the gravity waves excited by the tsunami of the main shock.

The second disturbance can be detected by satellites PRN03, PRN06, PRN14, and PRN22 around 17:30 UT. The disturbance results are shown in Fig. 12.9. The right side displays motion trajectories of PRN03 (blue), PRN06 (black), PRN14 (red), and PRN22 (green) from 16:30 to 19:00. In general, the amplitude of the second disturbance is about 0.2 TECU, which is slightly lower than the intensity of the first disturbance. By velocity fitting, the propagation speed of the disturbance is about 272 m/s, which is roughly similar to that of the first disturbance. Taking

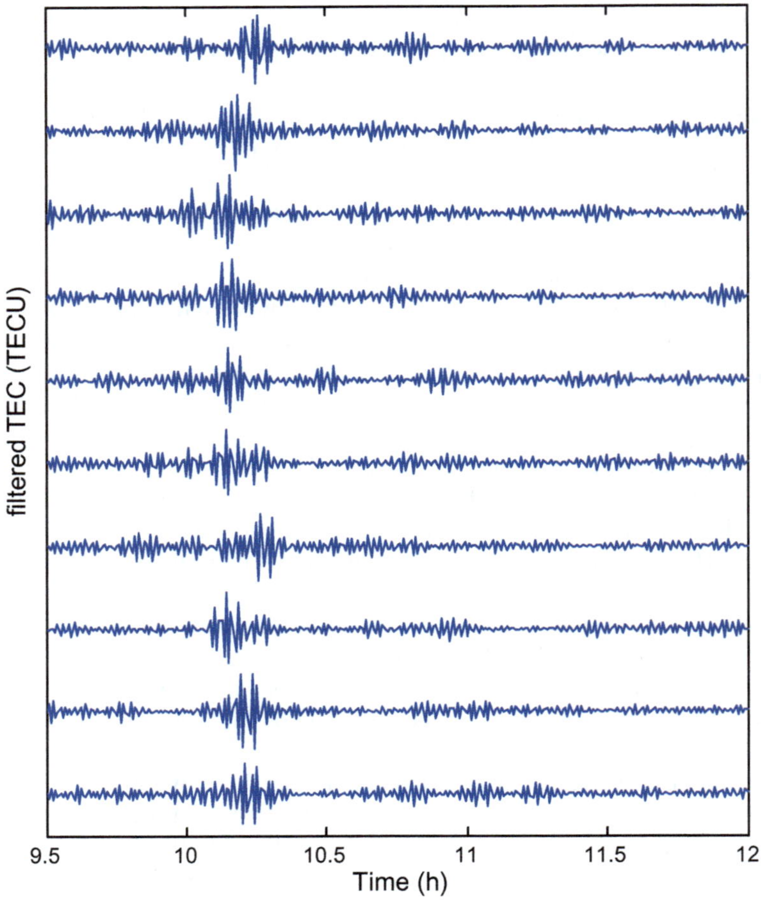

Fig. 12.7 Filtered TEC time series obtained from GPS stations in Hawaii after the 2011 Japan earthquake and tsunami

the disturbance detected at 6050 km from the epicenter at 17:30 as an example, it takes about 6.2 h from the epicenter to Hawaii at a speed of 272 m/s, and the time transmitting from the vicinity of the epicenter should be between 9:00 and 9:30 UT. This means the disturbance began to propagate from the epicenter about 3 h later after the main shock, so the disturbance was not caused by the tsunami accompanying the main shock. The excitation source of the disturbance should be the tsunami caused by the aftershock.

The far-field disturbance detected by the ionosphere over the Hawaiian Islands has only one mode compared to the near-field disturbance detected in the epicenter. According to the propagation speed of the mode, the disturbance can be caused by gravity waves. The far field cannot detect the disturbance with characteristics of the Rayleigh wave or infrasound wave throughout the day. This is because the energy

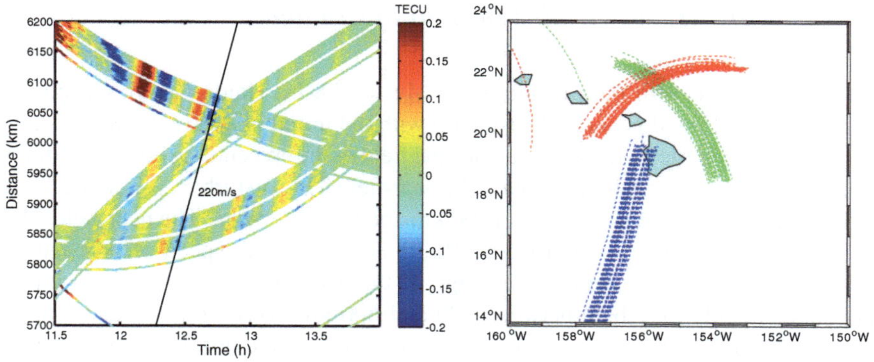

Fig. 12.8 The first response of the ionosphere over the Hawaiian Islands after the tsunami earthquake in Japan

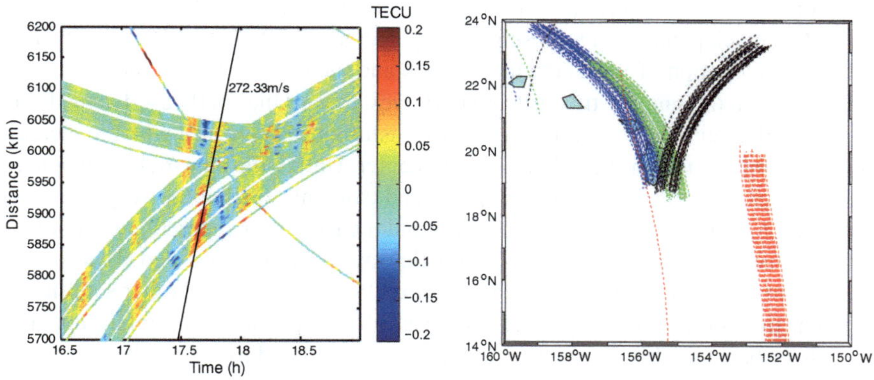

Fig. 12.9 The second response to the ionosphere over the Hawaiian Islands after the tsunami earthquake in Japan

of the Rayleigh wave or infrasound wave excited by the earthquake is exhausted within 1500 km and cannot be transmitted to long distance. From the duration of the disturbance, the disturbances detected in the vicinity of the epicenter and the far-field disturbances are consistent, which are both between 1 and 1.5 h. Although the upper atmosphere is a good medium for gravity wave propagation, the energy consumption during propagation is unavoidable. Therefore, from the amplitude of TEC disturbance, the intensity of near-field ionospheric disturbance (0.4 TECU) is higher than that of far-field disturbance (0.2–0.25 TECU). A difference can also be observed in the intensity of the two disturbances in the far field. The intensity of the first disturbance is slightly greater than the intensity of the second disturbance, which may be caused by the intensity of the first tsunami being greater than the intensity of the second tsunami.

12.3.2 2012 Haida Gwaii Tsunami in Canada

A big earthquake with the magnitude of Mw = 7.8 hit Haida Gwaii, Canada on October 28, 2012 and generated a tsunami with the initial height of 3.6 m. First, we analyzed ionospheric disturbances at the near field of the epicenter. About 400 continuous GPS stations were used to obtain the near-field TEC disturbances. Figure 12.10 shows the TEC map between 0 and 1200 km and the SIP tracks. The near-field TEC disturbances could be detected from PRN01 and the fitting speed of propagation is about 1754 m/s, which is in the range of the acoustic waves. The ionospheric anomalies mainly distribute between 200 and 600 km away from the epicenter as shown in the left panel of Fig. 12.10. The amplitude of the ionospheric disturbances is about 0.2 TECU as can be seen from the left panel of Fig. 12.10. The amplitude is smaller than the amplitude of the perturbations after the Japan earthquake and tsunami, which may result from the smaller magnitude of the earthquake. The right panel of Fig. 12.10 displays the SIP tracks of PRN01 from 3 o'clock to 5 o'clock in UTC. The black pentagram represents the epicenter (52.79°N, 132.10°W). The blue dots are the location of SIP at the main shock and the lines in red are the SIP tracks, indicating that the satellite travels from northwest toward the southeast. It can also be seen from the left panel, the amplitude of the disturbance decreased with the increase of the distance between the SIP and the epicenter. Unlike the near-field ionospheric disturbances of the 2011 Japan earthquake and tsunami, without the very fast mode generated by earth surface Rayleigh waves and the tsunami-triggered gravity waves, only the disturbance triggered by the acoustic waves could be detected about 10 min after the main shock southeast to the epicenter. We have analyzed other satellites in different directions, while only the southeast of the epicenter can obtain apparent TEC disturbances. This distribution feature may be related to the mechanism of the earthquake.

Figure 12.11 shows the filtered TEC time series obtained from GPS stations in the near field, Canada (left panel) and the far field, Hawaii (right panel) after the 2012 Haida Gwaii earthquake and tsunami. Figure 12.12 shows the ionospheric disturbances detected from GPS observations from Hawaii. The disturbances appeared at 8:20 UT and 3800 km away from the epicenter. The TEC disturbances could be found from five satellites (PRN04, PRN07, PRN08, PRN10, and PRN13). The SIP tracks of each satellite are displayed at the right panel of Fig. 12.12. The TEC disturbances of the near field are typical N-shape waves while the perturbations detected in Hawaii are shown in the form of quasi-cyclical fluctuations. The travel velocity of the TEC disturbances of Hawaii indicates that the ionospheric perturbations were related to the gravity waves while the near-field perturbations were triggered by the acoustic waves.

Figure 12.13 shows the spectrograms analysis of the 2012 Haida Gwaii, Canada earthquake. From up to bottom are spectrograms and filtered TEC and the distance time series obtained from station KFAP and KOSM for PRN04, station KTBW and P399 for PRN01. The first two cases are selected from the stations located in Hawaii and the last two cases are selected from near field of the epicenter as can be inferred

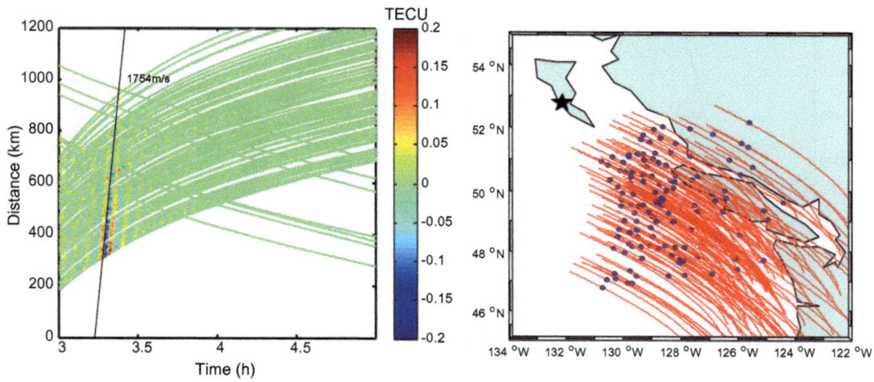

Fig. 12.10 TEC disturbances following the Haida Gwaii, Canada earthquake on October 28, 2012. The SIP tracks and the travel time diagram for PRN01. The blue dots indicate the SIP positions at the main shock. The oblique line is used for fitting the velocities of the TEC disturbances. The different colors show the amplitude of TEC disturbances

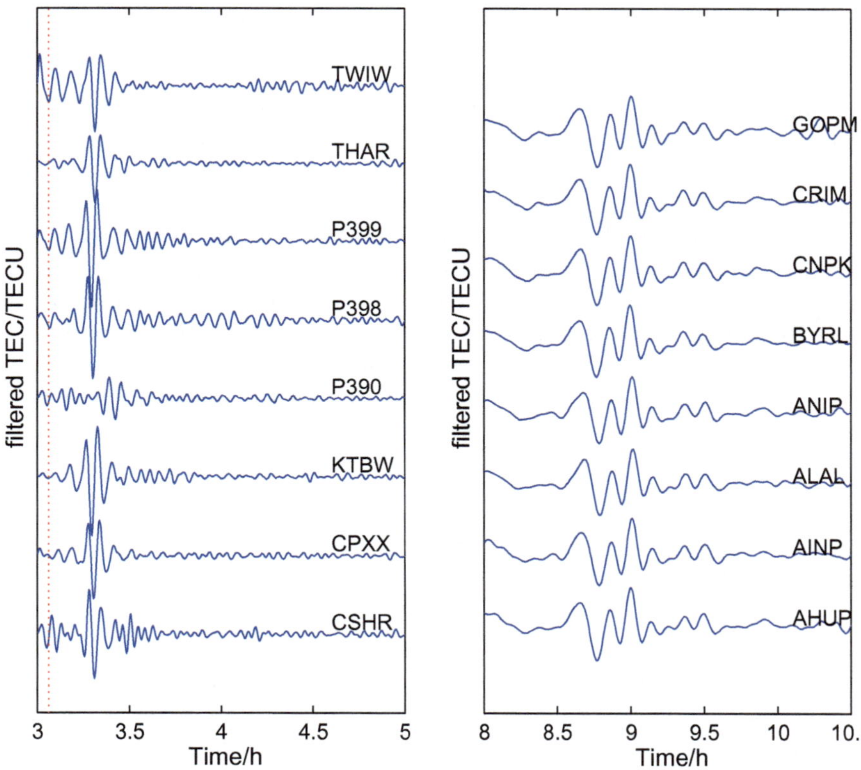

Fig. 12.11 Filtered TEC time series obtained from GPS stations in the near field, Canada (left panel) and the far field, Hawaii (right panel) after the 2012 Haida Gwaii earthquake and tsunami

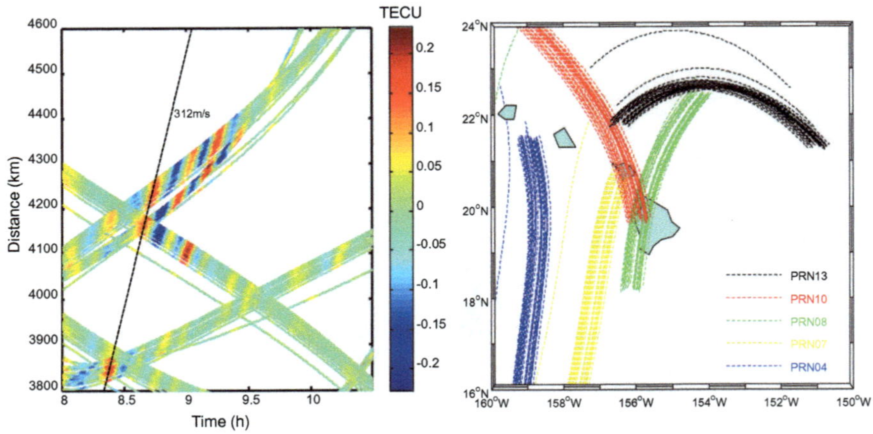

Fig. 12.12 TEC disturbances following the Haida Gwaii, Canada earthquake on October 28, 2012 obtained from GPS observations in Hawaii. The TEC travel time diagram and SIP tracks for PRN04, PRN07, PRN08, PRN10, and PRN13. The oblique line is used for fitting the velocities of the TEC disturbances. The different colors show the amplitude of TEC disturbances

from the distance between the SIP and the epicenter in green. The filtered TEC time series of each case indicates that the disturbances tend to be in the form of N-shape while quasi-cyclical fluctuations are in Hawaii. The disturbances in the near field lasted shorter than the disturbances 3800 km away from the epicenter. For example, the TEC anomalies of KFAP appeared at around 8:30 UT and disappeared at 9:30 UT, which lasted for an hour and the disturbance of KOSM also lasted for about half an hour. While for KTBW and P399, the ionospheric disturbances only lasted for about 10 minutes. There also exist different features in their frequency peaks. The frequency centers at around 2.0 mHz, which is in the range of gravity waves for the first two cases in Hawaii and especially for KOSM, and there tends to be two peak of the frequency. While for cases of the near field, the frequency centered at 3.7 mHz is in the range of the acoustic waves.

12.3.3 2010 Maule Tsunami in Chile

An earthquake with magnitude of Mw = 8.8 occurred in Maule, Chile at 06:34:14 UT on Feb 27, 2010. The location of the epicenter is 35.91°S and 72.73°W. This big earthquake generated a giant tsunami with the initial height of 4.5–5.0 m. Figure 12.14 shows the detected TEC disturbances near the epicenter. The left panel shows selected cases with apparent TEC perturbations and the station names are given in the end of each time series. The vertical red line indicates the main shock time. And the right panel shows the geometry of the stations and SIPs. The black triangle represents the location of the GPS stations, the blue dot indicates the location of the SIP at the main

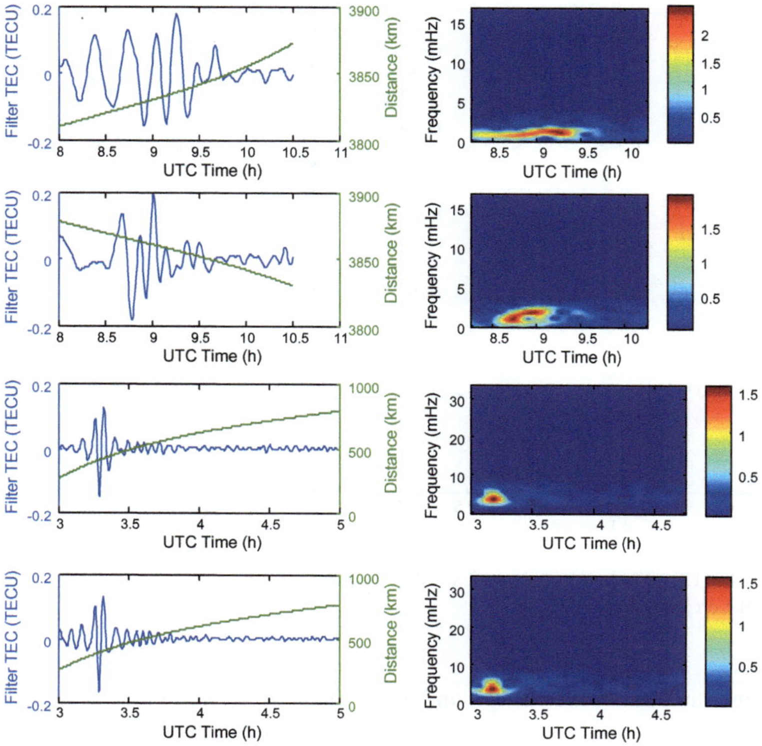

Fig. 12.13 Spectrograms for station KFAP and satellite PRN04, station KOSM and satellite PRN04. Spectrograms are for station KTBW and P399 of PRN01. The left panel displays the filtered TEC time series in blue and the distance between the SIP and the epicenter in green

shock point, and the red curve shows the SIP tracks. The TEC anomalies could be obtained about 10 min after the main shock, which is similar to the case of Haida Gwaii, Canada earthquake. Because of the lack of the GPS observations, we cannot get the TEC map and fit the travel velocity in the near field. While based on the shape of the disturbance, the disturbances can be proved to be generated by the acoustic waves just like the Canada earthquake.

Figure 12.15 shows the ionospheric disturbances detected from GPS observations from Hawaii. The disturbances appeared at 22:15 UT and 9500 km away from the epicenter. The TEC disturbances could be found from four satellites, as shown in the left of the upper panel, PRN03 in black, PRN09 in red, PRN17 in blue, and PRN27 in green. The right of the upper panel shows the distribution of the GPS stations in Hawaii and the red dot indicates the location of the GPS station. The lower panel shows the TEC map obtained from Hawaii, and the fitting propagation speed is 208 m/s, which belongs to the gravity waves range. The amplitude of the disturbances is about 1.0 TECU, which is much bigger than the Haida Gwaii, Canada earthquake. The big amplitude might result from the bigger intensity of the tsunami

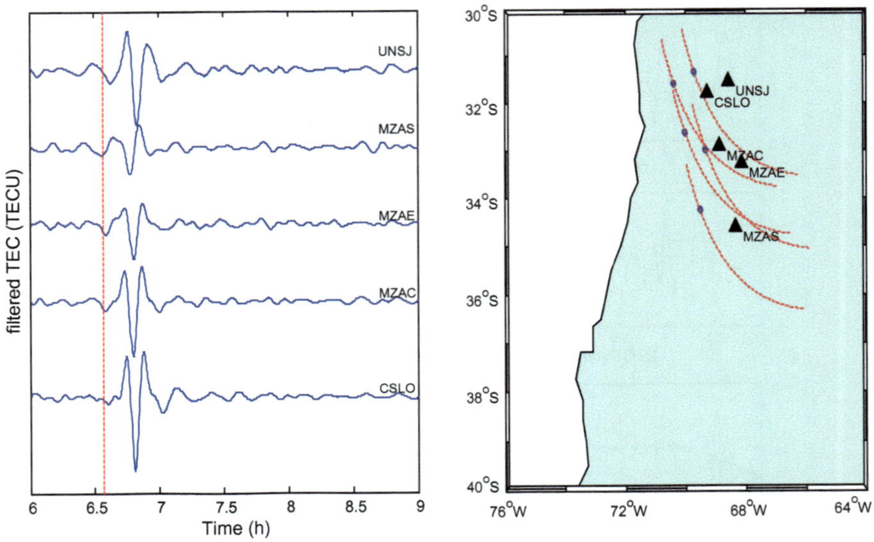

Fig. 12.14 TEC disturbances time series and corresponding SIP tracks following the Maule, Chile earthquake on February 27, 2010

and bigger amount of the background TEC. The TEC disturbances of the near field also tend to be typical N-shape waves while the perturbations detected in Hawaii are shown in the form of quasi-cyclical fluctuations.

12.3.4 2014 Iquique Tsunami in Chile

At 23:56:47 UT of April 1, 2014, the Mw = 8.3 earthquake struck northwestern Chile with a focal depth of 10.1 km. The epicenter was located at 19.61° south latitude and 70.77° west longitude, 99 km northwest of Iquique, Chile. After the earthquake broke out, there were many aftershocks. The aftershocks reached more than ten times after the main shock broke out only within 2 hours, while the magnitude of the largest aftershocks reached 6.2. After the earthquake, the US Pacific Tsunami Warning Center issued a tsunami warning to the countries of the Americas such as Chile, Peru, Mexico, and the United States. The tsunami predicted by the US National Oceanic and Atmospheric Administration not only affects the coastal countries near the epicenter but also the waves can cross the Pacific Ocean. It is expected to attack New Zealand after 15 h, and attack Australia and the Alaska region 18 h later. After 21 h, it will arrive at the coastal areas of Japan. An hour after the main shock, the first wave reached the northern part of Chile, with the highest wave reaching 2.1 m, causing serious damage to the coast (Fig. 12.16).

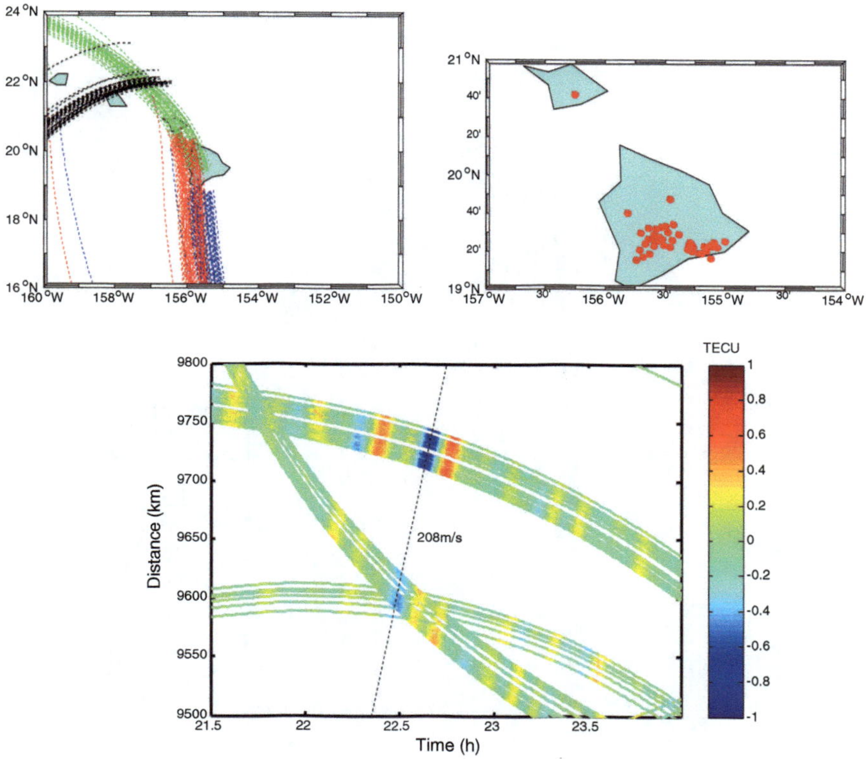

Fig. 12.15 The ionospheric disturbances obtained from GPS observations in Hawaii. The left of the upper panel shows the SIP tracks of PRN03, PRN09, PRN17, and PRN28 and the right panel shows the distribution of the GPS stations. The lower panel shows the TEC map

The energy released by the earthquake tsunami caused ionization disturbances and TEC anomalies. According to the tsunami propagation time prediction map, the waves will reach the coastal areas of New Zealand after 15 h. We use the ionosphere over New Zealand as an example to explore the far-field ionospheric response to the 2014 Iquique earthquake tsunami in Chile. In response to frequent earthquake activities, New Zealand has established an intensive GPS network consisting of more than 200 GPS stations with continuous observations. The station distribution is shown in Fig. 12.17. GPS stations are mainly located in the northeast of New Zealand, which are the most geological active area in New Zealand.

The response of the ionosphere over New Zealand to the Chilean Iquique earthquake tsunami is shown in Fig. 12.18. The disturbance is the result of PRN17. According to the satellite's trajectory, the disturbance is mainly distributed in the northeast of New Zealand. From the change of the TEC disturbance with time and distance, the maximum amplitude of the disturbance is about 0.3 TECU, which is lower than the amplitude of the near-field disturbance. The far-field disturbance is

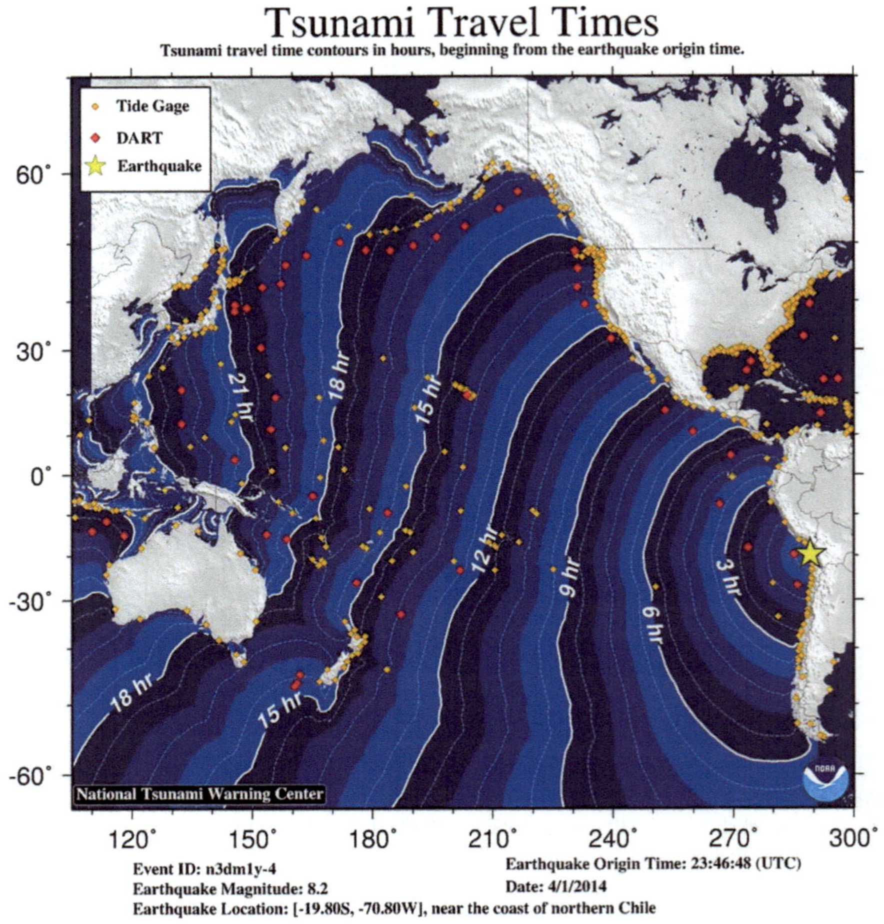

Fig. 12.16 Tsunami wave height and propagation time map in 2014, Chile

mainly concentrated in the range of 9300–9600 km from the epicenter. The most obvious disturbance is concentrated between 14:00 and 15:00 UT, and the TEC disturbance term has multiple extreme points. Through velocity fitting, the perturbation propagates at about 185 m/s in the ionosphere above New Zealand. The transmission time of the disturbance from the epicenter to the location of 9300 km takes about 14 h theoretically at a speed of 185 m/s. The disturbance at 14:30 is 9300 km and also consistent with the prediction given by US National Ocean and Atmospheric Administration. From the characteristics of the disturbance waveform, duration, and propagation speed, the disturbance is a far-field ionospheric anomaly caused by atmospheric gravity waves.

Through spectrum analysis, the near-field and far-field ionospheric responses after the 2014 Chile earthquake tsunami are similar to the 2011 earthquake tsunami in

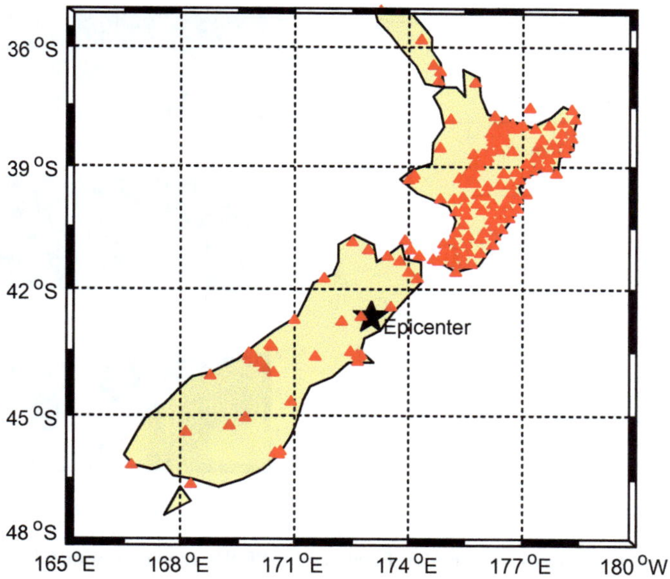

Fig. 12.17 Distribution of GPS stations in New Zealand

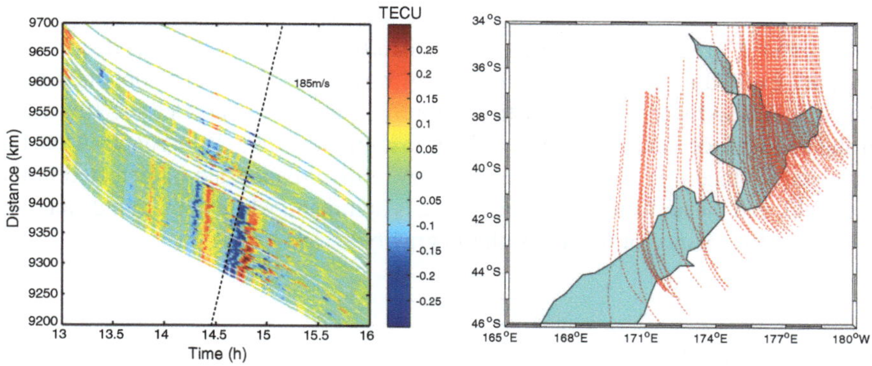

Fig. 12.18 Far-field ionospheric disturbances following the Iquique earthquake tsunami in Chile

Japan. Figure 12.19 shows the TEC disturbance and its frequency characteristics after the earthquake tsunami outbreak. The first set is the result of the station AUKT and satellite PRN16, which occurs in the range of 200–300 km from the epicenter, with an amplitude of 0.4 TECU and a characteristic frequency of 1.2 mHz, which is a disturbance of gravity wave excitation. The second set is derived from the observation of satellite PRN23 at station CJNT. The disturbance occurs at a position of 400–500 km from the epicenter, which belongs to the near-field disturbance. The peak frequency of the disturbance signal is between 3 and 4.5 mHz. Since the propagation velocity of the disturbance cannot be accurately fitted, the type of the disturbance cannot be

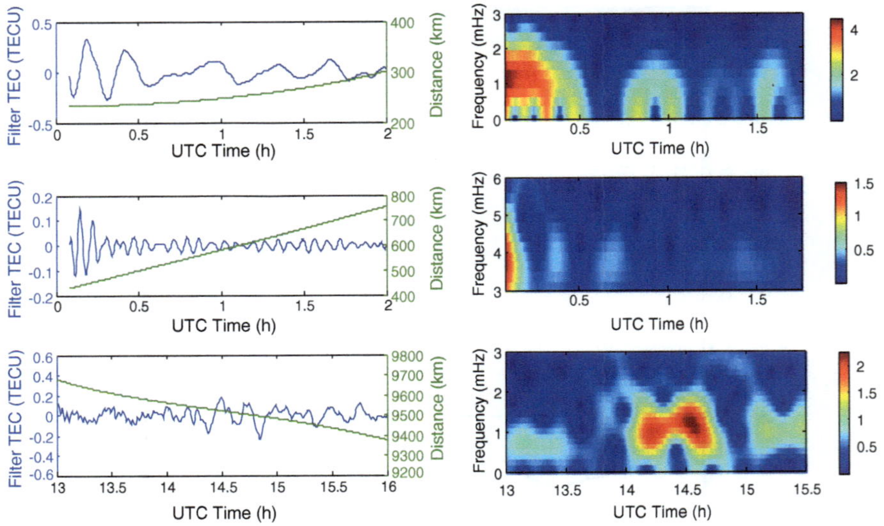

Fig. 12.19 Spectrograms for station AUKT and satellite PRN16 (upper), station CJNT and satellite PRN23 (middle), and station ATJN and satellite PRN20 (bottom). The left panel displays the filtered TEC time series in blue and the distance between the SIP and the epicenter in green

determined. Moreover, the intensity of the disturbance is lower than the disturbance intensity excited by the atmospheric gravity wave. The third set is obtained from the New Zealand station ATJN with satellite PRN20. The characteristic frequency of the disturbance is also 1.2 mHz, which is a gravity wave disturbance. After the earthquake tsunami in Chile, multiple modes of disturbance can be detected in the near field, and the atmospheric gravity wave is dominant, while the far field can only detect the disturbance of the atmospheric gravity wave.

12.4 Correlation Analysis

The activity of the ionosphere is affected by many factors. The main energy comes from solar radiation, so the solar activity has an important influence on the ionosphere. In addition, the geomagnetic field also affects the ionosphere. The charged particles in the ionosphere will be moved under the action of electromagnetic field (Kherani et al. 2012). Changes in the electromagnetic field can also cause ionospheric disturbances leading to changes in the local ionospheric electron density. These factors can be effectively removed by the Butterworth bandpass filter, and then ionospheric disturbances associated with the earthquake tsunami are estimated. However, the ionospheric disturbances associated with earthquake tsunami are still affected by many factors, such as the magnitude of the earthquake (Cahyadi and Heki 2015) and the wave height of the tsunami. The magnitude of the earthquake is related to the energy released by the geological activity, while the height of the tsunami represents the gravitational potential energy carried by the waves. Both are used to measure energy. The type of fluctuation induced by the earthquake tsunami directly determines the type of disturbance. Different types of disturbances have significant differences. The seismic infrasound has a fast propagation speed with short disturbance duration, and cannot be transmitted over long distances. The disturbance excited by atmospheric gravity wave is similar to the quasi-periodic simple harmonics, which has a low propagation speed and a low loss during propagation. In addition, the distance from the epicenter to the epicenter can affect the intensity of the disturbance. The intensity of the subsonic and Rayleigh wave decreases with the increase of the epicenter distance, while the disturbance excited by the atmospheric gravity wave occurs in relatively distant areas. During the earthquake explosion, the interaction between the faults can also affect the ionospheric disturbance. Furthermore, the ionospheric disturbances in the uplift zone and the subsidence zone have different polarities after the earthquake. The strike-slip earthquake can also stimulate the ionospheric disturbance with relatively small disturbance intensity. The direction of action between faults will affect the direction of propagation of shallow gravity waves, which in turn affects the direction of gravity waves in the atmosphere, causing different disturbance characteristics in different directions of the epicenter. For example, after the 2011 Japanese tsunami, ionospheric disturbances for near-field disturbances are mainly distributed in the northwest and south of the epicenter (Jin et al. 2014). Of course, the size of the total electron content of the background field directly affects the size of the TEC disturbance. Therefore, the perturbation percentage instead of the TEC disturbance value can more accurately reflect the relationship between the TEC disturbance intensity and other factors.

The following method is used to determine the percentage of TEC disturbances. Firstly, the cases where the maximum value of the TEC disturbance term is less than 0.02 TECU are removed, because 0.02 TECU is the normal range of TEC variations in the normal state. The perturbation percentage can be obtained by taking the quotient of the absolute value of the TEC disturbance term and the TEC value of the corresponding time. The disturbance intensity characteristics can be reflected to some extent by the average value of the maximum perturbation percentage, but the method of averaging cannot accurately represent the disturbance intensity as the range of disturbance changes is relatively large. Here, it uses a percentage interval to reflect the general characteristics of the disturbance intensity. This interval satisfies the following four conditions:

(1) The average of the percentage maxima should be included in the interval;
(2) The interval contains 60% of the TEC time series;
(3) The principle of minimum width, that is, the interval with the smallest width among all the intervals satisfies the above two conditions;
(4) Take the interval from the mean to the minimum squared sum of the boundary distances in all the intervals satisfying the first three terms.

Taking the ionospheric disturbance over the Hawaiian Islands after the 2011 Japanese tsunami as an example, the average value of the disturbance extremes is about 6%. Intervals [5, 8%] and [2.2, 7.1%] both satisfy the condition of 60%, but the former has a section length of 3% smaller than the latter section length of 4.9%, so the former is selected as the disturbance section for expressing the disturbance loudness. The third condition is to make the interval more concentrated, while the fourth condition is to make the value distribution in the interval closer to the normal distribution. If the interval satisfying the above four conditions is still not unique, it means that the distribution of other perturbation percentages is relatively uniform around the mean. In this case, condition 2 needs to be adjusted by increasing the TEC time series included in the interval. This processing strategy can not only reflect the overall characteristics of the disturbance intensity but also effectively reduce the impact of gross errors.

The following will explore the relationship between earthquake magnitude, tsunami wave height, and ionospheric disturbance intensity. The far-field ionospheric disturbances at thousands of kilometers away are selected as research objects. On the one hand, the 2014 Chilean tsunami and the 2012 Canadian tsunami cannot detect atmospheric gravity near the epicenter. The perturbation of the wave excitation and the type of approach disturbance will interfere with the relationship between atmospheric gravity wave disturbance and other factors. At the same time, significant ionospheric disturbances can be detected after four far-off tsunamis. The disturbances are caused by atmospheric gravity waves excited by the tsunami.

Figure 12.20 shows the relationship between the ionospheric disturbance intensity and the magnitude of the earthquake. The abscissa indicates the magnitude and the ordinate indicates the intensity of the TEC disturbance. As the magnitude of the earthquake changes, the ionospheric disturbance intensity changes significantly. The magnitudes of the earthquake from the low to the high are 7.8 for the Canadian earthquake in 2012, 8.2 for the Chilean earthquake in 2014, 8.8 for the Chilean earthquake in 2010, and 9.0 for the Japanese earthquake in 2011. The disturbance intensity intervals are 0.2–0.3 and 0.3–0.5%, 1–3.5 and 2–8%. The difference between the strongest disturbance and the weakest disturbance intensity is tens of times. The ionospheric disturbance intensity is related to the magnitude of the relevant earthquake. It can be seen from Fig. 12.20 that the ionospheric disturbance intensity increases significantly with the increase of the magnitude, showing a positive correlation trend.

Figure 12.21 shows the variation of the disturbance intensity with the tsunami wave height. The abscissa is the tsunami wave height and the ordinate is the disturbance intensity. The blue line and the red line represent the tsunami wave height range and the ionospheric disturbance dynamic intensity interval, respectively. The disturbance intensity is also positively correlated with the tsunami wave height. The disturbance intensity increases with the increase of the magnitude and tsunami wave height, which is related to the magnitude of energy propagation. The larger the magnitude, the higher the energy carried by the wave.

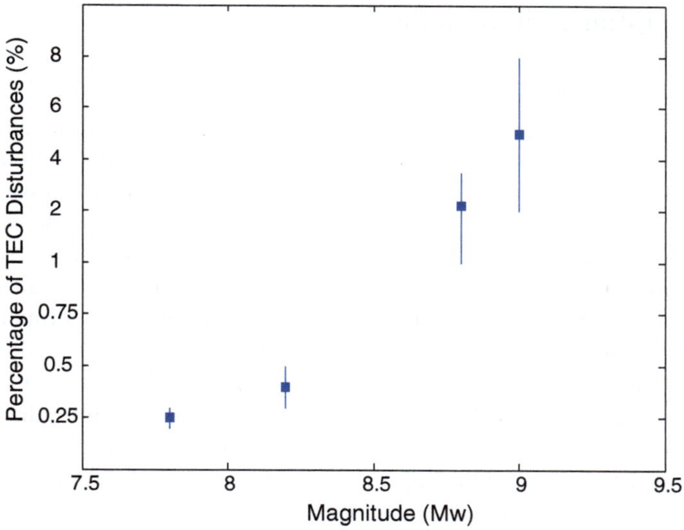

Fig. 12.20 Relationship between ionospheric disturbance intensity and earthquake magnitude

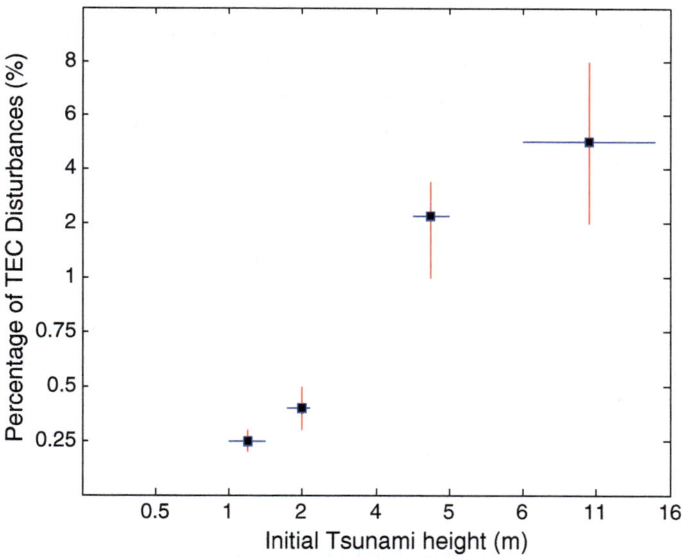

Fig. 12.21 Relationship between ionospheric disturbance intensity and tsunami wave height

12.5 Coupling and Modeling

Fluctuations caused by the earthquake and tsunami include Rayleigh waves, infrasound waves, and gravity waves. In our study, we mainly focused on atmospheric gravity waves. The kinematic definition of gravitational waves is that internal gravitational waves are linear perturbations of the carrier medium and will propagate freely along their rays once they are generated. The nonlinear effects and the background will only slowly change their properties. Nonlinear effects (such as wave rupture, etc.) are only very localized events in time and space. Atmospheric gravity waves exist in the structural system of lightweight and heavyweight. The formation of this structure by the influence of gravity in the atmosphere under certain conditions, such as volcanic eruptions, earthquakes, tsunamis, nuclear explosions, and lee waves, can generate gravity waves. Tsunami is an important source of atmospheric gravity waves. When a tsunami occurs, the vibration of water molecules on the surface of the ocean surface will cause the upper air molecules to deviate from the oscillation of the equilibrium position to induce atmospheric gravity waves.

The propagation of gravity waves has always been a hot research topic. In the vertical direction, gravity waves can cause the coupling of the bottom neutral atmosphere with the ionosphere. In the horizontal direction, gravity waves can generate thousands of kilometers of energy transmit. The heating intensity of the middle and upper atmosphere is several hundred times that of the solar ultraviolet radiation. Hines (1960) deduced the dispersion relation of gravity waves propagating as monochromatic plane waves under the condition of ignoring viscous and thermal conduction, assuming atmospheric isothermal conditions:

$$N = m^2 = \frac{\omega_b^2 - \omega_a^2}{C^2} - k^2\left(1 - \frac{B_f^2}{\omega_b^2}\right) \tag{12.1}$$

where N is atmosphere refraction index, m is vertical wavenumber, ω_a is the sonic cutoff frequency, ω_b is the Eigen frequency, k is the horizontal wavenumber, C is the velocity of voice, and B_f is the Brunt frequency:

$$B_f^2 = \left(\frac{g}{\theta}\right)\left(\frac{d\theta}{dz}\right) = \frac{g}{T}\left(\frac{\partial T}{\partial z} + \frac{g}{c_p}\right) \tag{12.2}$$

where g is the gravity acceleration, θ is the potential temperature, T is the atmospheric temperature, and c_p is the constant pressure heat capacity.

12.5.1 *Neutral Particle and Ion Coupling Process*

After the tsunami erupted, disturbances could be detected in the ionosphere. Hines (1972) believed that the disturbances detected in the ionosphere were related to the atmospheric gravity waves excited by the tsunami. The atmospheric gravity waves transmitting upward into the ionospheric plasma will affect the coupling process of the neutrality particles and ions, destroying the steady-state balance, causing local electron density changes, and finally disturbing the total electron content. Kelly and Heelis (2007) quantitatively analyzed the momentum exchange between neutral particles and ions in the plasma in consideration of the geomagnetic field by the magnetic fluid beam equation. The magnetic fluid equation is as follows:

$$\rho_i \frac{d\vec{v}_i}{dt} = -\nabla p_i + \rho_i \vec{g} + n_i q_i (\vec{E} + \vec{v}_i \times \vec{B}) - \rho_i \mu_{in} (\vec{v}_i - \vec{v}_n) \quad (12.3)$$

where i represents the ion type, ρ_i represents the electron density after perturbation, p_i represents the ion pressure, v_i represents the ion perturbation velocity, which is zero in the absence of perturbation, v_n represents the velocity of the neutral particle, q_i represents the ion charge amount, n_i represents the ion perturbation number density, E is the electric field strength, B is the geomagnetic field strength, B represents the collision frequency between the neutral particles and ions, and g is the gravitational acceleration. The velocity of the atmospheric gravity wave in the ionosphere does not change relatively much, so the acceleration of the gravity wave on the right side of the above formula can be neglected.

$$k = -\nabla p_i + \rho_i g + n_i q_i E \quad (12.4)$$

$$\Omega = \frac{q_i n_i B}{\rho_i} \quad (12.5)$$

where k represents a term irrelevant to the change of the magnetic field and Ω represents the magnetic rotational frequency of the ion, so the ion perturbation velocity is

$$\vec{v}_i = \vec{v}_n + \frac{\vec{k}}{\rho_i \mu_{in}} + \frac{n_i q_i \vec{v}_i \times \vec{B}}{\rho_i \mu_{in}} + \frac{\Omega (\vec{v}_i \times \vec{B})}{\mu_{in} B} \quad (12.6)$$

The component of the ion perturbation velocity is

$$v_{iz} = v_{nz} + \frac{k_{iz}}{\rho_i \mu_{in}} \quad (12.7)$$

$$v_{ix} = v_{nx} + \frac{\Omega_i}{\mu_{in}} v_{iy} + \frac{k_{ix}}{\rho_i \mu_{in}} \quad (12.8)$$

$$v_{iy} = v_{ny} - \frac{\Omega_i}{\mu_{in}} v_{ix} + \frac{k_{iy}}{\rho_i \mu_{in}} \quad (12.9)$$

where v_{iz} represents the direction parallel to the magnetic field B, v_{ix} and v_{iy} represent the direction perpendicular to the direction of the magnetic field, and v_{ix} and v_{iy} are components affected by the earth's magnetic field.

Occhipinti et al. (2008) found that the momentum transfer of neutral molecules and ions has a maximum in the horizontal direction near the geomagnetic equator by studying the relationship between the momentum transport of ionospheric ions and neutral molecules and the geomagnetic latitude. With the increase of geomagnetic latitude, the momentum transmission is mainly in the vertical direction when entering the high magnetic latitude and area. It can be seen that in the process of coupling the atmospheric gravity wave with the ionosphere, the atmospheric gravity wave component parallel to the direction of the magnetic field has an important influence on the coupling process. The influence of geomagnetic field on momentum transmission varies with altitude.

The influence on momentum transmission is expressed as $\frac{\Omega_i}{\mu_{in}}$, which increases with height. Therefore, this effect is relatively small in the ionosphere E-layer. The E-layer coupling is mainly affected by the amplitude of atmospheric gravity wave vibration. However, after reaching a certain height, the ion number density, the collision frequency, and the amplitude of the atmospheric gravity wave decrease, and the effect of the magnetic field on the momentum transmission also decreases. In fact, the ionospheric disturbances after the tsunami are mainly concentrated in the F-layer, the region most affected by the geomagnetic field.

The perturbation of the ion density of the layer is mainly due to the breaking of the steady state of the ion balance. The atmospheric gravity wave causes the ionospheric disturbance by affecting the rate and loss rate of the ion chemical reaction. The ionospheric ion chemical reaction equation (Schunk and Sojka 1996) and the chemical reaction rate (Rees 1989) are shown in Tables 12.2 and 12.3.

12.5.2 Coupling Model

The background wind field has an important influence on the gravity wave propagation in the atmosphere. Studies have shown that the wind field can suppress the propagation of atmospheric gravity waves, affecting the transmission of energy, when the horizontal phase velocity of the gravity wave is consistent with the direction of the background wind field. When directions of gravity wave and wind field are opposite, the wavelength of the atmospheric gravity wave becomes longer, propagating upward better. In addition, the inhomogeneous atmospheric molecular viscosity in the vertical direction also hinders the upward propagation of atmospheric gravity waves. In order to fully consider the influence of background wind field and atmospheric molecular viscosity, we use a full wave model to simulate the propagation of atmospheric gravity waves. The full wave model proposed by Hickey et al. (1997) is a Navier–Stokes atmospheric equation of motion that satisfies the boundary conditions and simulates the propagation of atmospheric gravity waves in the ionosphere

Table 12.2 Ionic reaction formula and reaction rate

	Ionic reaction formula	Reaction rate ($cm^3\ s^{-1}$)
1	$O^+ + N_2 \rightarrow NO^+ + N$	$k_1 = 5 \times 10^{-13}$
2	$O^+ + O_2 \rightarrow O_2^+ + O$	$k_2 = 2 \times 10^{-11}(300/T)^{0.4}$
3	$O^+ + NO \rightarrow NO^+ + O$	$k_3 = 8 \times 10^{-13}$
4	$O^+ + e \rightarrow O(^5P) + hv_{1356}$	$k_4 = 7.3 \times 10^{-13}$
5	$O_2^+ + N_2 \rightarrow NO^+ + NO$	$k_5 = 5 \times 10^{-16}$
6	$O_2^+ + NO \rightarrow NO^+ + O_2$	$k_6 = 4.4 \times 10^{-10}$
7	$O_2^+ + e \rightarrow O^+ + O$	$k_7 = 1.6 \times 10^{-7}(300/T)^{0.55}$
8	$N_2^+ + O \rightarrow NO^+ + N$	$k_8 = 1.4 \times 10^{-10}(300/T)^{0.44}$
9	$N_2^+ + O \rightarrow O^+ + N_2$	$k_9 = 1 \times 10^{-11}(300/T)^{0.23}$
10	$N_2^+ + O \rightarrow O^+ + N_2$	$k_{10} = 5 \times 10^{-11}(300/T)^{-1}$
11	$N_2^+ + NO \rightarrow NO^+ + N_2$	$k_{11} = 3.3 \times 10^{-10}$
12	$N_2^+ + e \rightarrow N + N$	$k_{12} = 1.8 \times 10^{-7}(300/T)^{0.39}$
13	$NO^+ + e \rightarrow N + O$	$k_{13} = 4.2 \times 10^{-7}(300/T)^{0.85}$

Table 12.3 Loss rate and production rate of major ions

	Loss rate L (s^{-1})	Production rate P ($cm^3\ s^{-1}$)
O^+	$k_1N_2 + k_2O_2 + k_3NO + k_4e$	$k_9ON_2^+$
O_2^+	$k_5N_2 + k_6NO + k_7e$	$k_2O_2O^+ + k_{10}O_2O_2^+$
N_2^+	$(k_8 + k_9)O + k_{10}O_2 + k_{11}NO + k_{12}e$	
NO^+	$k_{11}e$	$k_1N_2O^+ + k_3NOO^+$ $+k_5N_2O_2^+ + k_6NOO_2^+$ $+k_8ON_2^+ + k_{11}NON_2^+$

within 500 km, taking into account the background wind field and molecular viscous conditions. The full wave model is as follows:

$$\frac{D\rho}{Dt} + \rho\vec{\nabla}\cdot\vec{v} = 0 \tag{12.10}$$

$$\rho\frac{D\vec{v}}{Dt} + \vec{\nabla}p - \rho\vec{g} - \vec{\nabla}\cdot\left(\rho k\vec{\nabla}\vec{v}\right) - \vec{\nabla}\left(\rho\eta_e\vec{\nabla}\vec{v}\right) + \rho\mu_{in}(\vec{v} - \vec{v_i}) + \rho K_R\vec{v} = \rho\omega\xi \tag{12.11}$$

$$\rho c_v\frac{DT}{Dt} + p\vec{\nabla}\vec{v} - \vec{\nabla}\left(\lambda_m\vec{\nabla}T\right) - \frac{c_pT}{\theta}\vec{\nabla}\left(\rho\kappa_e\vec{\nabla}\theta\right) + c_v\rho K_NT = 0 \tag{12.12}$$

$$p = \frac{\rho R_0 T}{M} \tag{12.13}$$

$$\theta = T\left(\frac{p_0}{p}\right)^{R/c_p} \tag{12.14}$$

Equation 12.10 represents the continuity equation, where $\frac{D}{Dt} = \frac{\partial}{\partial t} + \vec{U} \cdot \vec{\nabla}, \vec{U}$ is the background wind speed, ρ is the density of the neutral atmosphere, and $\vec{v}$ is the velocity of the atmospheric gravity wave. Equation 12.11 is the momentum conservation equation, p stands for atmospheric pressure, $\vec{g}$ is gravity acceleration, k stands for molecular motion viscosity coefficient, η_e stands for vortex momentum diffusion coefficient, μ_{in} is the collision frequency of ions and neutral particles, $\vec{v_i}$ represents ion perturbation rate, K_{R} is Rayleigh friction coefficient, ω is the initial wave frequency, ξ is the displacement in the maximum vertical direction of the sea surface during tsunami, and $\omega\xi$ is the vertical velocity amplitude, which is the excitation source of the full wave model. Equation 12.12 is an energy conservation equation, where c_v is the constant volume heat capacity, c_p is the constant pressure specific heat capacity, λ_m represents the molecular thermal conductivity, κ_e is the vortex thermal diffusivity, K_N is the Newton cooling coefficient, and K_R is used to simulate the boundary condition, and its value varies with height. Equation 12.13 is the equation of state of the ideal gas, where R_0 is the general gas constant and M is the molecular average mass. Equation 12.14 is the bit temperature equation and p_0 is 1000 mbar.

The disturbance of the ionospheric wave excited by the tsunami is divided into dynamic process and chemical process. Based on the dynamic process and chemical process, the coupling model of atmospheric gravity wave and ionosphere are established as follows (Yan 2016):

$$\frac{\partial n_i}{\partial t} + w_i \frac{dn_i}{dz} + n_{i0}\vec{\nabla}\vec{v}_i = P_i - n_{i0}L_i - n_i L_{i0} \tag{12.15}$$

$$\vec{v}_i = (\Omega_i^2 + \mu_{in}^2)^{-1}\{\mu_{in}\Omega_i\vec{v}_n \times \vec{B} + \Omega_i^2(\vec{v}_n \times \vec{B})\vec{B} + \mu_{in}^2\vec{v}_n\} \tag{12.16}$$

$$n_e = \sum_{i=1}^{3} n_i \tag{12.17}$$

Equation 12.15 represents the chemical process of the coupling of gravity waves and ionosphere ions, where $i = 1, 2, 3$ represents three ion types, n_{i0} represents the number density of ions in the homogeneous state, n_i represents the perturbation number density of ions, and v_i is the disturbance velocity of an ion. Disturbance velocity, w_i is the component of the ion perturbation velocity in the vertical direction, L_i is the chemical loss rate of the ion under the disturbance condition, and L_{i0} is the ion loss rate in the homogeneous state. Equation 12.16 shows the dynamics process, where Ω_i is the magnetic rotation frequency, $\Omega_i = q_i n_i B/\rho_i$, $\mu_{in} = 2.6 \times 10^{-15}(n_n + n_i)M^{-12}$, indicating the ion and the collision frequency of a neutral molecule varies with height, where n_n represents the neutral particle number density, M is the average molecular mass of the neutral particle, v_n represents the velocity of the neutral particle, and B is the magnetic field strength. Equation 12.17 represents the principle of electrical neutrality.

In the absence of atmospheric gravity waves and other influencing factors, the electron number density can be obtained by the Chapman layer function, while the total electron content is the integral of the electron number density in height. The Chapman layer is defined as follows (Chapman 1950):

$$n_{eo}(z) = N_{max}\exp\left\{0.5\left[1 - \frac{(z - z_{max})}{H_0} - \exp\left(-\frac{z - z_{max}}{H_0}\right)\right]\right\} \tag{12.18}$$

In Eq. 12.18, N_{max} and z_{max} represent the maximum electron density of the ionosphere E- and F-layers and the corresponding height, and H_0 represents the atmospheric elevation. Taking the Japanese tsunami in 2011 as an example, in the vicinity of the epicenter, the maximum electron number density of the ionosphere E′ layer is $1.54 \times 10^7\ \text{m}^{-3}$, which appears at 105 km, while the maximum electron number density of the F-layer is $1.2 \times 10^{12}\ \text{m}^{-3}$, appearing at 300 km. the TEC near the Japanese tsunami epicenter is about 20.92 TECU (Yan 2016) by electronic number density integration.

In order to simplify the problem, four chemical reaction formulas are mainly considered. The first two are free recombination processes, while the latter two are charge exchange processes. The reaction formula is shown in Table 12.4.

Table 12.5 is the ion loss rate and production rate based on the reaction formula and reaction rate table.

Firstly, we use the dynamic process equation to calculate the perturbation velocity of each ion using the dynamic theory of Macleod (1966) without considering diffuse and polarized electric fields. θ represents the tilt angle of the magnetic field,

Table 12.4 Main ion reaction formula and reaction rate

Ion reaction formula	Reaction rate ($\text{cm}^3\ \text{s}^{-1}$)
$O_2^+ + e \to O + O$	$\alpha_1 = 1.6 \times 10^{-7}(300/T)^{0.55}$
$NO^+ + e \to N + O$	$\alpha_2 = 14.2 \times 10^{-7}(300/T)^{0.85}$
$O^+ + O_2 \to O_2^+ + O$	$\beta_1 = 2 \times 10^{-11}(T/300)^{-0.4}$
$O^+ + NO \to NO^+ + O$	$\beta_2 = 8 \times 10^{-13}$

Table 12.5 Loss rate and production rate of major ions

	Loss rate L (s^{-1})	Production rate P ($\text{cm}^3\ \text{s}^{-1}$)
O^+	$\beta_1 O_2 + \beta_2 NO$	
O_2^+	$\alpha_1 e$	$\beta_1 O_2 O^+$
NO^+	$\alpha_2 e$	$\beta_2 NO O^+$

$$\frac{\vec{B}}{B} = i\cos\theta + k\sin\theta \tag{12.19}$$

Assuming that the magnetic field is a dipole field, the relationship between the magnetic field dip angle and the latitude is

$$s\sin\theta = -2\sin\varphi/\sqrt{1+3\sin^2\varphi} \tag{12.20}$$

$$\cos\theta = -\cos\varphi/\sqrt{1+\sin^2\varphi} \tag{12.21}$$

The wave velocity of the atmospheric gravity wave obtained by the full wave model simulation is the neutral particle velocity, $v_n = u_n i + v_n j + w_n k$, where i represents the south direction, j represents the east direction, and k represents the vertical direction. The direction of B is taken northward in the northern hemisphere and the dynamic equation can be converted to

$$\begin{aligned}\left(\Omega_i^2 + \mu_{in}^2\right)v_i = {} & \left\{\mu_{in}\Omega_i v_n \sin\theta + \Omega_i^2(u_n\cos\theta + w_n\sin\theta)\cos\theta + \mu_{in}^2 u_n\right\}i \\ & + \left\{\mu_{in}\Omega_i(w_n\cos\theta - u_n\sin\theta) + \mu_{in}^2 v_n\right\} \\ & + \{-\mu_{in}\Omega_i v_n\cos\theta + \Omega_i^2(u_n\cos\theta + w_n\sin\theta)\sin\theta + \mu_{in}^2 w_n\}k\end{aligned} \tag{12.22}$$

In the kinetic equation, $\vec{\nabla}\vec{v}_i$ s is the divergence of the ion perturbation velocity. If the perturbation velocity is assumed to be obtained by the propagation wave numbers k_x and k_y of the plane wave in the horizontal direction,

$$\vec{\nabla}\vec{v}_i = -\vec{i}k_x n_i - \vec{j}k_y v_i + \partial w_i/\partial z \tag{12.23}$$

Then, the kinetic equation is used to solve the perturbation number density of each ion. Finally, the electric neutrality principle is used to solve the electron number density under the disturbance state (Walterscheid and Schubert 1987).

12.6 Concluding Remarks

GPS TEC was used to detect the ionospheric disturbances following the four earthquakes tsunamis in the near and far fields. The disturbance caused by the tsunami-induced gravity waves could be detected in the far-field ionosphere thousands of kilometers away. The perturbation propagation speed is between 150 and 300 m/s and the duration is usually greater than 1 hour. The near-field disturbance is more complicated. For the 2011 Japanese tsunami and the 2014 Chilean tsunami, three types of disturbances can be detected in the near field, and the disturbances excited by atmospheric gravity waves are dominant. For the 2010 Chilean tsunami and the

2012 Canadian earthquake tsunami, it can detect disturbances caused by earthquakes. The difference between near-field disturbance and far-field disturbance is caused by the different propagation characteristics of the Rayleigh wave or infrasound wave excited by the earthquake and the atmospheric gravity wave excited by the tsunami. The infrasound wave propagates faster in the atmosphere, but the energy loss is not fast. Rayleigh waves are usually transmitted on the surface, while atmospheric gravity waves travel at a slower rate, but can spread energy thousands of kilometers away.

Different types of disturbances have different frequency characteristics. Combined with other characteristics of the disturbance, the spectrum analysis of the TEC disturbance time series can accurately determine the type of disturbance and distinguish the ionospheric anomalies excited by the earthquake and the tsunami. The ionospheric disturbance is affected by many factors. The disturbance intensity can be described by a perturbation percentage interval that satisfies certain conditions. Studies have shown that the intensity of ionospheric disturbances induced by tsunami-induced atmospheric gravity waves is related to earthquake magnitude and tsunami wave height, and increases with magnitude and tsunami wave height.

References

Cahyadi, M. N., & Heki, K. (2015). Coseismic ionospheric disturbance of the large strike-slip earthquakes in north Sumatra in 2012: Mw dependence of the disturbance amplitudes. *Geophysical Journal International, 200*(1), 116–129.

Calais, E., & Minster, J. B. (1995). GPS detection of ionospheric perturbations following the January 17, 1994, Northridge earthquake. *Geophysical Research Letters, 22*(9), 1045–1048.

Chapman, S. (1950). Upper atmospheric nomenclature. *Journal of Atmospheric and Terrestrial Physics, 55*(4), 395–399.

Coisson, P., Lognonné, P., Walwer, D., et al. (2015). First tsunami gravity wave detection in ionospheric radio occultation data. *Earth & Space Science, 2*(5), 125–133.

Crowley, G., Azeem, I., Reynolds, A., Duly, T. M., Mcbride, P., Winkler, C., et al. (2016). Analysis of traveling ionospheric disturbances (TIDs) in GPS TEC launched by the 2011 Tohoku earthquake. *Radio Science, 51*(5), 507–514.

Daniels, F. B. (1952). Acoustical energy generated by the ocean waves. *Journal of the Acoustical Society of America, 24*(1).

Galvan, D. A., Komjathy, A., Hickey, M. P., Stephens, P., Snively, J., Tony Song, Y., et al. (2012). Ionospheric signatures of Tohoku-Oki tsunami of March 11, 2011: Model comparisons near the epicenter. *Radio Science, 47*(4), RS4003.

Garcia, R. F., Doornbos, E., Bruinsma, S., et al. (2015). Atmospheric gravity waves due to the Tohoku-Oki tsunami observed in the thermosphere by GOCE. *Journal of Geophysical Research Atmospheres, 119*(8), 4498–4506.

Heki, K., Otsuka, Y., Choosakul. N., et al. (2006). Detection of ruptures of Andaman fault segments in the 2004 great Sumatra earthquake with coseismic ionospheric disturbances. *Journal of Geophysical Research Solid Earth, 111,* B09313. https://doi.org/10.1029/2005JB004202.

Hickey, M. P., Walterscheid, R. L., Taylor, M. J., et al. (1997). Numerical simulations of gravity waves imaged over Arecibo during the 10-day January 1993 campaign. *Journal of Geophysical Research Space Physics, 102*(A6), 11475–11489.

Hickey, M. P., Schubert, G., & Walterscheid, R. L. (2009) Propagation of tsunami-driven gravity waves into the thermosphere and ionosphere. *Journal of Geophysical Research Space Physics,* A08304. https://doi.org/10.1029/2009ja014105.

Hines, C. O. (1960). Internal atmospheric gravity waves at ionospheric height. *Canadian Journal of Physics, 38*(7), 1424–1427.

Hines, C. O. (1972). Gravity waves in the atmosphere. *Nature, 239*(239), 73–78.

Jin, S. G., Jin, R., & Li, J. H. (2014). Pattern and evolution of seismo-ionospheric disturbances following the 2011 Tohoku earthquakes from GPS observations. *Journal of Geophysical Research: Space Physics, 119*(9), 7914–7927. https://doi.org/10.1002/2014JA019825.

Jin, S. G., Occhipinti, G., & Jin, R. (2015). GNSS ionospheric seismology: Recent observation evidences and characteristics. *Earth Science Reviews, 147,* 54–64. https://doi.org/10.1016/j.earscirev.2015.05.003.

Kelley, M. C., & Heelis, R. A. (2007). The earth's ionosphere: Plasma physics and electrodynamics. *Science, 1*(3), 206–210.

Kherani, E. A., Lognonné, P., Hébert, H., Rolland, L., Astafyeva, E., Occhipinti, G., et al. (2012). Modelling of the total electronic content and magnetic field anomalies generated by the 2011 Tohoku-Oki tsunami and associated acoustic-gravity waves. *Geophysical Journal International, 191*(3), 1049–1066.

Macleod, M. A. (1966). Sporadic E Theory. I. Collision-geomagnetic equilibrium. *Journal of the Atmospheric Sciences, 23*(1), 96–109.

Meng, X., Komjathy, A., Verkhoglyadova, O. P., Yang, Y., Deng, Y., & Mannucci, A. J. (2015). A new physics-based modeling approach for tsunami-ionosphere coupling. *Geophysical Research Letters, 42*(12), 4736–4744.

Occhipinti, G., Kherani, E. A., & Lognonné, P. (2008). Geomagnetic dependence of ionospheric disturbances induced by tsunamigenic internal gravity waves. *Geophysical Journal International, 173*(3), 753–765.

Occhipinti, G., Lognonné, P., Kherani, E. A., et al. (2006). Three-dimensional waveform modeling of ionospheric signature induced by the 2004 Sumatra tsunami. *Geophysical Research Letters, 33*(20), 382–385.

Rees, M. H. (1989). *Physics and chemistry of the upper atmosphere*. Cambridge University Press.

Rolland, L. M., Occhipinti, G., Lognonné, P., & Loevenbruck, A. (2010). Ionospheric gravity waves detected offshore Hawaii after tsunamis. *Geophysical Research Letters, 37*(17), 122–132.

Schunk, R. W., & Sojka, J. J. (1996). USU model of the global ionosphere. Step Handbook of Ionospheric Models.

Walterscheid, R. L., & Schubert, G. (1987). A dynamic-chemical model of tidally fluctuations in the OH nightlow. *Journal of Geophysics Research Atmosphere, 92*(A8), 8775–8780.

Yan, Z. (2016). *Ionospheric disturbance response of atmospheric gravity wave induced by tsunamis*. Ph.D. Thesis.

Part III
Volcano Atmospheric Disturbances

Chapter 13
Volcano Atmospheric Disturbances

13.1 Background

Large volcanic eruptions can trigger acoustic resonance between the Earth surface and the atmosphere. Some of the resonance waves leak upward into the ionosphere and trigger ionospheric disturbances (Afraimovich et al. 2010; Caudron et al. 2015; Nakashima et al. 2016; Shah and Jin 2015; Jin et al. 2010, 2011, 2014, 2017; Wang et al. 2009). The atmospheric–ionospheric anomalies can be detected by the atmospheric temperatures and ionospheric total electron content (TEC). A detailed method using GPS TEC was introduced by Jin et al. (2015) to study the 2008 Wenchuan earthquake and the 2011 Japan earthquake. The atmospheric–ionospheric disturbance triggered by volcanic eruptions can be observed as temperature and TEC anomalies. For example, The SO_2 and temperature anomalies can be detected following the volcanic eruption (Okazaki and Heki 2012). Distinct TEC disturbances with an amplitude of 0.03–0.16 TECU were detected by GPS after 12 min of the eruption of the Asama volcano in Japan on September 1, 2004 and the velocity of the ionospheric anomalies was around 1.1 km/s (Heki 2006). The propagation speed of the anomalies was similar to the velocity of acoustic waves in the ionosphere.

The ionospheric anomalies triggered by strong earthquakes or large volcano eruptions were found in the form of a typical N-shape wave at first. While the mechanism and processes of earthquake or volcano are complicated and the atmospheric environment change is varying. The TEC disturbance triggered by strong earthquakes or large volcano eruptions often contains several modes. The Tohoku earthquake in Japan on March 11, 2011 was a massive earthquake with the magnitude of 9. The seismic Rayleigh waves (2–3 km/s), acoustic waves (0.3–1.5 km/s), and gravity waves (0.1–0.3 km/s) were detected after the earthquake from GPS observations (Jin et al. 2014; Rolland et al. 2011; Saito et al. 2011). For the volcano, different ionospheric disturbance modes were also found after the explosion. The detected TEC disturbances following the eruption of the Kelud volcano in 2014 traveled with two different speeds, 0.3 km/s and 4.0 km/s, which were triggered by gravity waves and

S. Jin et al., *GNSS Atmospheric Seismology*,
https://doi.org/10.1007/978-981-10-3178-6_13

acoustic waves, respectively (Caudron et al. 2015). Furthermore, the spectral analysis of the Soufriere Hill Volcano (SHV) in Montserrat, Lesser Antilles erupted on 13 July 2003 showed two peak frequencies at 1 and 4 mHz, indicating the ionospheric anomalies were associated with both the gravity and acoustic waves (Kanamori et al. 1994).

It is widely recognized that the big earthquakes, large explosions as well as tsunamis can trigger pressure waves in the atmosphere and cause the ionospheric disturbance. Meanwhile, the volcano explosion can trigger seismic waves in the solid Earth due to the lithosphere–atmosphere coupling (Dautermann et al. 2010). The ionospheric disturbances with high spatial and temporal resolution obtained from GPS observations and seismic data recorded by seismometer were combined together to analyze the Kelud volcano in February 2014. Both the spectra analyses had two peaks at the frequency of 3.7 and 4.8 mHz (Nakashima et al. 2016), which indicated the harmonic ground motions were excited by the coupling between the solid Earth and the lower atmosphere. In order to study the coupling mechanism between the ionosphere and volcano, the TEC disturbances were initially modeled. The PREM model for the solid Earth and MSIS-E-90 atmosphere model was used to model the TEC disturbance caused by the SHV, and the modeled TEC anomalies matched well with the observed TEC anomalies from GPS observations in the duration time, waveform characteristics and the arrival time (Dautermann et al. 2009b). However, volcanic ionospheric disturbances are complex for different volcanoes, e.g., the 2015 eruptions of Chile's Calbuco volcano (Shults et al. 2016).

In this chapter, detailed methods to detect atmospheric disturbances following the volcano are presented as well as case studies and theoretical modeling. For example, the atmospheric–ionospheric disturbances following the Calbuco volcano eruption on 22 and April 23, 2015 are investigated by GPS, including the amplitude of TEC disturbances and the duration. Furthermore, SO_2 and temperature anomalies obtained from Aura/Ozone Monitoring Instrument (OMI) and COSMIC vertical atmospheric profiles are further used to study atmospheric and ionospheric responses to the eruption of the volcano.

13.2 Observation Methods and Theory

13.2.1 Ionospheric Disturbance Estimation

The dual-frequency phase and code observations can estimate the ionospheric delay and TEC (Jin et al. 2004, 2008, 2015; Li and Jin 2017). In order to detect ionospheric disturbances following the volcano, the relative TEC instead of absolute TEC from GNSS measurements is used (Jin et al. 2013). The carrier phase measurement was used because we concentrate more on the disturbances of TEC rather than the absolute value of TEC while the code measurement contains larger noise than phase observations. Before estimating the STEC, the cycle slips must be detected and repaired to get clean carrier phase observation (Wu et al. 2010). Here a second-order and time-difference phase ionospheric residual (STPIR) algorithm was used (Cai et al. 2013). The instrument bias, the ambiguity of the carrier phase and residuals are regarded as constants for one continuous arc. After obtaining the TEC series, the variations of TEC series can be derived by smoothing the original TEC series or calculating the gradient of the TEC series. In our work, the four-order zero-phase Butterworth band-pass filter with cutoffs at 3 and 8 mHz was applied to isolate the acoustic component of TEC perturbation (Dautermann et al. 2009b).

In order to study the distribution of the ionospheric disturbance, each filtered TEC time series is associated with a sub-ionospheric point (SIP). The SIP is defined as the ground projection of the intersection between the satellite–receiver line of sight (LOS) and the altitude of the peak electron density. Here we assume the altitude of the peak electron density is 300 km for the first eruption and 400 km for the second eruption.

13.2.2 Theoretical Modeling

In order to better monitor and understand the temporal change of the solid Earth and atmospheric coupling, the normal modes which are eigenfunctions to the equations of motion for the solid Earth plus atmosphere system are used to simulate displacements in the atmosphere and solid Earth (Lognonne et al. 1998). The mass density for the solid Earth is from the PREM model and for atmosphere under the altitude of 500 km is from the MSIS-E-90 atmosphere model. Also, the solar flux index is taken into account. The gravitational potential (V) is obtained from Eq. (13.1) and the perturbation displacement u is obtained from Eq. (13.2) (Dautermann et al. 2009a)

$$\nabla^2 V = 4\pi G\rho(\vec{x}, t) \tag{13.1}$$

$$\rho \frac{d^2}{dt^2} u_i(\vec{x}, t) = \rho(\vec{x}, t) g_i + \frac{\partial \tau_{ij}}{\partial x_j} \tag{13.2}$$

where $i = 1, 2, 3$, ρ is the mass density, g is the acceleration, u is the perturbation displacement, and x is the equilibrium position. Equations (13.1) and (13.2) are simultaneously integrated with a free boundary condition at the top of the model domain to obtain the displacement, distribution of the mass density and gravitational potential. The mapping algorithms developed by Lognonne et al. (1998) were applied to deal with the energy leakage out of the domain defined by a half-space boundary condition.

This method was used by Dautermann et al. (2009b) to model the 2003 explosive eruption of the Soufriere Hills Volcano, Montserrat and found that modes $n = 3$, 5, 7 carry the majority of the energy released by the explosion in the atmosphere.

The results were in accordance with the frequency content and other atmospheric explosions.

13.2.3 Source Energy

The explosion energy of volcanic eruption is needed to calculate the displacement. The released energy by the explosion of volcano includes acoustic energy and seismic energy. The acoustic energy was first obtained by Heki (2006) with the function to fit the filtered TEC time series:

$$f(t) = -at\exp\left(\frac{-t^2}{2\sigma^2}\right) \tag{13.3}$$

where a represents the amplitude and σ represents the period. The function gets its maximum and minimum at $t = -\sigma$ and $t = \sigma$, respectively. The energy released by a volcanic eruption could be obtained from the following equation:

$$E_{acoustic} = \frac{2\pi r^2}{\rho_{atmos} c_{atmos}} \int \Delta P(t)^2 dt \tag{13.4}$$

where r is the distance from the volcano, ρ_{atmos} and c_{atmos} are density and the velocity of the acoustic wave in the atmosphere, respectively. The parameter ΔP is the pressure perturbation that is obtained from the TEC disturbance relative to the background as

$$\Delta P = \frac{P \times \Delta TEC}{TEC} \tag{13.5}$$

The energy obtained from the method introduced above is only a small portion. This method works better on eruptions in form of N-shape perturbations while many eruptions are in the form of quasi-periodic waves. The total energy released by the explosion of volcano includes acoustic and seismic energy. Because of the uncertainty of Heki's (2006) method, Dautermann et al. (2009b) introduced a new method to calculate the acoustic from GPS data and the estimated acoustic energy released by SHV was 1.53×10^{10} J. The seismic energy was obtained by Kanamori et al. (1993) from the following equation:

$$\mathrm{E}_{Seismic} = 4\pi r^2 \rho c Q(r)^2 \int v^2(t) dt \tag{13.6}$$

where r is the distance from the explosion source, ρ is the average density, c is the velocity of the S-wave and $Q(r)$ represents the attenuation function from Kanamori et al. (1993), and $v(t)$ is calculated from the strain data based on the following equation:

$$v(t) = \frac{\Delta L}{\Delta t} = \frac{\sqrt[3]{V_0}}{3\Delta t} \times Volumetric\ Strain \tag{13.7}$$

where V is the volume of the strainmeter. Dautermann et al. (2009b) found that the seismic energy was two orders of the magnitude smaller than the acoustic energy. The seismic energy was neglected when calculating the source moment and the source moment of the SHV was 1.02×10^{13} Nm in Dautermann et al. (2009b).

13.2.4 Volcanic Ionospheric Coupling

All calculated modes are summed as (Dautermann et al. 2009a):

$$\vec{u}(\vec{x}, t) = \sum_{i,n,l,m} \mathrm{Real}[\varepsilon_{n,l,m,i,i}(\vec{x}_s) M_{ii} \vec{u}_{n,l,m}(\vec{x}) \frac{1 - e^{i\delta_{n,l}t}}{\delta^2_{n,l}}] \tag{13.8}$$

where l, m, n represent angular, azimuthal and radial orders of the normal modes, respectively, M is moment tensor and ε is the strain component of a certain mode and σ is the eigenfrequency (Lognonne 1991). The electron density perturbations are computed by assuming that the ionospheric charged particles follow the motion of the neutral gas through collision interactions. The collision interactions are calculated from the Navier–Stokes equation of magnetohydronamics and the electron plasma velocity could be obtained from following equation:

$$\rho_e \frac{d\vec{v}_e}{dt} = -\nabla p + \rho_e \vec{g} + Ne(\vec{v}_e \times \vec{B}) - \rho_e \nu_{en}(\vec{v}_e - \frac{d\vec{u}}{dt}) \tag{13.9}$$

The equation for the charge density from the eruption time to the observation time is integrated as (Dautermann et al. 2009a):

$$\delta \mathrm{N}(\vec{r}_l, t) = -\frac{\partial N}{\partial z} \int_{t_{explosion}}^{t} v_{e,z}(\vec{r}_l, \hat{t}) d\hat{t} - N(z) \int_{t_{explosion}}^{t} \nabla \cdot v_e(r_l, \hat{t}) d\hat{t} \tag{13.10}$$

and in order to obtain the TEC disturbance following the explosion, the electron density are integrated along the line of sight.

13.3 Atmospheric Disturbances and Characteristics

13.3.1 Volcano and Data

The Calbuco volcano (41.33°S, 72.618°W) in 2015 is a stratovolcano in southern Chile, which locates at the southeast of Llanquihue Lake and the northwest of Chapo Lake. The Calbuco volcano erupted explosively at 21:04 UT (18:04 LT) on April 22, 2015 and ended at 4:00 UT (1:00 LT) on April 23, 2015. The ash plume of the first eruption reached an altitude of more than 15 km above the crater, where it drifted primarily ENE. The second eruption lasted nearly 6 hours and generated ash plume, which rose higher than 15 km and drifted in NNE and East. The GPS data used in our work were obtained from 50 continuous GPS stations. The GPS stations belong to IGS and South America GPS networks (RAMSAC and UNAVCO). All GPS observations with 15 s interval or 30 s interval were applied to calculate the GPS TEC time series in our work. While the ionospheric anomalies detected from 15 s interval GPS observations were more distinct when compared with results from 30 s'. Figure 13.1 shows the distribution of GPS stations used here. The red pentagram represents the location of the 2015 Calbuco volcano and the blue points represent the location of the GPS stations. The bottom right panel shows the topography of the near field of the volcano. In addition, SO_2 and temperature anomalies are obtained from Aura/Ozone Monitoring Instrument (OMI) and COSMIC vertical profiles.

13.3.2 Temperature Anomalies

The intensity of the pumice on April 22 and later on April 23, 2015 was declared the most intensive. After the eruption, the local atmospheric situations varied due to the volcanic tremors. In this study, the temperature profiles from COSMIC are interpolated at latitude 41.3°S and longitude 72.2°W, which is shown in Fig. 13.2 that the temperature profile is retreating the lower confidential interval at 12–15 km altitude by a significant value of 5 °C. To further detect the authenticity in the temperature profile, the data points at 200 hPa are interpolated around the eruption regime in Fig. 13.3 (Liu et al. 2017). A temperature anomaly of 8 °C surrounds the geographical location of the Calbuco volcano on 23 April in Fig. 13.3c, which is very significant from the rest of the observations. The anomalous zone in Fig. 13.3b is very similar to the real size of volcanic plume.

13.3.3 SO_2 Anomalies

The SO_2 data have been interpolated using the spline function at the significance level of $p = 0.1$ from April 21 to April 24, 2015. Additionally, the SO_2 data have

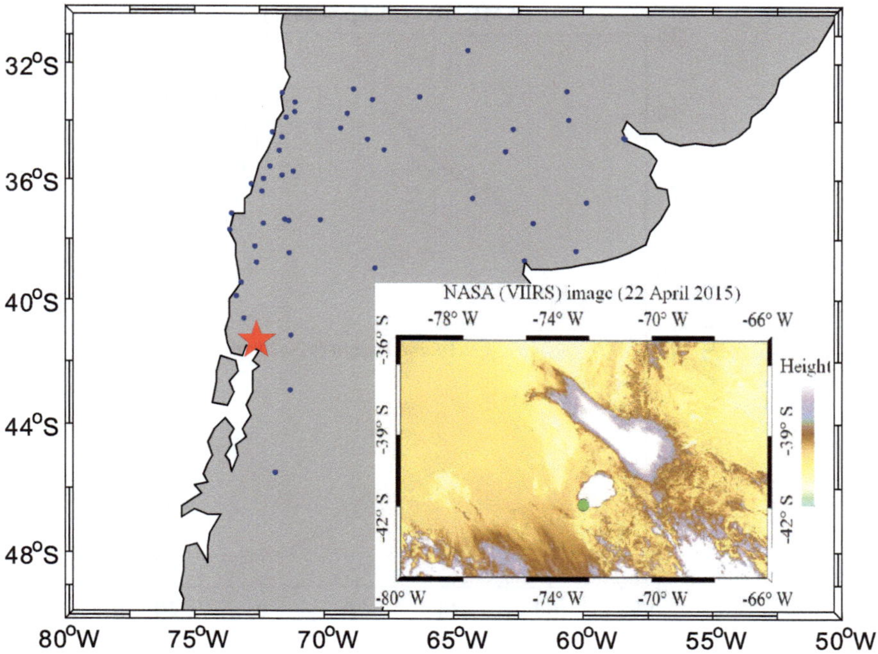

Fig. 13.1 The distribution of GPS stations around the volcano, and the blue point represents GPS stations. Geographical location of the Calbuco Volcano, Chile is displayed at the lower right corner. The pentagram and the green filled circle show the eruption place

been utilized for contiguity of atmospheric anomalies with the volcanic eruption. The daily SO_2 is measured by the Aura/Ozone Monitoring Instrument (OMI) with 1° spatial resolution. The unit for the SO_2 concentration in the atmosphere is Dobson (DU). To depict the vulnerability of the plume, the differential map of SO_2 is used. The concentration of SO_2 in the atmosphere after the eruption day is shown in Fig. 13.4. The differential maps of SO_2 showed a high value of 0.5 DU around the Calbuco location in Fig. 13.4c. The abnormal concentration of the SO_2 after the Mt. Pinatubo eruption in the surrounding areas has been examined by McCormick et al. (1995) by deriving the data from different ground and space-born observations. The atmospheric variations after the eruption at low altitude are distinctive from the rest of the regime in our observations too. All these results unanimously agree upon the induction of volcanic ashes and gas-particle in the lower atmosphere (Liu et al. 2017).

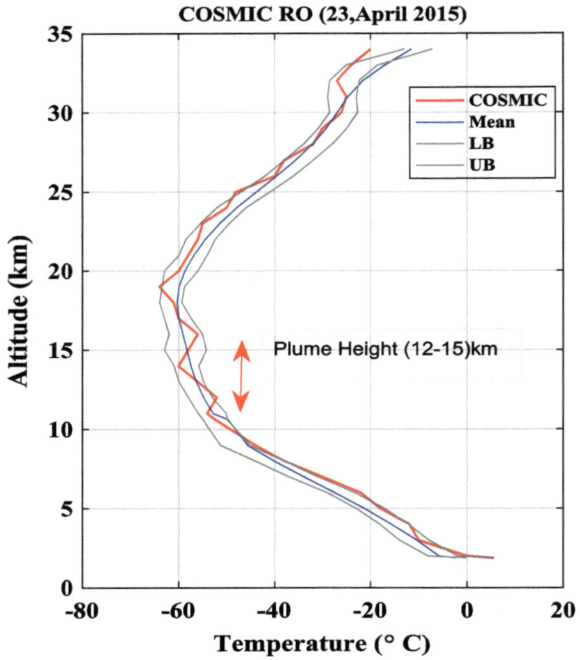

Fig. 13.2 Temperature profiles interpolated over the Calbuco Volcano from COSMIC GPS RO with the altitude in km. The legends show the necessary information

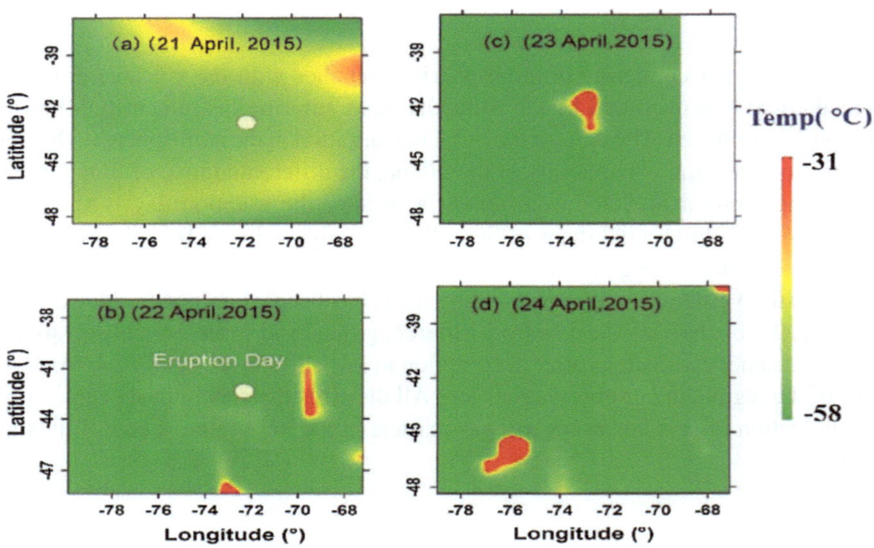

Fig. 13.3 COSMIC temperature profile at 200 hPa is interpolated from 21 to April 24, 2015. The white circle is showing the geographical location of the Volcano

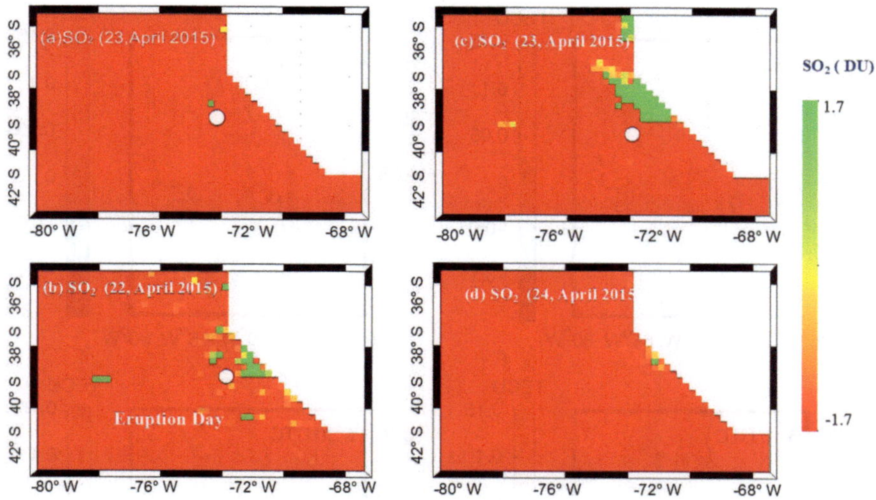

Fig. 13.4 The distribution of SO_2 concentration around the eruption day from OMI

13.4 Volcanic Ionospheric Disturbances

13.4.1 Co-volcanic Ionospheric Anomalies

The ionospheric responses to the Calbuco volcano eruption on 22 and 23 April, 2015 are investigated from GPS observations (Liu et al. 2017). Figure 13.5 shows the TEC disturbances after the volcano on 23 April from 4:45 to 5:30 UTC with a 15 min interval. The red pentagram represents the location of the volcano and the dots are corresponding to the position of SIP. The TEC disturbances or the filtered TEC time series are processed with a Butterworth filter with a 3–8 mHz window and displayed by different colors in Fig. 13.5. The purpose of the filter is to remove the trend of SIP's motion and ionospheric background changes. The detected TEC disturbance is mainly related to the volcano eruption. The ionospheric disturbance could be detected after 1 hour of the eruption at around 5:00 UTC. Then the disturbance became distinct at 5:15 and both positive and negative disturbances are located in the North of the volcano.

Figures 13.6 and 13.7 show the detailed information of the first eruption on April 22, 2015 and the second eruption on April 23, 2015, respectively. For the first eruption, there were about 10 observable satellites (satellite 1, 3, 4, 9, 11, 14, 17, 19, 23, 31, and 32), while the apparent ionospheric disturbances can only be detected from PRN03 and PRN23. In the left upper panel of Fig. 13.6, the dashed red line displays the SIP tracks of PRN03 from 21:00 to 23:00 UTC. The blue dot is the location of SIP when the volcano erupted. The right upper panel is the travel time diagram of PRN03. The dotted vertical line indicates the point when volcano erupted and the colors represent different amplitudes of filtered TEC in unit of TECU.

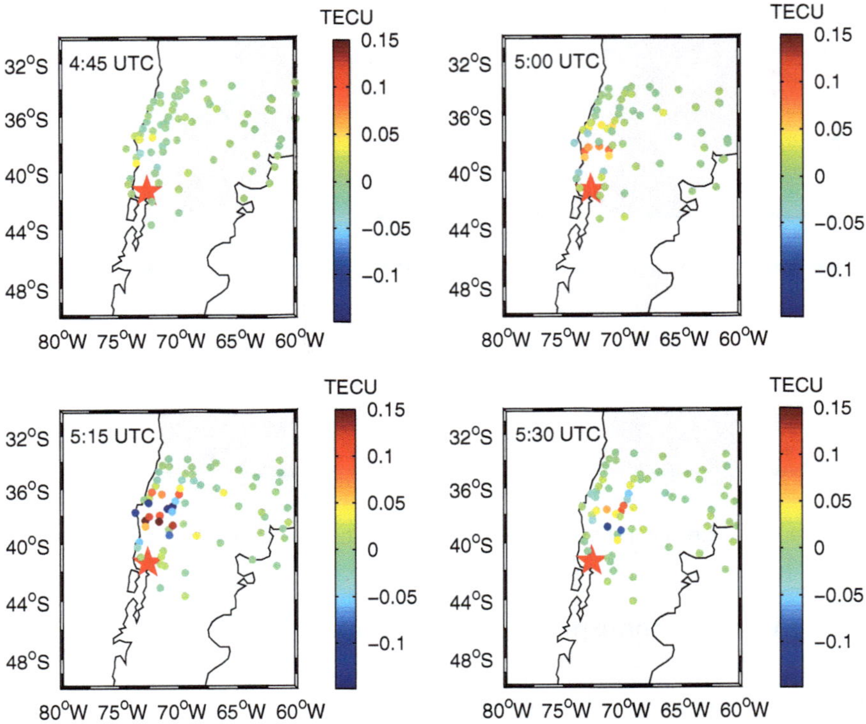

Fig. 13.5 Detrended TEC variations following the April 23, 2015 eruption of the Calbuco Volcano, Chile with a 15 min interval

The ionospheric anomalies are mainly obtained from the 15's carrier phase measurement. The detected TEC disturbances are in the form of quasi-periodic waves instead of the typical N-shape wave. The ionospheric anomalies could be divided into two distinct wave chains. The first wave chain appeared at around 21:30 UTC and the second occurred half an hour later. According to the travel time diagram, the second wave chain had larger anomaly amplitude and lasted longer when compared with the first wave chain. The propagation speed of TEC disturbances obtained from PRN03 was about 0.79 km/s. The lower panel shows the SIP tracks and travel time diagram for PRN23. The SIP tracks for PRN23 were similar to PRN03 and both of them are mainly located in the north of the volcano. The ionospheric disturbance can also be divided into several wave chains like the anomalies for PRN03. But the first wave chain appeared at 22:10 UTC and the wave chain at 22:45 UTC was more distinct when compared with results for PRN03. These different features of waveform for different satellites are caused by the movement direction of SIP and the positional relationship between SIP and volcano.

Although both SIP tracks of PRN03 and PRN23 are located in the north of the volcano, the travel direction is quite different. SIP tracks were gradually approaching the volcano while the SIP for PRN03 propagated from the west to the east and

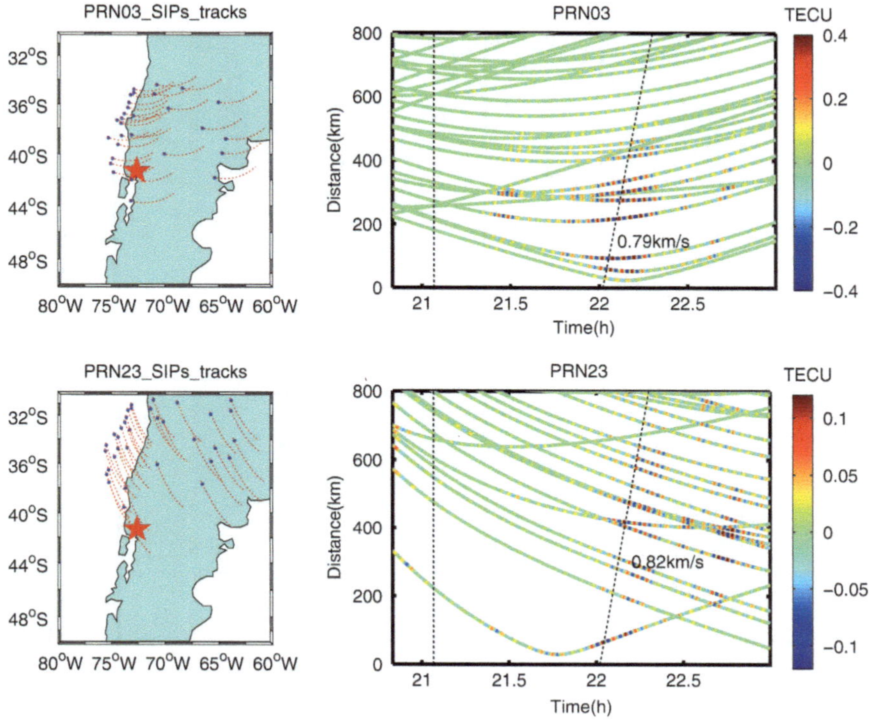

Fig. 13.6 TEC responses to the eruption on April 22, 2015. The upper panel is SIP tracks and the travel time diagram for PRN03 while the lower panel is for PRN23. The blue dots indicate the SIP tracks when the volcano erupted. The vertical dotted lines indicate the time of eruption of volcano and the oblique dotted lines are used for fitting the velocities of the TEC disturbances. The different colors show the amplitude of TEC disturbances

became far away from the volcano. The reason why there was no disturbance at 21:30 UTC for PRN23 was that the SIP tracks were out of the range of the volcanic affected areas. As for the distinct wave chain at 22:45, the SIP tracks for PRN23 are located in the affected area. Figure 13.7 displays the same information as Fig. 13.6 of the ionospheric disturbance following the second volcano eruption. For the second eruption, the observable satellites for most stations were satellite PRN05, 07, 10, 13, 15, 20, 28, and 30, the distinct ionospheric disturbance can be obtained from PRN13 and PRN28. Compared with the first volcano, the ionospheric disturbances were also detected at the north of the volcano, while the ionospheric anomalies only lasted about half an hour, which was shorter than the first eruption. Besides the above, the amplitude of most filtered TEC was smaller about 0.15 TECU. The velocity of the disturbance was around 0.90 km/s. There were also different features for PRN13 and PRN28.

According to previous research results, the amplitude of the TEC disturbance is affected by the intensity of the volcano, the background TEC and the distance

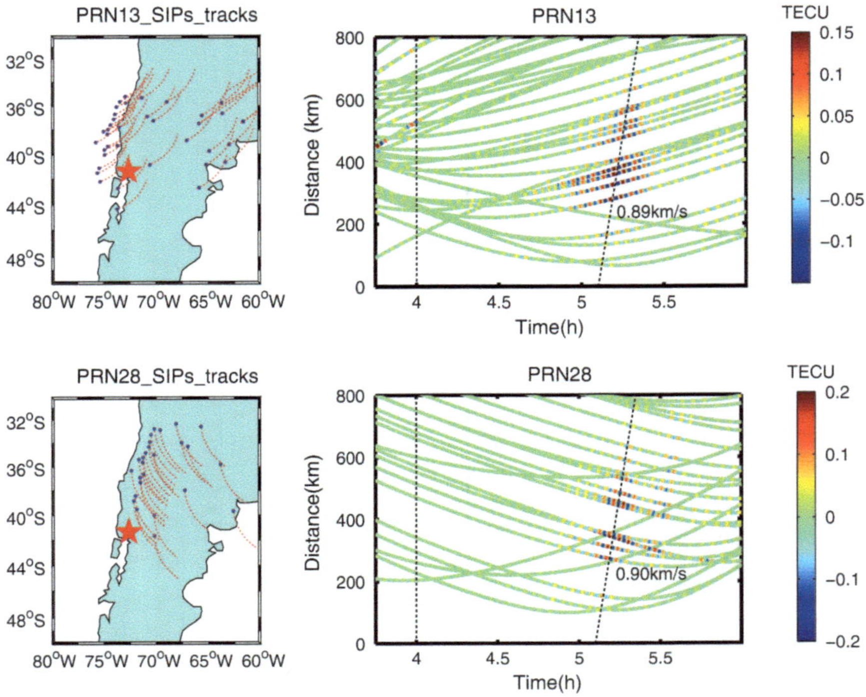

Fig. 13.7 The same as Fig. 13.2 for the eruption on April 23, 2015, while the satellites are PRN13 and PRN28

between the volcano and the influenced area. The second eruption was more intense than the first eruption, while the background TEC for the first eruption was three times as much as for the second eruption (Shults et al. 2016). That is why we find that, in general, the amplitude of TEC disturbance of the first eruption was much bigger than the second one. Seven stations with different distances were analyzed. As shown in Fig. 13.8, the left panel shows the location of volcano (red pentagram), the location of each selected station (black triangle) and the SIP tracks (blue curve). And the right panel is the filtered TEC time series of selected GPS stations. The distance was between the volcano and the SIP where the maximal TEC disturbance could be detected. The filtered TEC time series indicate that the amplitude of the disturbance changed with the distance. Firstly, the amplitude increased with the distance and got maximum among the selected cases at station LAJA with a distance of about 480 km away from volcano. Then the amplitude decreased with the distance as the trend of LAJA, NIHU, RMQE and MELP and finally the disturbance disappeared when the SIP reached a distance of 900 km away from the volcano. The changing trend of the amplitude of the TEC disturbance indicates that the perturbations did not travel in the vertical direction.

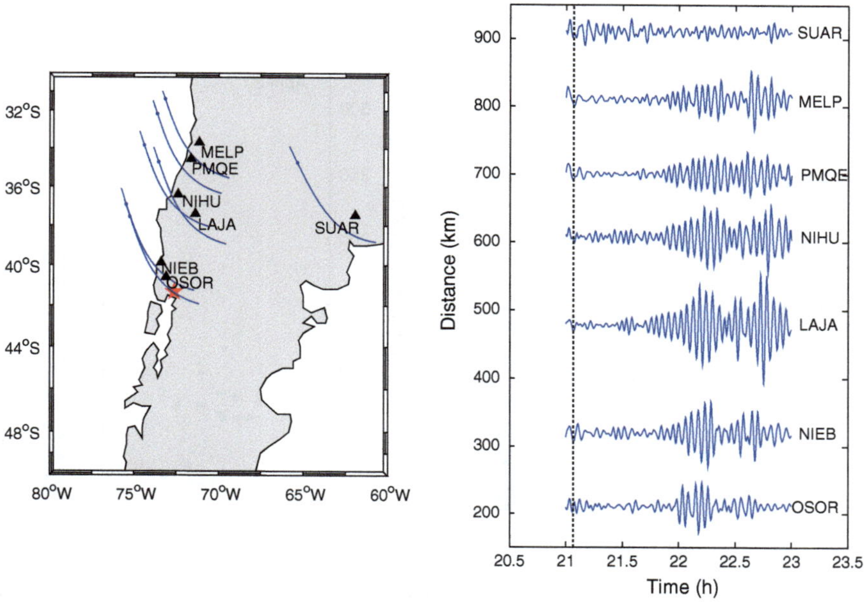

Fig. 13.8 Filtered TEC time series after the eruption on April 22, 2015. The stations' name is shown on the right. The left panel shows the location of each station analyzed and the SIP tracks for PRN23

13.4.2 Amplitude and Frequency

The amplitude of filtered TEC is considered to be influenced by the background TEC, the geometrical factor and the distance from away the volcano. But for the same satellite, the amplitude mostly relates to the distance. In our work, we applied a method to obtain the maximum disturbance point (MDP) by maximum disturbance distance (MDD) and Maximum Disturbance Azimuth (MDA). In our method, the volcano was considered as a point source and the anomalies traveled upward and reached a certain point, which was the MDP, of the ionosphere shell, and then the MDP acted as a point source to affect the surrounding area. The MDD is the distance between the MDP and the volcano and the MDA is the SIP Volcanic azimuth from MDA with the volcano. And SIP Volcanic azimuth and corresponding LOS satellite elevation angles distribution for maximum filtered TEC in each GPS observation arc are displayed in Fig. 13.9. When calculating the MDD, the maximum of each filtered TEC time series was regarded as the weights of the corresponding distance and the MDD could be obtained by the following equation:

$$MDD = \frac{\sum_{i=1}^{n} \text{distance}_i * TEC_i}{\sum_{i=1}^{n} TEC_i} \tag{13.11}$$

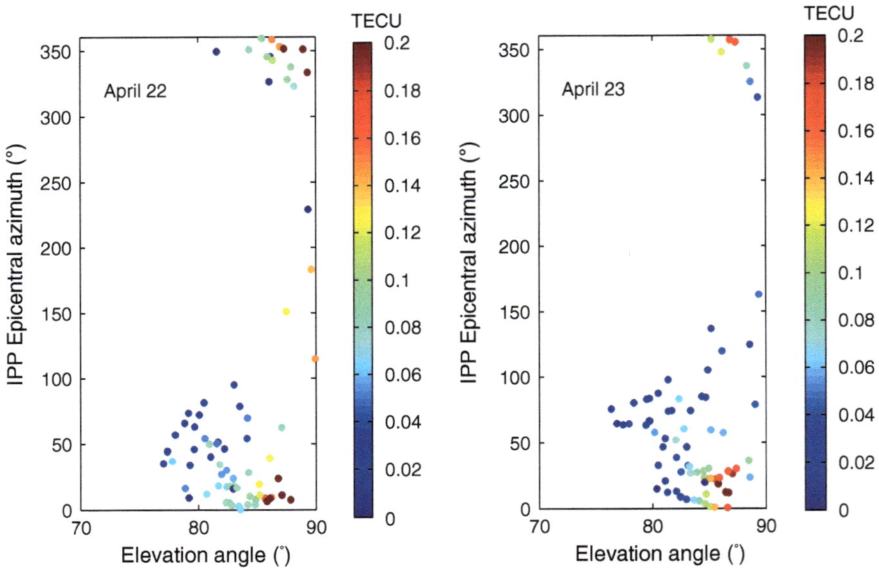

Fig. 13.9 Distribution of the IPP epicentral azimuth and corresponding LOS satellite elevation angles for maximum filtered TEC in each GPS observation arc for the first eruption (left panel) and the second eruption (right panel). Here the undisturbed arcs (maximum filtered TEC < 0.01 TECU) are not considered

Also for calculating MDA:

$$MDA = \frac{\sum_{i=1}^{n} \text{azimuth}_i * TEC_i}{\sum_{i=1}^{n} TEC_i} \tag{13.12}$$

The filtered TEC time series for PRN03 and PRN23 for the first volcano was studied in details based on this method without considering the cases which had the maximum filtered TEC less than 0.04 TECU (Liu et al. 2017). The calculated MDD was 350 km and the MDA was 5°. This result agreed with what could be found in Fig. 13.5, the nearer with the maximum disturbance point, the bigger of the TEC anomalies. Figure 13.6 also shows that the TEC disturbance could be detected from PRN03 at 21:30 UTC while there were no distinct TEC anomalies for PRN23 at that point. This difference could also be found in the spectrograms of filtered TEC for two satellites.

Figure 13.10 displays the spectrogram of the filtered TEC for some selected stations and satellites. The spectrogram associated with TEC time series were calculated using the short-time Fourier transform with the window of 30 min. From the up to the bottom, the first and second are PRN03 for station LAJA and PRN23 for station MHIN which obtained from the first eruption. And the third and the fourth are PRN13 for station PANE and PRN28 for station CAUQ that obtained from the second eruption. The left panel shows the filtered TEC time series in red and distance time series in blue, respectively. And the right panel displays the spectrograms of each selected

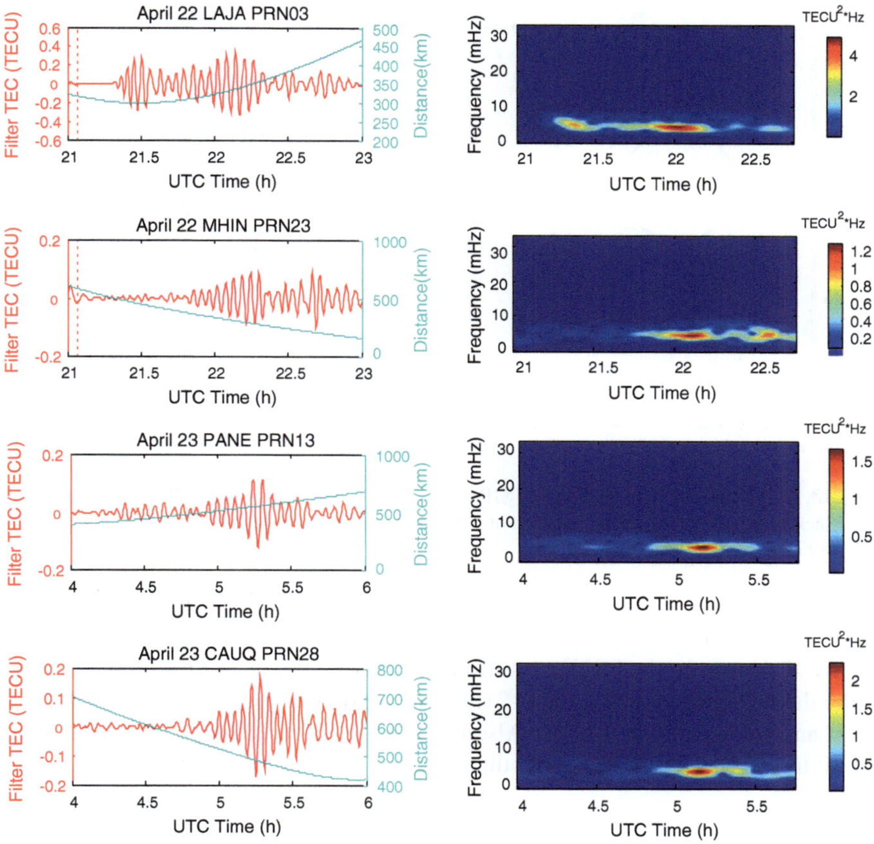

Fig. 13.10 Spectrograms for station LAJA and satellite PRN03, station MHIN and satellite PRN23 for the first eruption in the two top panels, and station PANE and satellite PRN13 as well as the station CAUQ and satellite PRN28 for the second eruption in the two bottom panels. The left panel displays the filtered TEC time series in red and the distances between the volcano and the IPP of corresponding stations and satellites

case. The frequency centered at about 4.0 and 3.8 mHz for the two eruptions, which was in the frequency range of infrasonic wave. The results are in agreement with the travel time diagram and the frequency for PRN03 has a peak at near 21:30 UTC while there was no peak for PRN23 at that point.

Figure 13.11 shows the position relationship between SIP tracks and MDP for PRN03 and PRN 23, respectively (Liu et al. 2017). The black triangle represents the location of MDP and the black circles indicate the influenced area with a radius of 300 km. The left panel, which is for PRN03, shows that the SIP tracks for most GPS stations traveled from the west to the east and also were in the range of the influenced area. While the right panel shows that the most SIP tracks traveled from the northwest to the southeast and most of them were out of the range of the influenced area at

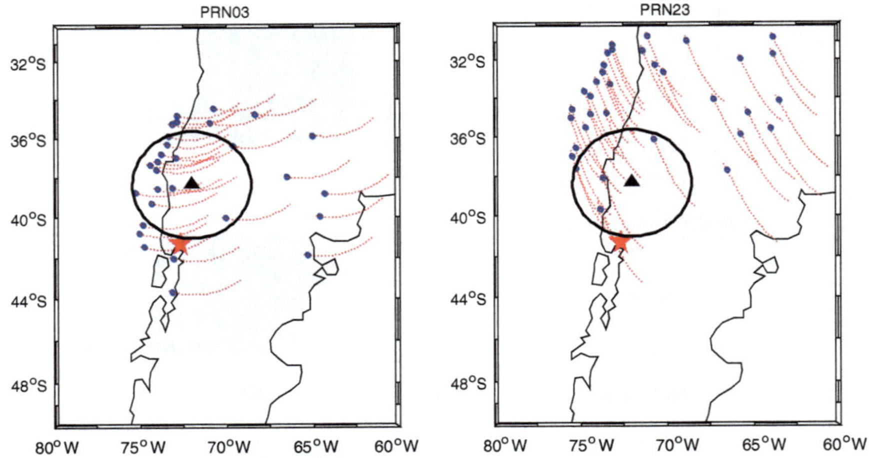

Fig. 13.11 The position relationship of SIP tracks and MDP for PRN03 and PRN23, respectively, the black triangle represents the location of MDP and the black circle indicates the influenced area with a radius of 300 km

first. And then they traveled into the influenced area and gradually approached to the MDP. However, the PRN23 is different from PRN03. Its most SIP tracks did not enter the influenced area at around 21:30 which explains why there were no detected TEC anomalies at around 21:30. Also because the SIP tracks were approaching the MDP in the selected time period, the largest TEC disturbance appeared in the last period.

13.5 Coupling Modeling and Discussions

Dautermann et al. (2009a) modeled the lithosphere–atmosphere–ionosphere coupling after the 2003 explosive eruption of the Souferiere Hills Volcano, Montserrat and their modeled signals match well with the signals recorded from strainmeters and GPS TEC signals in duration, waveform features and arrival times. Figure 13.12 displays the comparison of the synthetic and observed volumetric strain and TEC signals. Top three panels show the modeled signals (in red) and signals recorded at three strainmeter sites (AIRS, TRANTS, GERD). For both observed seismic signals and synthetic, the disturbance occurs at around 3:35 UT and falls off after 4:10 UT. Besides for AIRS and TRANTS, the modeled signal shows a distinct double wave train compared with the observed seismic wave. While there still exist mismatches between the modeled results and the observations, for example, the larger amplitude initial peak in the model and the simulated amplitude at site TRANTS is 50 times lower than observed. Taking TRANTS a coastal site into account, the mismatches might result from the tsunami.

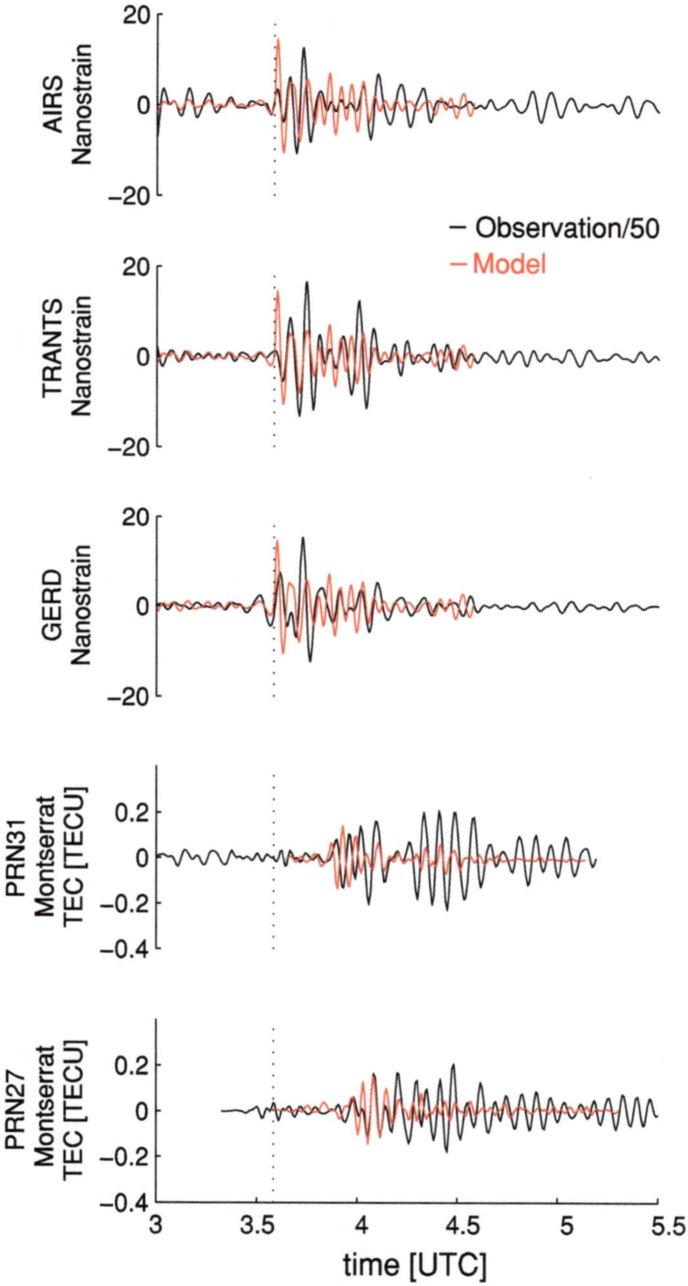

Fig. 13.12 Data recorded at the three strainmeter sites (top three panels, black) and stacked TEC time series using the Montserrat GPS sites for satellites PRN31 and PRN27 (bottom two panels, black) compared to model synthetics (red). Note that the observed strain signal amplitude at site TRANTS has been divided by 50 to match the synthetics (Dautermann et al. 2009a)

A similar double-wave train was recorded at borehole strainmeters, which indicated that part of the strain perturbation was due to energy leakage from the atmosphere to the solid Earth. Furthermore, the synthetic strain was split into its atmospheric and seismic mode components, which were used to compute an atmospheric source at the actual height of the SHV and for a source located 100 m below the ground surface. Figure 13.13 showed the model volumetric strain generated for an isotropic point source located within the atmosphere at 1 km altitude

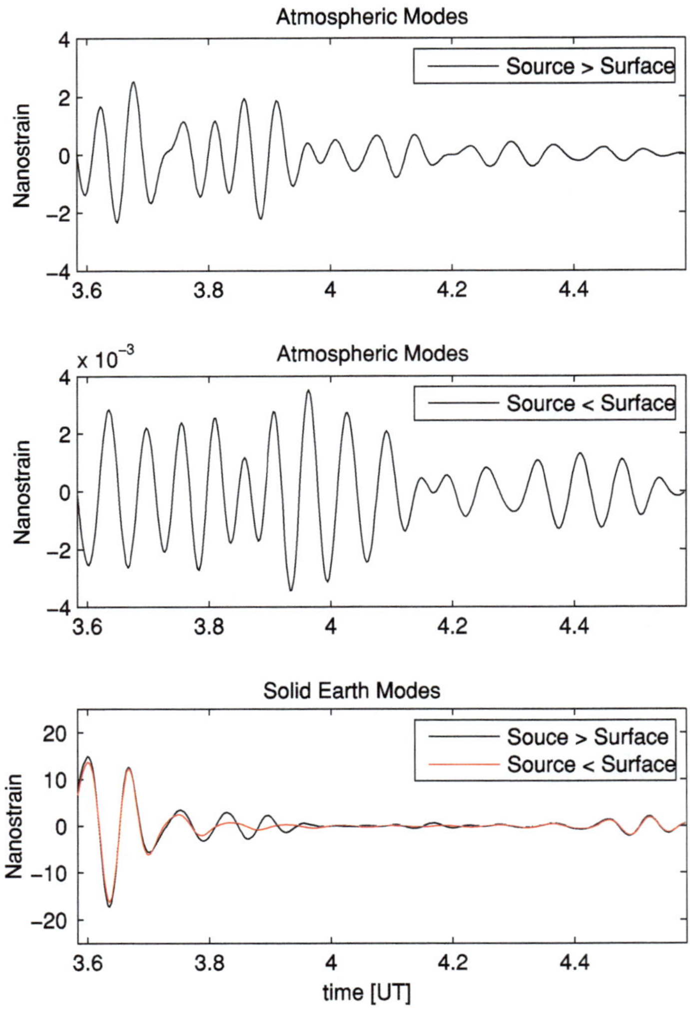

Fig. 13.13 Model volumetric strain generated for an isotropic point source located within the atmosphere at 1 km altitude (top and bottom red traces), and the same point source located 100 m below the surface (middle and bottom black traces) (Dautermann et al. 2009a)

(top and bottom red traces), and the same point source located 100 m below the surface (middle and bottom black traces) (Dautermann et al. 2009a). It can be seen that solid Earth modes were excited similarly by the two sources and resulted in a strain waveform with shorter duration and absence of a double wave train. Atmospheric modes showed the double wave train for a source located in the atmosphere, but not for an underground source, which resulted in strain amplitude 1000 times smaller.

According to the bottom two panels in Fig. 13.12, the arrival time and the amplitude of the first wave train of modeled signal matches well with the GPS TEC signal. For most selected cases, there exist three wave trains for the TEC time series obtained from GPS observations, while the modeled signal in the near field has two wave trains and one peak far from the SHV, and the amplitude for the second wave train of the synthetic is smaller than the GPS TEC signal. These mismatches might result from that the winds as well as temperature and deviations from the MSIS-E-90 are not taken into account.

13.6 Summary

The ionospheric anomalies following the April 2015 Calbuco volcano are investigated using GPS TEC and COSMIC GPS Radio Occultation. Significant TEC disturbances with the amplitude of 0.1–0.4 TECU are found about one hour after the eruptions and the disturbance lasts more than one hour for the first eruption and about half an hour for the second eruption. The propagation velocities from the travel time diagram were around 800 m/s and 900 m/s for the two eruptions, respectively. The spectrogram of filtered TEC time series is centered at 3.7 mHz, which is in the frequency range of infrasonic waves. The distribution and characteristics of the TEC anomalies are closely associated with the position relationship between SIP tracks and volcano. Meanwhile, the temperature anomaly from COSMIC vertical profiles over the Calbuco volcano is also detected, which is further confirmed by the spatial interpolation of the 200 hPa pressure level. The volume of the temperature anomaly is almost similar to the real-time image by the NASA VIIRS. The difference between anomalous and non-anomalous zones is about 8 °C, which may be the post-volcanic plumes perturbations. The observation of SO_2 also shows anomalous enhancement on the April 23, 2015, immediately after the eruption day. Based on our observations, the April 23, 2015 is declared the anomalous day for observing the volcanic generated temperature anomalies. All these results suggest the influence of volcanic eruption on the lower atmosphere, but further observations and studies are needed in the future.

References

Afraimovich, E., Ding, F., Kiryushkin, V., Astafyeva, E., Jin, S. G., & Sankov, V. (2010). TEC response to the 2008 Wenchuan earthquake in comparison with other strong earthquakes. *International Journal of Remote Sensing, 31*(13), 3601–3613. https://doi.org/10.1080/01431161003727747.

Cai, C., Liu, Z., Xia, P., & Dai, W. (2013). Cycle slip detection and repair for undifferenced GPS observations under high ionospheric activity. *GPS Solutions, 17*(2), 247–260.

Caudron, C., Taisne, B., Garcés, M., Alexis, L. P., & Mialle, P. (2015). On the use of remote infrasound and seismic stations to constrain the eruptive sequence and intensity for the 2014 Kelud eruption. *Geophysical Research Letters, 42*(16), 6614–6621.

Dautermann, T., Calais, E., Lognonne, P., & Mattioli, G. S. (2009a). Lithosphere—atmosphere—ionosphere coupling after the 2003 explosive eruption of the Soufriere Hills volcano, Montserrat. *Geophysical Journal International, 179*(3), 1537–1546.

Dautermann, T., Calais, E., & Mattioli, G. S. (2009b). Global positioning system detection and energy estimation of the ionospheric wave caused by the 13 July 2003 explosion of the Soufriere Hills volcano, Montserrat. *Journal of Geophysical Research Solid Earth, 114*(B2), 311–320.

Dautermann, T., Calais, E., Lognonné, P., & Mattioli, G. S. (2010). Lithosphere—atmosphere—ionosphere coupling after the 2003 explosive eruption of the Soufriere Hills volcano, Montserrat. *Geophysical Journal International, 179*(3), 1537–1546.

Heki, K. (2006). Explosion energy of the 2004 eruption of the Asama volcano, central Japan, inferred from ionospheric disturbances. *Geophysical Research Letters, 33*(14), L34303. https://doi.org/10.1029/2006GL026249.

Jin, S. G., van Dam, T., & Wdowinski, S. (2013). Observing and understanding the Earth system variations from space geodesy. *Journal of Geodynamics, 72,* 1–10. https://doi.org/10.1016/j.jog.2013.08.001.

Jin, S. G., Han, L., & Cho, J. (2011). Lower atmospheric anomalies following the 2008 Wenchuan Earthquake observed by GPS measurements. *The Journal of Atmospheric and Solar-Terrestrial Physics, 73*(7–8), 810–814. https://doi.org/10.1016/j.jastp.2011.01.023.

Jin, S. G., Jin, R., & Li, J. H. (2014). Pattern and evolution of seismo-ionospheric disturbances following the 2011 Tohoku earthquakes from GPS observations. *Journal of Geophysical Research: Space Physics, 119*(9), 7914–7927. https://doi.org/10.1002/2014JA019825.

Jin, S. G., Jin, R., & Li, D. (2017). GPS detection of ionospheric Rayleigh wave and its source following the 2012 Haida Gwaii earthquake. *Journal of Geophysical Research: Space Physics.* https://doi.org/10.1002/2016ja023727.

Jin, S. G., Luo, O. F., & Park, P. H. (2008). GPS observations of the ionospheric F2-layer behavior during the 20th November 2003 geomagnetic storm over South Korea. *Journal of Geodesy, 82*(12), 883–892. https://doi.org/10.1007/s00190-008-0217-x.

Jin, S. G., Occhipinti, G., & Jin, R. (2015). GNSS ionospheric seismology: Recent observation evidences and characteristics. *Earth Science Reviews, 147,* 54–64. https://doi.org/10.1016/j.earscirev.2015.05.003.

Jin, S. G., Wang, J., Zhang, H., & Zhu, W. Y. (2004). Real-time monitoring and prediction of the total ionospheric electron content by means of GPS observations. *Chinese Journal of Astronomy and Astrophysics, 28*(3), 331–337. https://doi.org/10.1016/j.chinastron.2004.07.008.

Jin, S. G., Zhu, W. Y., & Afraimovich, E. (2010). Co-seismic ionospheric and deformation signals on the 2008 magnitude 8.0 Wenchuan Earthquake from GPS observations. *International Journal of Remote Sensing, 31*(13), 3535–3543. https://doi.org/10.1080/01431161003727739.

Kanamori, H., Mori, J., & Harkrider, D. G. (1994). Excitation of atmospheric oscillations by volcanic eruptions. *Journal Geophysical Research, 99,* 21947–21961. https://doi.org/10.1029/94JB01475.

Kanamori, H., Mori, J., Hauksson, E., Heaton, T. H., Hutton, L. K., & Jones, L. M. (1993). Determination of earthquake energy release and ML using TERRAscope. *Bulletin of the Seismological Society of America, 83*(2), 330–346.

Li, J. H., & Jin, S. G. (2017). High-order ionospheric effects on electron density estimation from Fengyun-3C GPS radio occultation. *Annales Geophysicae, 35*(3), 403–411. https://doi.org/10.5194/angeo-35-403-2017.

Liu, X., Zhang, Q., Shah, M., Hong, Z. (2017). Atmospheric-ionospheric disturbances following the April 2015 Calbuco volcano from GPS and OMI observations. *Advances in Space Research, 60*(12), 2836–2846.

Lognonne, P. (1991). Normal modes and seismograms in an anelastic rotating Earth. *Journal Geophysical Research, 96,* 20309–20319.

Lognonne, P., Clevede, E., & Kanamori, H. (1998). Computation of seismograms and atmospheric oscillations by normal-mode summation for a spherical earth model with realistic atmosphere. *Geophysical Journal International, 135,* 388–406.

McCormick, M. P., Thomason, L. W., & Trepte, C. R. (1995). Atmospheric effects of the Mt. Pinatubo eruption. *Nature, 373,* 399–404.

Nakashima, Y., Heki, K., Takeo, A., Cahyadi, M. N., Aditiya, A., & Yoshizawa, K. (2016). Atmospheric resonant oscillations by the 2014 eruption of the Kelud volcano, Indonesia, observed with the ionospheric total electron contents and seismic signals. *Earth & Planetary Science Letters, 434,* 112–116.

Okazaki, I., & Heki, K. (2012). Atmospheric temperature changes by volcanic eruptions: GPS radio occultation observations in the 2010 Icelandic and 2011 Chilean cases. *Journal of Volcanology and Geothermal Research, 245–246,* 123–127.

Rolland, L. M., Lognonné, P., Astafyeva, E., Kherani, E. A., Kobayashi, N., Mann, M., et al. (2011). The resonant response of the ionosphere imaged after the 2011 off the pacific coast of Tohoku earthquake. *Earth, Planets and Space, 63*(7), 853–857.

Saito, A., Tsugawa, T., Otsuka, Y., Nishioka, M., Iyemori, T., Matsumura, M., et al. (2011). Acoustic resonance and plasma depletion detected by GPS total electron content observation after the 2011 off the pacific coast of Tohoku earthquake. *Earth, Planets and Space, 63*(7), 863–867.

Shah, M., & Jin, S. G. (2015). Statistical characteristics of seismo-ionospheric GPS TEC disturbances prior to global Mw $\geq$ 5.0 earthquakes (1998–2014). *Journal of Geodynamics, 92,* 42–49. https://doi.org/10.1016/j.jog.2015.10.002.

Shults, K., Astafyeva, E., & Adourian, S. (2016). Ionospheric detection and localization of volcano eruptions on the example of the April 2015 Calbuco events. *Journal of Geophysical Research: Space Physics, 121,* 10303–10315. https://doi.org/10.1002/2016JA023382.

Wang, K.-Y., Lin, S.-C., & Lee, L.-C. (2009). Immediate impact of the Mt Chaiten eruption on atmosphere from FORMOSAT-3/COSMIC constellation. *Geophysical Research Letters, 36,* L03808. https://doi.org/10.1029/2008GL036802.

Wu, Y., Jin, S. G., Wang, Z., & Liu, J. (2010). Cycle slip detection using multi-frequency GPS carrier phase observations: A simulation study. *Advances in Space Research, 46*(2), 144–149. https://doi.org/10.1016/j.asr.2009.11.007.

Chapter 14
Volcanic Plumes Detection from GNSS SNR

14.1 Volcanic Plumes

The erupting volcano will eject a large amount of rock mass, volcanic ash, volcanic gas, and other substances, which can seriously affect the integrity of buildings, damage the safety of land or air traffic, endanger the growth of plants, and affect the healthy life and safety of human beings. Therefore, it is important to monitor and forecast volcanic ash. A variety of ground or air sensors are used to monitor volcanic eruptions around the world. These two styles of devices have their own advantages and disadvantages. Ground-based equipment can provide relatively high precision and spatial-temporal coverage with limited coverage. Conversely, satellite observations can provide a global coverage with low spatial resolution and accuracy. In order to better reduce the damage of volcanic plumes, related monitoring and forecasting systems such as Istituto Nazionale di Geofisica e Vulcanologia, Osservatorio Etneo (INGV-OE), Light Detection And Ranging (LiDAR), Doppler radar, etc. have gradually been developed and used (Scollo et al. 2009, 2014; Gouhier et al. 2012).

In addition, volcanic plumes may also affect GNSS satellite signals. Houlie et al. (2005) found the anomaly of the GPS linear combination during the eruption of Miyake Jima in 2000 and proposed that these anomalies are due to effects of the hot volcanic plume on GPS path delay. Aranzulla et al. (2013) applied the robust cross-statistics to assess the plume detection by GPS signals and suggested that the GPS network at Etna was able to detect the volcanic plume formed during the eruption of 4–5 September 2007. Grapenthin et al. (2013) found systematic spikes in subdaily positioning solutions to phase delays for station–satellite pairs that cross dense parts of volcanic plumes, showing that the phase residuals along the sky tracks of satellites provide easy access to plume azimuths. All the above methods used the standard geodetic processing techniques for position and treated the plume signal as an atypical atmospheric effect. These technically challenging methods require

---

The original version of this chapter was revised: Belated corrections have been incorporated. The correction to this chapter is available at https://doi.org/10.1007/978-981-10-3178-6_16

S. Jin et al., *GNSS Atmospheric Seismology*,
https://doi.org/10.1007/978-981-10-3178-6_14

analysis of thousands of GPS carrier phase ranges and estimation of the effects of station coordinates, orbits, relativity, clocks, atmospheric delaysdelays, and phase ambiguities, while both the plume and the receiver's positions are time-varying phenomena. Furthermore, because of the application of the least squares estimation, any mis-modeling of GNSS errors is contributed to other estimated parameters and satellite observations that do not cross the plume.

Recently, GNSS signal-to-noise ratio (SNR) observations are related to surface characteristics and volcanic ash (Jin and Najibi 2014; Jin et al. 2011, 2016, 2017; Larson 2013). Larson (2013) proposed a new method to detect volcanic plumes using GPS signal strength of signal-to-noise ratio (SNR) data, instead of the GPS carrier phase data. The eruptions of Mt. Redoubt in 2009 and Mt. Okmok in 2008 were taken as examples and SNR data were very sensitive to large ash particles, which can be modeled without estimating position so as to provide a clearer picture of a volcanic plume. Fournier and Jolly (2014) computed the GPS LC phase residuals and SNR data at the time of the eruption of the Tongariro Te Maari crater, New Zealand in 2012, showing that several processes successively affected the signal at the closest GPS station. Therefore, GPS LC phase residuals and SNR data have the potential to monitor and understand eruptive sequences when visual observations are limited. In this chapter, SNR data are used to monitor volcanic plume variations for recent Aso volcanic eruption as well as Etna Volcano and Okmok Volcano.

14.2 Theory and Method

14.2.1 GPS SNR Data

The GPS signal is a modulated wave for navigation and positioning by the wide users. It includes carrier, range code, and navigation message in the L-band. The commonly used carrier waves are L1 with carrier frequency f1 equal to 1575.42 MHz and wavelength $\lambda 1$ equal to 19.032 cm, and L2 with carrier frequency f2 equal to 1227.60 MHz and wavelength $\lambda 2$ equal to 24.42 cm. The latest carrier wave is L5 with carrier frequency f5 equal to 1176.45 MHz and wavelength $\lambda 5$ equal to 25.480 cm. There are two types of range code,namely, C/A code and P code. The former can only be modulated on L1 with code rate equal to 1.023 MHz and the latter's code rate is equal to 10.23 MHz.

All GPS receivers routinely calculate and distribute a SNR value because it provides information about how well the tracking algorithms are working inside the receiver. SNR data correspond to carrier-to-noise-density ratio (C/N0), the ratio of signal power to the noise power spectral density. SNR is related to C/N0 through the noise bandwidth (B) as in SNR = (C/N0)/B in units of decibels. GPS receiver manufacturers primarily report this quantity assuming a 1-Hz bandwidth, or dB-Hz. Corresponding to L1, L2, and L5, the type of SNR are S1, S2, and S5. SNR data contribute nothing to the distance between the satellite transmitting the signal and the antenna on the ground receiving the signal, and make no direct contribution to navigation and positioning. As a result, geodesists and geophysicists always ignore SNR data. However, SNR data remains valuable with the feature of directly measuring signal blockages. For example, an ash-laden volcanic plume crosses a GPS signal and parts of that signal are attenuated and scattered, which means that the signal arriving at the GPS receiver has less power than the normal. In contrast, water in the atmosphere does not significantly reduce GPS signal power. The lack of sensitivity to water is the same reason that L-band radars are being developed to observe ash clouds (Aranzulla et al. 2013). Previous GPS plume studies used carrier phase data, which is sensitive to water vapor, temperature, and ash, while the remarkable feature only influenced by ash is why SNR data apply to volcanic research (Ohta and Iguchi 2015).

SNR data rely heavily on the GNSS satellite elevation angle. When GNSS satellite elevation angles increase from 5° to 65°, SNR values will slowly increase, mainly due to antenna gain type. When elevation angles less than 25° L1 SNR data have larger high-frequency variations than high-elevation L2P SNR data, ascribing to the fact that the C/A code is much shorter than L2P or L2C and suffers from cross-channel interference. Nevertheless, because the receiver cannot use the encrypted code in its retrieval, L2P has much lower SNR values than C/A or L2C, but users have no access to L2C data unless they request it.

When elevation angles are below 25°, ground reflection will cause the obvious oscillations of GNSS SNR data and the frequency is related to surface characteristics of GNSS antenna surroundings, e.g., soil moisture content, snow depth, and sea level height (Jin and Komjathy 2010; Larson et al. 2010; Jin et al. 2014). Observed modulations exist because the reflected signal travels a longer distance than the direct signal and interferes with the direct signal. In order to eliminate the disturbance of ground reflection, we only use the GNSS SNR data for volcano plume detection with high-elevation angle satellites of over 25°.

Besides, we should verify whether GNSS SNR data are influenced by ionospheric scintillations, since ionospheric activities also produce significant changes in SNR data. It is worth mentioning that ionospheric scintillation often occurs above high latitudes and low latitudes but rarely occurs above the middle latitudes. Furthermore, to get more full analyses in the plume detection based on GNSS SNR data, we compare our results with other instrument measurements, such as scatter radar. Here, the size of ash particle is not considered and will be considered in the future work. In fact, volcanic eruptions contain volcanic ash with solid particles, water vapor, and volcanic gases with sulfide. In higher elevations, water vapor will separate from volcanic ash because water vapor will keep enough density and volume in higher elevations, but volcanic ash will settle and dispel along horizontal direction. As a result, along the vertical and horizontal direction, the farther the distance to volcano, the lower the density of volcanic ash is. Hence, the distance between the GNSS station and volcano is farther, the impact on GNSS SNR data is smaller. It means that the ash density, dispersion direction and distance play important roles in GNSS SNR attenuation.

14.2.2 *Detection Method*

First, we preprocess the GNSS SNR data in which we abandon the data whose elevation angles are below 25° to avoid ground reflection. Then the SNR data with the elevation angle are checked if SNR data increase with the increase of elevation angles. Second, we filter the direct components in GNSS SNR data and remain the interference components. Here, we get the SNR correction profile by the four-order polynomial fitting. These modeled SNR data can be subtracted from the observed SNR values from each GNSS station on someday, and then we can obtain the differenced SNR time series. After the abnormal variation of SNR data is obtained, we can roughly estimate the volcanic eruption time.

In addition to using the SNR data for a correction profile, the standard deviation of these data is also calculated. Here, a running average standard deviation of 9 points (4.5 min) is used to characterize the residual. Also, we set a threshold value of 1 dB-Hz as an auxiliary condition to judge the SNR anomaly.

14.3 GPS SNR Anomalies Following Volcanoes

14.3.1 *Mt. Aso Volcano*

Mount Aso is the largest active volcano in Japan and also one of the largest volcanoes in the world. It locates in Aso Kujū National Park in Kumamoto Prefecture, the island of Kyushu. Its peak is 1,592 m above mean sea level. Japan Meteorological Agency (JMA) reported that explosive eruptions at Asosan's Nakadake Crater occurred at 01:46 UTC, 8 October 2016, after a period of deformation was detected. Inclement weather prevented visual observations, although satellite images showed the 8 October 2016 ash plume rising to an altitude of 11 km (36,100 ft). The alert level was raised to 3 (scale of 1–5). Ash deposits extended as far as 1.6 km on the NW flank and 1 km on the SE flank, and were abundant on the NE flank.

The GNSS SNR data are used to monitor volcanic plumes of the 2016 Mt. Aso volcano from GEONET stations established by the Geographical Information Authority of Japan (GSI). The GNSS data are sampled at 1-s or 30-s intervals, which depend on the situation of each station, e.g., the type of GNSS receiver and the telemetry system. According to Grapenthin et al. (2013), a sampling rate of 15-s or 30-s can analyze the signal across the volcanic plume. All the stations record GPS dual-frequency data. In our study, the sampling rate of the 30-s data is used. We choose 9 GEONET stations around Mount Aso volcano in the range of 30 km. Figure 14.1 shows the satellite track, the position of selected GEONET stations and the location of Mount Aso volcano. We extract GNSS SNR data from GEONET stations around Mount Aso volcano during 16:00–18:00 on 8 October 2016 (UTC). The changing relationship between SNR data with time or satellite elevations is analyzed and the differenced

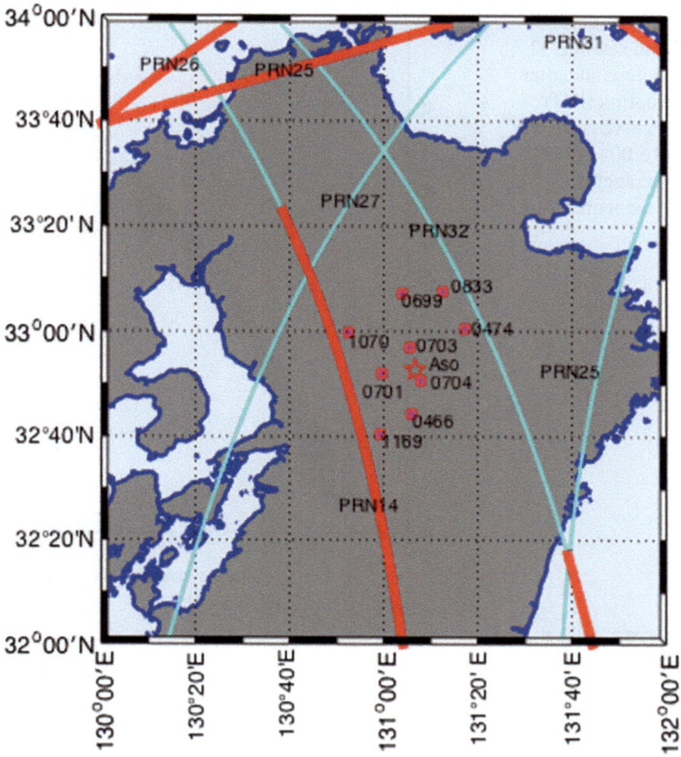

Fig. 14.1 Magenta circles represent the location of GPS sites. The 2016 Mount Aso volcano is shown by the red star. The cyan lines symbolize the shadow of satellite traces during the whole day in latitude and longitude, and the red lines show the satellite trace between 16:00 and 18:00 UTC on DOY 281, 2016

SNR value of each station by binomial fitting method is calculated, which show the occurring probability from the differenced SNR value.

Figure 14.2 provides a typical example of L2 SNR data collected from the station 0703 around the 2016 Mount Aso volcano. SNR data collected during the eruption day and 2 days before and after the eruption (from day of year (DOY) 279 to 283) are shown. We can see that the SNR value increases following the increase of the satellite angle and the profiles are similar except the DOY 281 (Zhang 2018). The 3 dB in SNR data drop on DOY 281 when the elevation angle is about 45°, which are mostly related to the antenna gain. Figure 14.3 shows a typical example of L5 SNR data from station 0703, similar to L2 SNR. The change in SNR values is consistent with the behavior of satellite that is setting with the exception of the outliers on DOY 281. The rest of eight stations have no anomalies. L5 SNR value is 15 dB larger than L2 SNR and has a drop of 2 dB at the elevation angle of 45°. The occurrence of the drop in SNR value may be related to plume dispersion and the signal path between satellite and station.

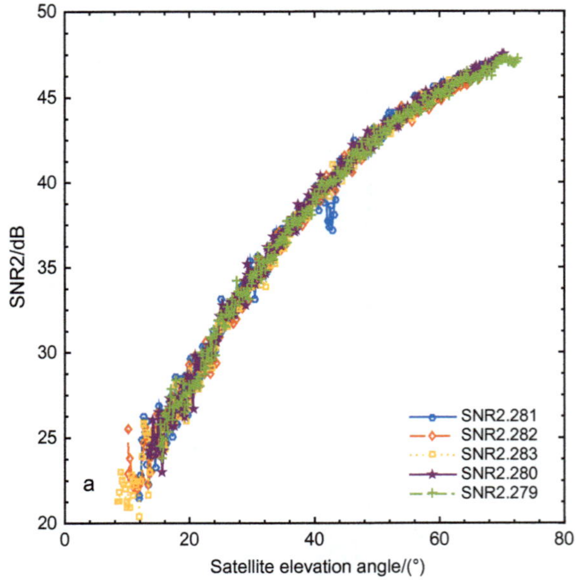

Fig. 14.2 L2 SNR data of station 0703 and satellite 32 for two days before and after Mount Aso eruption (DOY 281). A rising curve is shown for the hours 16:00–18:00 UTC. Elevation angles decrease with the increase of time

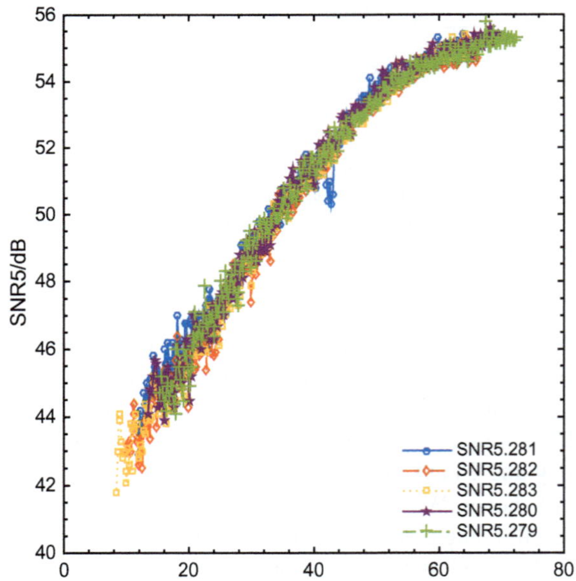

Fig. 14.3 L5 SNR data of station 0703 and satellite 32 for two days before and after Mount Aso eruption (DOY 281). A rising curve is shown for the hours 16:00–18:00 UTC. Elevation angles decrease with the increase of time

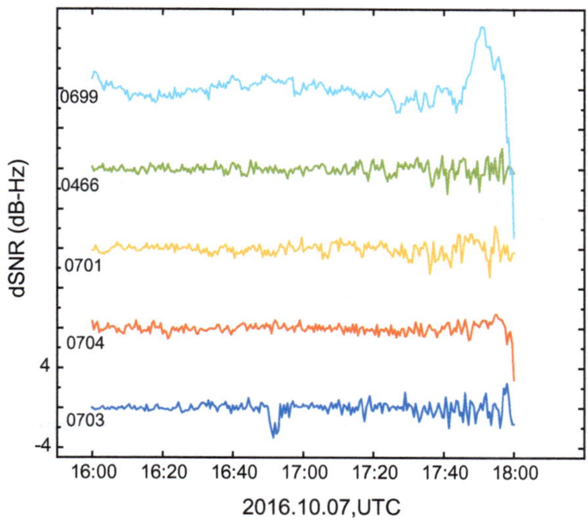

Fig. 14.4 Differenced SNR time series for the sites around Mount Aso volcano

Figure 14.4 summarizes the differenced SNR time series for the sites around Mount Aso volcano, which clearly shows that only differenced SNR calculated by L2 SNR data from station 0703 has obvious anomalies between 16:40 and 17:00 UTC when the eruption occurred at 16:46 UTC. In order to further research the relationship between differenced SNR and volcanic plume, differenced SNR time series and satellite–station pair path time series are taken into account. Figure 14.5a shows the differenced SNR data after removing the correction profile. There is a sharp drop during 16:45–17:00 UTC. We can see that the residuals are slightly bigger at the end of the arc than at the ending, probably due to that this satellite is setting. In Fig. 14.5b, c, horizontal and vertical projections are plotted to intuitively display the relative geometric information between each receiver–satellite line of sight and Mount Aso volcano. Satellite PRN32's signal crosses the ring region with Mount Aso volcano as the center and a radius of 2–4 m and the lowest crossing altitude is 2 km.

Here, we analyze the detection capability of volcanic plume with a circle in a 1-km radius from 6 sites within 20 km of the summit over a 24-hour period. The potential plume detection is assessed for two regions with 0–2 and 2–4 km above the volcanic vent. Figure 14.6 shows detection probability and contour map of volcanic ash cylindrical plume with a radius of 1 km (red dotted circle) over the volcanic vent (red star) of Mount Aso volcano on 7 October 2016. The closer the GPS site to the volcanic vent, the easier the plume is monitored. The contour line for volcanic ash cylindrical plume detection with the high probability of detection is over 600 m.

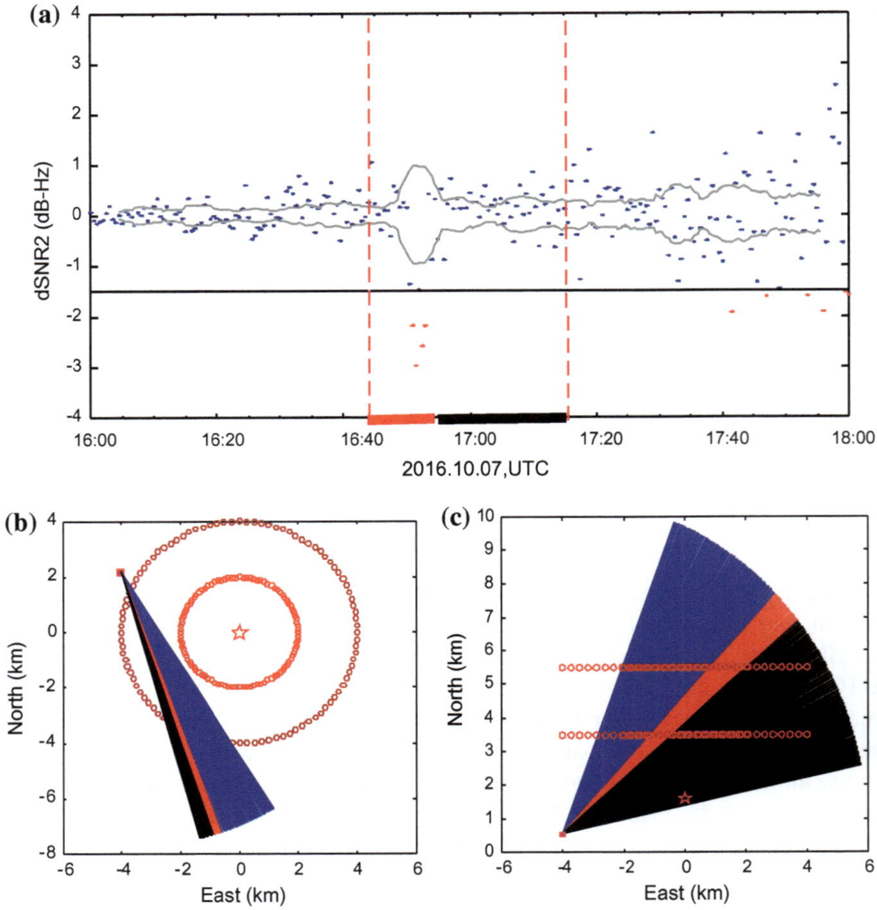

Fig. 14.5 **a** Differenced SNR observations (blue dots) with one standard deviation in gray using 4.5 min smoothing for GNSS station 0703 and satellite PRN32. The horizontal black line represents the threshold value. SNR values less than −1.5 dB-Hz are red dots. The vertical red dashed line represents the start time and the end time of the eruptive. The red thick line shows the time interval with SNR drop. The black thick line shows the time interval without SNR drop. **b** The geometry of each satellite line of sight is depicted in top view, the blue line represents the satellite–station pair path before the eruption, the red line shows the satellite–station pair path during the eruption, and the black line represents the satellite–station pair path after the eruption. **c** The geometry of each satellite line of sight is depicted in side view

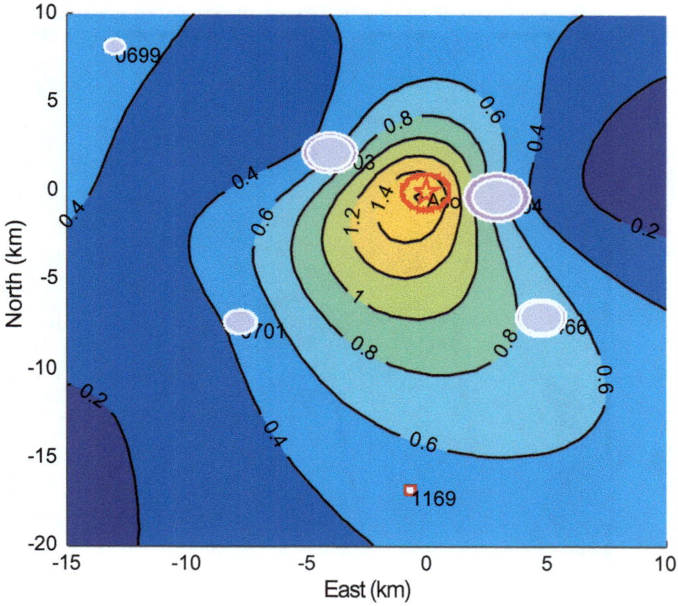

Fig. 14.6 Detection probability and contour map of volcanic ash cylindrical plume with a radius of 1 km (red dotted circle) over the volcanic vent (red star) of Mount Aso volcano on October 7, 2016. The circles show the 10 times probability for the mean line of sight signal crossing above the volcanic vent within 0–2 km (Darker purple circles) and 2–4 km (lighter purple circles). The red circle represents a top view of the volcanic plume with a radius of 1 km from the volcano

14.3.2 Mt. Etna Volcano

Mount Etna is an active stratovolcano on the east coast of Sicily, Italy. It lies above the convergent plate margin between the African Plate and the Eurasian Plate. It is the highest active volcano in Europe outside the Caucasus. It is currently 3329 m in altitude, though this varies with maximum eruptions. INGV reported that after a progressive intensification of activity during the evening of 2 December 2015, an eruption at Etna's Voragine Crater erupted between 03:30 and 04:10UT on 3 December 2015. During the peak period, the sustained lava fountains rose over 1 km above the crater with some jets of hot material rising 3 km. An ash plume rose several kilometers and drifted NE, causing ashfall in Linguaglossa, Francavilla di Sicilia, Milazzo, Messina, and Reggio Calabria.

The SNR data are used to monitor volcano plume of Mt. Etna volcano from dense Etnanet stations established by INGV-OE. The GPS data were sampled at 30-s intervals. In order to avoid data shift and obtain the same satellite geometry, we choose the data from 1 December 2015 to 5 December 2015 during 2:00–4:00 UTC from stations around Mount Etna in the range of 10 km. Figure 14.7 shows the satellite trace and the location of Mount Etna volcano and Etnanet GPS stations. We

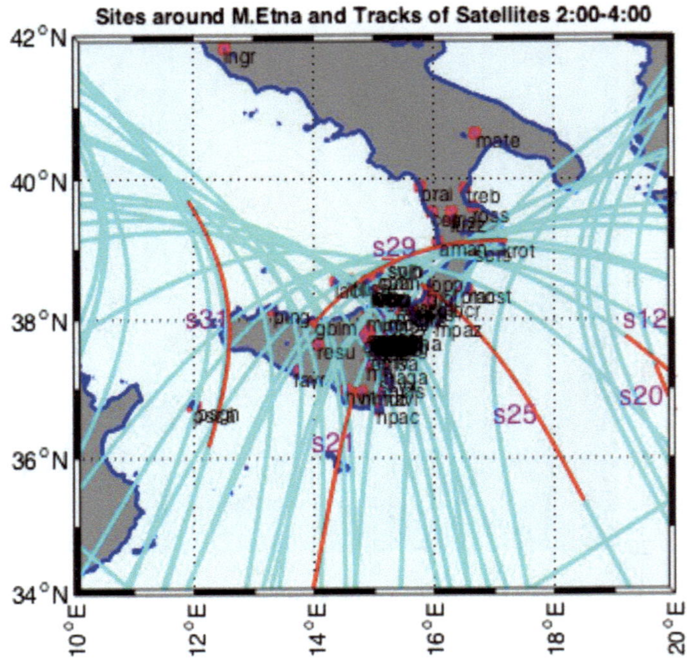

Fig. 14.7 The location of GPS sites (the magenta circle) and Mount Etna volcano (the red star). The cyan line represents the shadow of satellite traces during the whole day in latitude and longitude and the red line show the satellite trace between 2:00 and 4:00 UTC on DOY 337

can see that the traces of satellite PRN07, PRN13, PRN21, and PRN27 are closer to the Mount Etna volcano and the orientation of satellite antenna can be estimated. Considering the plume diffusion direction and many trials, we finally find that the station ECNE for satellite PRN21 in the NE direction of the Mount Etna volcano was badly affected because the station ECNE locates at the plume diffusion direction with 1276 m from the Mount Etna volcano. Figure 14.8a, b show a typical example of L1 and L2 SNR data collected from the station ECNE with satellite PRN21, respectively and Fig. 14.8c is a typical example of L1 SNR data collected from the station EPLU with satellite PRN21. A rising curve is shown for 2:00–4:00 UTC and the elevation angle of satellite increases with the increase of time (Fig. 14.8). For station ECNE, we can see that the L1 SNR value increases following the increase of satellite elevation angle and almost keeps unchanged when the elevation angle is larger than 60°and the profiles are similar except DOY 337 (Fig. 14.8a). Significant SNR difference with about 6 dB for satellite PRN21 is found for two elevation angle periods, 30°–35° and 45°–50°, corresponding to the two time periods, 2:27–2:47UTC and 3:05–3:15 UTC. The L2 SNR value at station ECNE increases following the increase of satellite elevation angles and the two periods of drops keep same with L1 SNR, but the drop value of L2 SNR is 4 dB larger than L1 SNR (Fig. 14.8b). The L1 SNR value at EPLU increases following the increase of satellite elevation angles and almost keeps

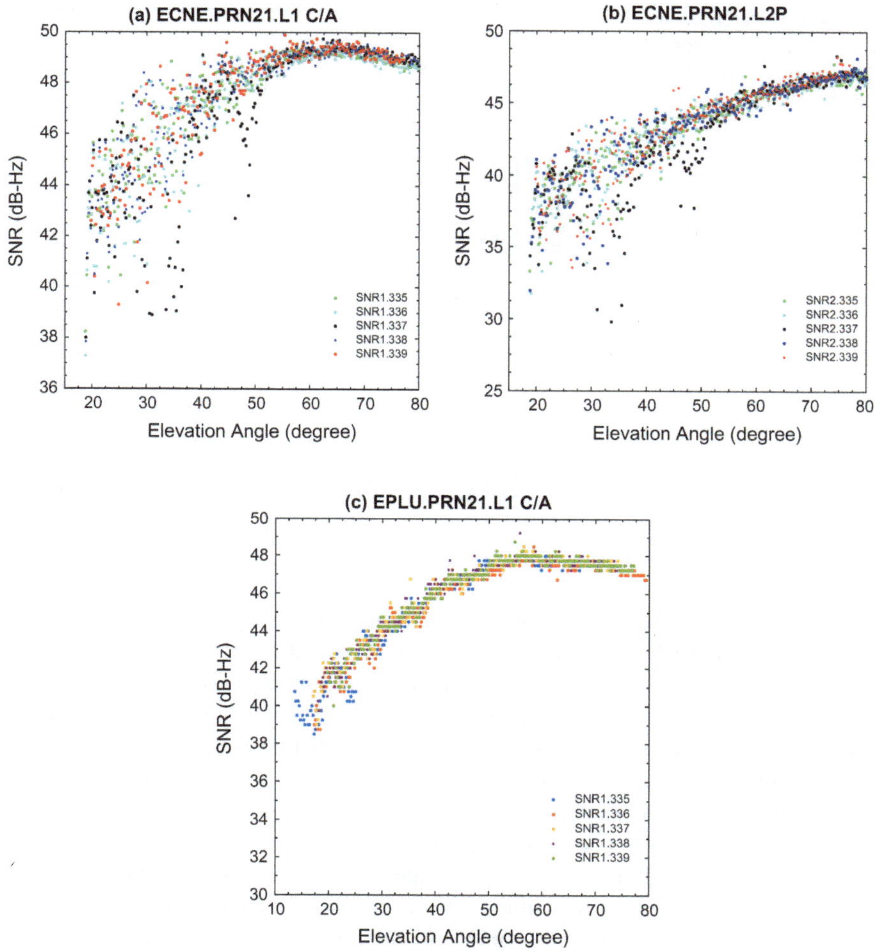

Fig. 14.8 SNR data for two days before and after Mount Etna eruption (DOY 337). **a** L1 SNR for station ECNE and satellite PRN21, **b** L2 SNR for station ECNE and satellite PRN21, and **c** L1 SNR for station EPLU and satellite PRN21

unchanged when the elevation angle is larger than 60° and the profile is similar for all five days (Fig. 14.8c).

Figure 14.9 shows differenced SNR time series for GPS stations around Mount Etna volcano. It is clearly seen that only differenced SNR time series calculated by L2 SNR data from station ECNE have obvious anomalies during two periods, 2:27–2:47 UTC and 3:05–3:15 UTC. The differenced SNR series at station EMCN have a 3 dB drop at the elevation angle of 27°, which may be caused by the circumstance such as ground reflection because the same drop is found on DOY 336 and 338. In order to further study the relationship between differenced SNR and volcanic plume, the time series of differenced SNR and satellite–station pair path are taken into account.

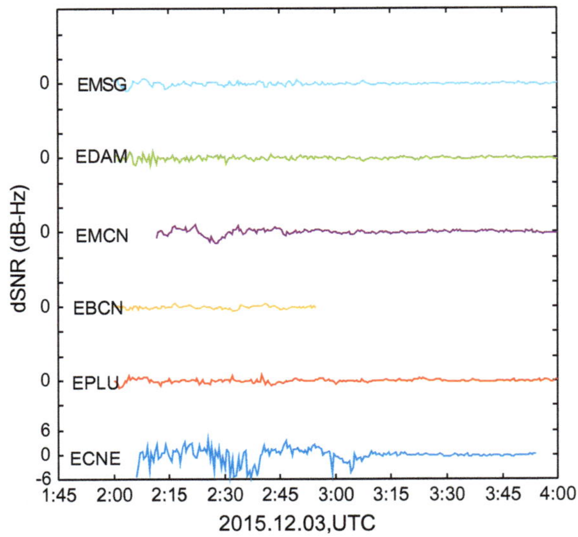

Fig. 14.9 Differenced GNSS SNR time series for GPS stations around Mount Etna volcano

Figure 14.10a shows the differenced SNR data after removing the correction profile. There are two sharp drops during 2:27–2:47 UTC and 3:05–3:15 UTC. The anomaly is slightly bigger at the end of the arc than at the beginning, probably because this satellite is setting. In Fig. 14.10b, c, horizontal and vertical projections are plotted to intuitively display the relative geometric information between each receiver–satellite line of sight and Mount Etna volcano. The signal of satellite PRN21 just crosses the vent of Mount Etna volcano with the highest altitude of 8 km.

Figure 14.11 shows the detection capability for a 1-km radius plume at 12 stations within 10 km of the summit over a 24-hour period. The circles show the probability for the mean line of sight to be within 0–2 km above the volcanic vent (Darker purple circles) and 2–4 km above the volcanic vent (lighter purple circles). The red square represents the location of GPS stations and the red circle represents a top view of the volcanic plume with a radius of 1 km from the volcano. The potential plume detection is assessed for two regions with 0–2 and 2–4 km above the volcanic vent. It is clearly seen that the closer the GPS site to the volcanic vent, the easier the plume is monitored. The contour line with the high probability of volcanic plume detection is over 2 km.

14.3.3 Mt. Okmok Volcano

Mount Okmok is the highest point on the rim of Okmok Caldera on the northeastern part of Umnak Island in the eastern Aleutian Islands of Alaska. Alaska Volcano Observatory (AVO) reported that a strong explosive eruption on Okmok began abruptly at 19:43 UTC on 12 July 2008, and ash and gas from the initial explosions

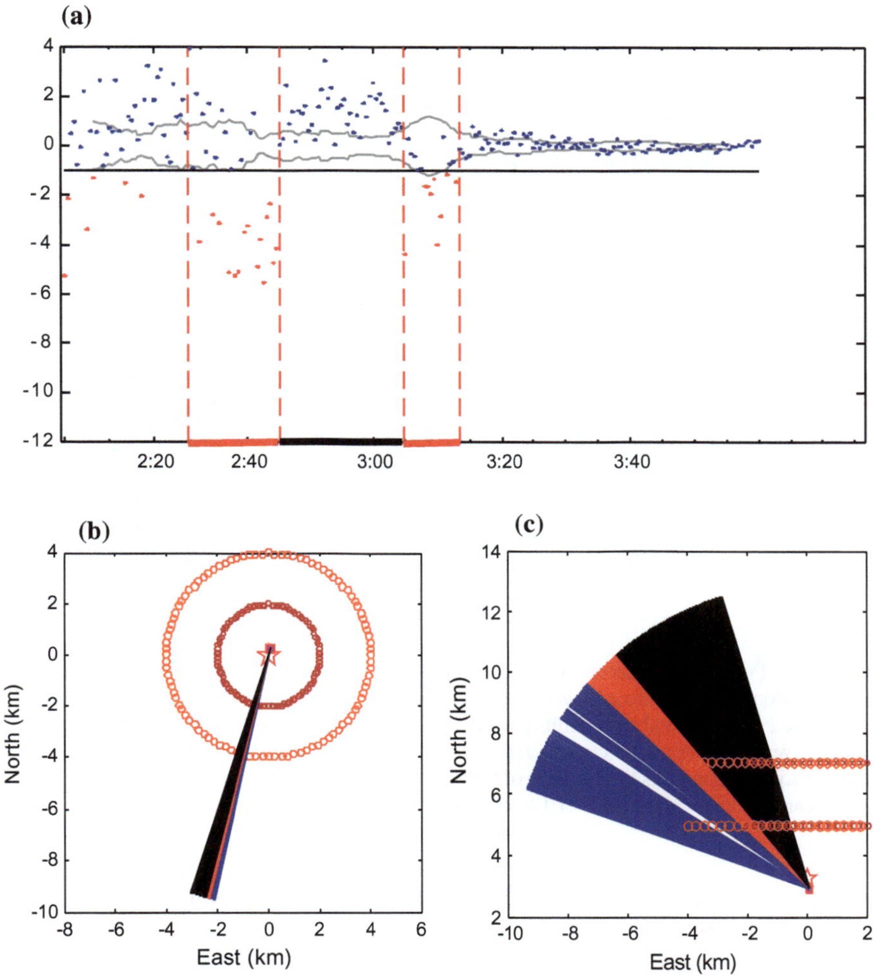

Fig. 14.10 **a** Differenced SNR observations (blue dots) with one standard deviation in gray using 4.5-min smoothing for GPS station ECNE and satellite PRN21. The horizontal black line represents the threshold value. SNR values less than −1 dB-Hz are red dots. The vertical red dashed line shows the start time and the end time of the eruption. The red thick line shows the time interval with SNR drop. The black thick line shows the time interval without SNR drop. **b** The geometry of each satellite line of sight is shown in top view, the blue line represents the satellite–station pair path before the eruption, the red line represents the satellite–station pair path during the eruption, and the black line represents the satellite–station pair path after the eruption. **c** The geometry of each satellite line of sight is shown in side view

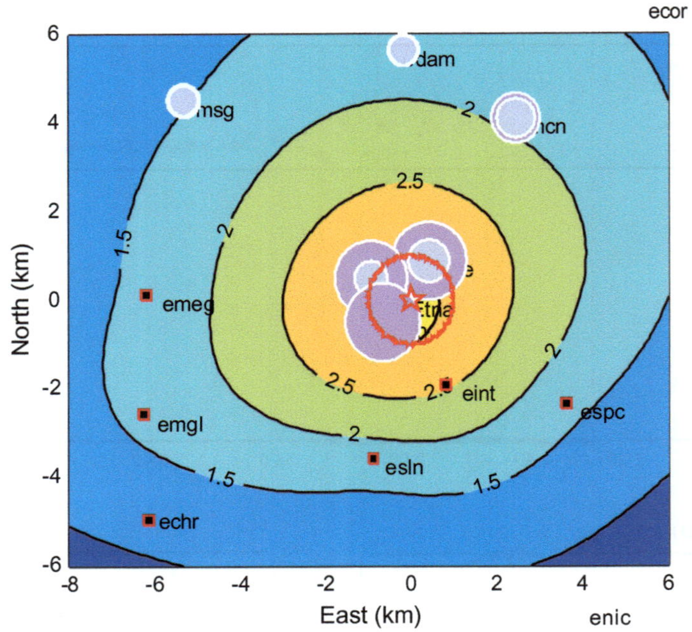

Fig. 14.11 Probability of detection and contour map of volcanic ash cylindrical plume with radius 1 km (red dotted circle) centered over the vent (red star) of Mount Etna volcano during 03 December 2015. The circles show the probability for the mean line of sight to be within 0–2 km above the volcanic vent (Darker purple circles) and 2–4 km above the volcanic vent (lighter purple circles). The red square represents the location of GPS stations and the red circle represents a top view of the volcanic plume with a radius of 1 km from the volcano

reached at least 15 km on 12 July 2008 and drifted as a large cloud above the North Pacific. Two GPS receivers were operating on the island at the time of the first seismic activity with station OKFG and OKSO located within 15 km of the caldera. The GPS data from the Alaska Volcano Observatory are sampled at 30-s intervals. In order to avoid data shift and obtain the same satellite geometry, we choose the data from 11 July 2008 to 16 July 2008 during 19:00–21:00 UTC from stations OKFG and OKSO. Figure 14.12 shows the trace of satellites and the location of Mount Okmok volcano and two GPS stations. The traces of satellite PRN08, PRN25, and PRN27 in red curve line are closer to Mount Okmok volcano and the orientation of satellite antenna can be estimated. Following the plume diffusion direction, the signals from satellite PRN08, PRN25, and PRN27 may have been influenced by volcanic plume. The differenced SNR series show anomalies of station OKFG and satellite PRN08. For example, Fig. 14.13 shows differenced SNR time series at stations OKFG and OKSO around Mount Okmok volcano on 12 July 2008 (DOY 194). We can see that SNR has no anomalies at OKSO within $[-1, 1]$, while there are two significant large variations at OKFG above 2 dB between 20:36–20:57 UTC.

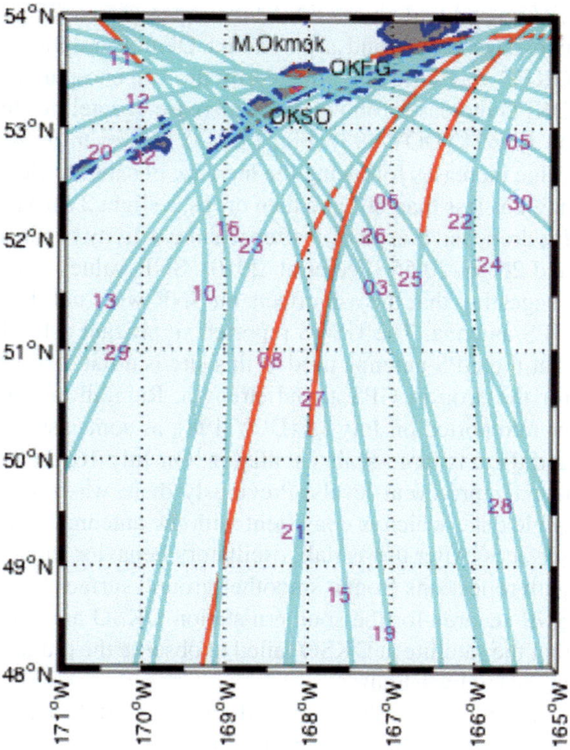

Fig. 14.12 The location of Okmok volcano (red star), the satellite trace (red and green curve line) and GPS stations OKFG and OKSO (magenta circle). The green and red curve lines show the trace of satellite PRN08 between 19:00 and 21:00 UTC on DOY 194

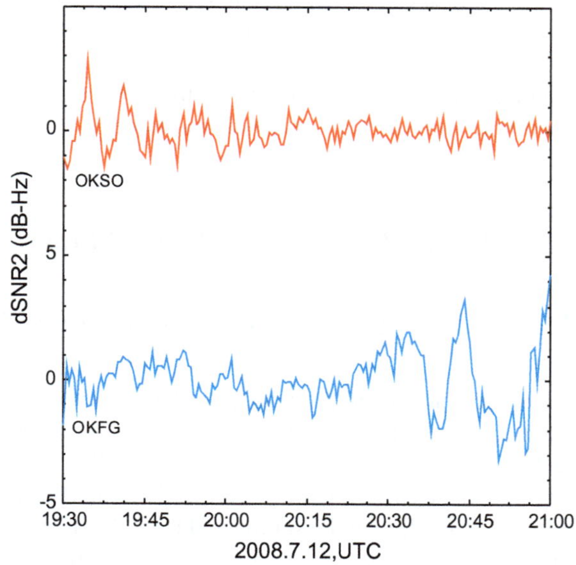

Fig. 14.13 Differenced SNR time series at stations OKFG and OKSO around Mount Okmok volcano on July 12, 2008 (DOY 194)

Figure 14.14 shows SNR time series at two GPS stations OKFG and OKSO for two days before and after Mount Okmok eruption (DOY 194). The GPS station OKFG was seriously affected by the volcano plume. The L1 SNR value at station OKFG increases following the increase of satellite elevation angle without any drop on July 11 (DOY 193) (Fig. 14.14a). However, on July 12 (DOY 194) the L1 SNR value increases following the increase of satellite elevation angle when the elevation angle is less than 60° and then drops, a slight 2 dB-Hz increase and a dramatic 7 dB-Hz drop, and it is just 1 h after seismic activity began, corresponding to 20:35–20:40 and 20:46–20:55 (Fee et al. 2010). SNR values remained low for the next 2 days, suggesting that the ash effects on SNR were not due to a plume, but sitting on the GPS antenna. The USGS reported significant ash fall (10 cm) near OKFG. Given that the GPS antenna used at this site is a flat disk, this is the simplest explanation for the drop in GPS signal strength. Rainfall and heavy winds on Umnak Island were reported on July 15 (DOY 198), as some ash is still remaining on the antenna and there is large-scale oscillation. On July 16, the OKFG SNR data have returned to their pre-event levels. Previously there was no oscillation in the low elevation angle data, which is consistent with the antenna being surrounded by rough ground surfaces. After the ashfall, oscillatory behavior can be observed, which is consistent with reflections from a smoother ground surface (Larson 2013). In Fig. 14.14b, the SNR records for the southern station OKSO are consistent for all days, indicating that the satellite at OKSO failed to observe the plume effect and the antenna was not impacted by ashfall.

Figure 14.15a shows the differenced SNR data after removing the correction profile. There are two drops during 20:35–20:40 and 20:46–20:55 UTC. In Fig. 14.15b, c, horizontal and vertical projections are plotted to intuitively display the relative geometric information between each receiver–satellite line of sight and Mount Okmok volcano. Though the signal path is a little far away from Mount Okmok volcano, the plume diffusion is far enough (15 km), so the signal of satellite PRN08 still crosses the plume with the highest crossing altitude of about 8 km.

14.4 Characteristics of Volcanic Plumes

14.4.1 Propagation Velocity of Volcanic Plumes

When the volcanic eruption happened, the average propagation velocity of volcanic plumes can be calculated using the distance from the volcanic plume and the intersection point of the station and satellite as well as the required time from the volcanic eruption to the changing time of the SNR. When the SNR begins to change, the angle α from the satellite to the station and from the intersection point to the volcanic vent can be calculated using the known variation time dt, satellite elevation ele, the distance s of the station and the volcanic vent, and the angle β from the estimated intersection to the volcanic vent and the plumb line (Fig. 14.16). Thereafter, the

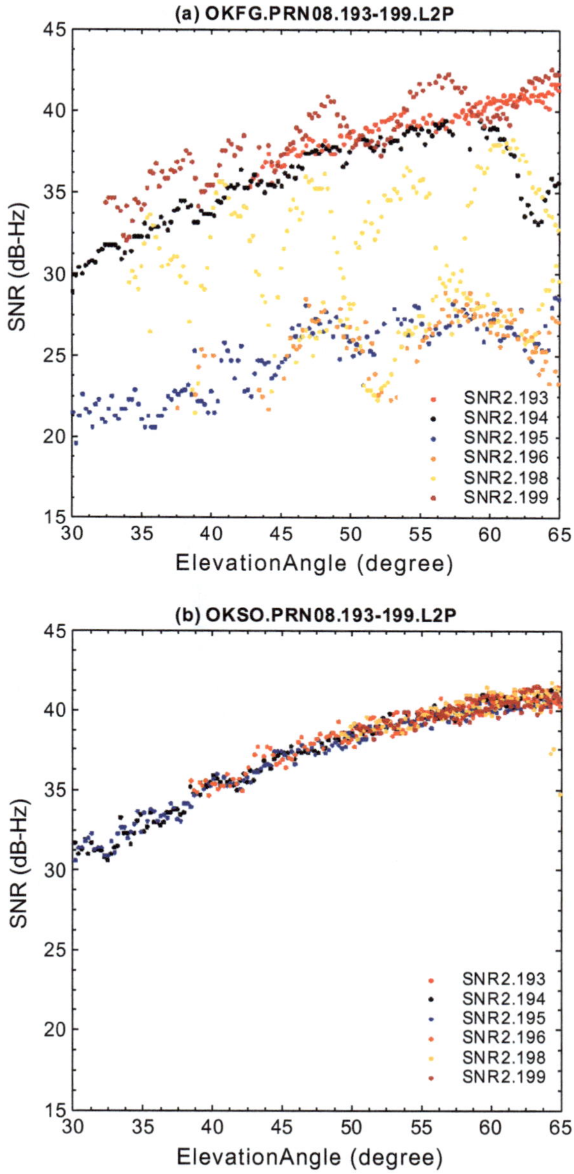

Fig. 14.14 SNR data at two GPS stations OKFG and OKSO for two days before and after Mount Okmok eruption (DOY 194). **a** L1 SNR of station OKFG and satellite PRN08, and **b** L1 SNR of station OKSO and satellite PRN08

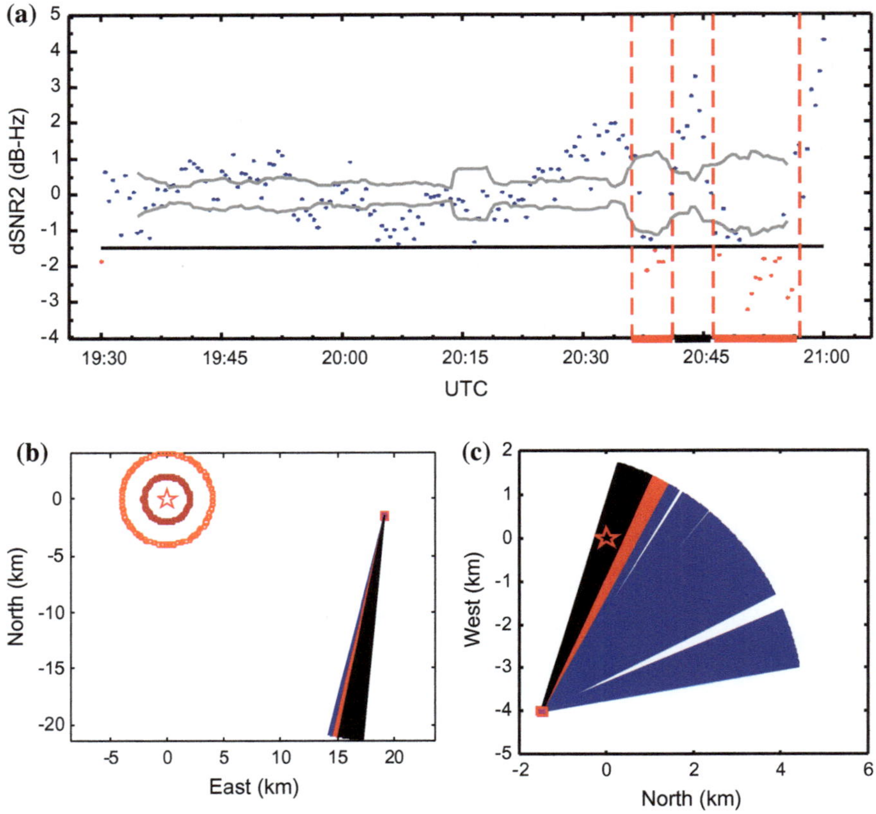

Fig. 14.15 **a** Differenced SNR observations (blue dots) with one standard deviation in gray using 4.5-min smoothing for GPS station OKFG and satellite PRN08. The horizontal black line represents the threshold value. SNR values of less than −1.5 dB-Hz are red dots. The vertical red dashed lines represent the start time and the end time of the eruption. The red thick line shows the time interval with SNR drop. The black thick line shows the time interval without SNR drop. **b** The geometry of each satellite line of sight is depicted in top view. The blue line represents the satellite–station pair path before the eruption, the red line shows the satellite–station pair path during the eruption, and the black line represents the satellite–station pair path after the eruption. **c** The geometry of each satellite line of sight is depicted in side view

distance ρ from the intersection point to the volcanic vent can be acquired according to the sine law and then the average propagation velocity of volcanic plumes can be estimated at approximately 30 m/s (Zhang 2018).

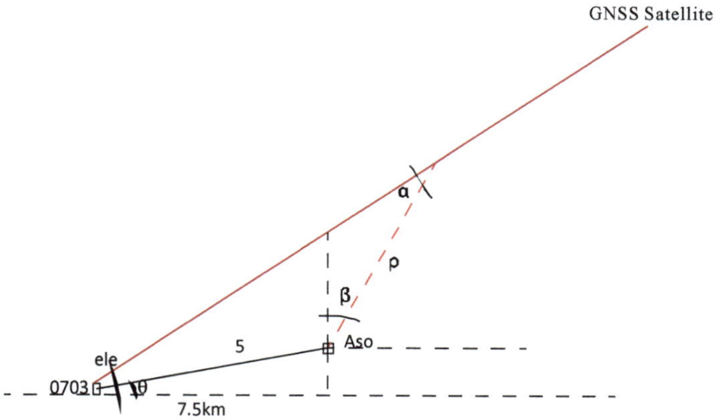

Fig. 14.16 Geometric map of the station, volcanic vent and intersection point

14.4.2 Relationship of SNR Anomaly and Volcano Magnitude

As mentioned earlier, from a qualitative point of view, the amplitude of the SNR anomaly is related to the intensity of volcanic eruption, concentration of the volcanic plumes, migratory direction of the volcanic plumes and the relative position of the station–satellite pair. The amplitude of the SNR anomaly will be bigger with the stronger intensity and concentration when the path of the satellite–station pair crosses the volcanic plumes. Here we investigate the relationship between the amplitude of the SNR anomaly, the intensity of the volcanic eruption and concentration of the volcanic plumes from a quantitative perspective.

The following volcanoes are analyzed, including Okmok volcano in July 2008, Etna volcano in December 2015, Aso volcano in October 2016, Redoubt volcano in March 2009, Te Maari volcano in August 2012, Etna volcano in November 2013 and Sakurajima volcano in July 2012 as the examples. The possible relationships are investigated for the Volcanic Explosivity Index (VEI) and the maximum vertical penetration of the volcano with the amplitude of the SNR anomaly. A nearly approximate negative linear relationship is found (Zhang 2018). Figures 14.17 and 14.18 show the nearly linear relationship of the amplitude of the SNR anomaly and the VEI grades of the volcanos or the maximum vertical penetration, indicating that the bigger the Volcanic Explosivity Index (VEI) or the maximum vertical penetration of the volcano is, the larger the effect of volcanic plume on the SNR anomaly is

14.4.3 Relationship of SNR Anomaly and Change Interval

As shown in Fig. 14.19, the correlation coefficient of the SNR anomaly amplitude and anomalous SNR change interval is 0.64, which indicates significant (Zhang 2018).

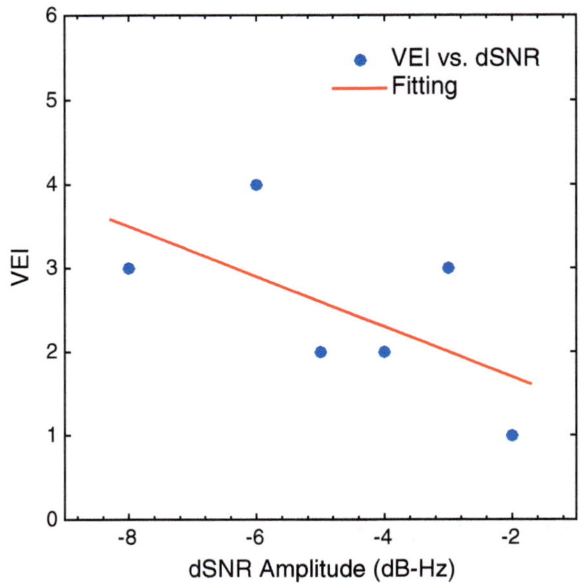

Fig. 14.17 Relationship of the volcanic explosivity index (VEI) and the amplitude of the SNR anomaly

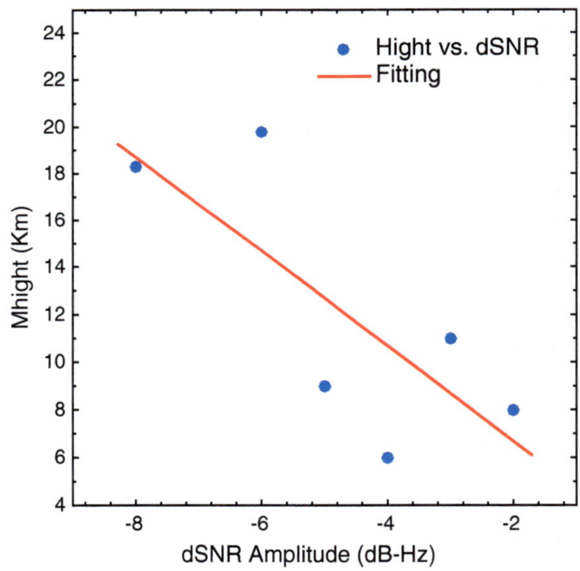

Fig. 14.18 Relationship of the maximum vertical penetration of the volcano and the amplitude of the SNR anomaly

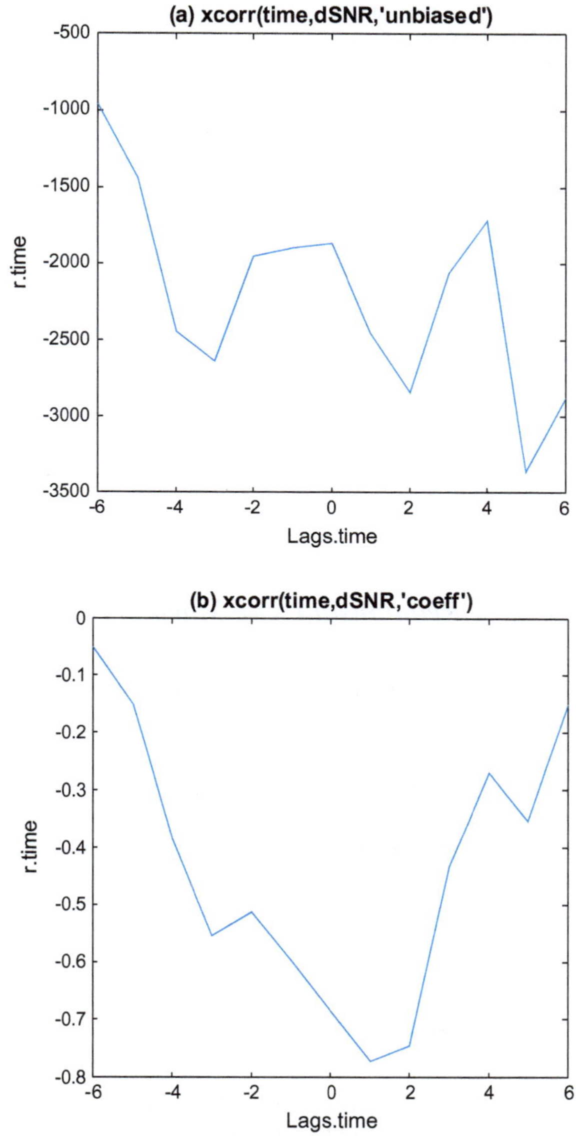

Fig. 14.19 Relationship between the SNR anomaly amplitude and SNR change interval with cross-correlation function (**a**) and cross-correlation coefficients (**b**)

According to the anomalous SNR change interval, we can infer that the path of the station and satellite and the volcanic plumes are changing with the time. The SNR anomaly exists when the path of the station and satellite crosses the volcanic plume with a certain concentration. When the concentration is not enough or the path does not cross the volcanic plume, the SNR is normal without anomaly.

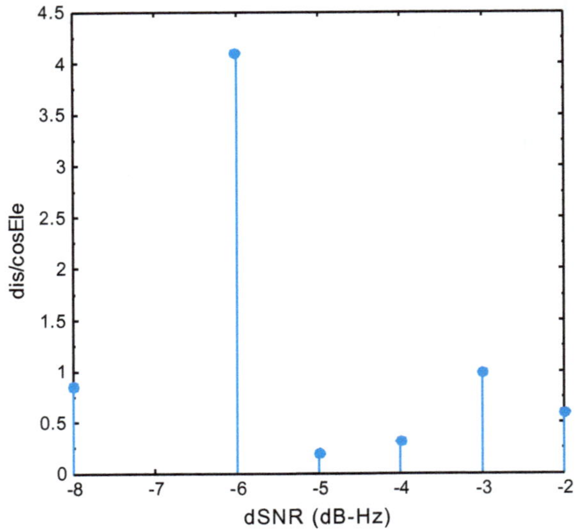

Fig. 14.20 Relationship between the SNR anomaly amplitude and satellite elevation or the distance to the volcano

14.4.4 Relationship with Volcano Distance

The SNR anomaly amplitude is complicated and affected by many factors. In the perspective of the satellite elevation and distance of the station and the volcanic vent, we can not find the same height and distance. Hence, we need to analyze in combination of multi-influencing factors. In this section, we analyze the cosine value of the satellite elevation and the distance. From Fig. 14.20, we can see that the ordinate values are not more than 1 (distance in unit of 10 km), which demonstrates that the SNR anomaly amplitude is also affected by any other factors such as the magnitude and the concentration of the volcano (Zhang 2018).

14.5 Conclusion

The volcanic plume is successfully detected by GPS SNR during eruptions at Mount Aso, Mount Etna, and Mount Okmok volcanoes. For Mount Aso volcano, only GPS stations with less than 5 km from the volcano vent can detect volcanic plume using the SNR technique, but for Mount Etna volcano only stations with less than 3 km from the vent are successful. However, there will be no satellite crossing the plume during the eruption sometimes. In the future, more high-rate GPS SNR data (1 s) and multi-signals from GLONASS, Galileo and BeiDou will increase the probability of volcanic plume detection. Furthermore, because water and water vapor are not known in signal attenuation during volcanic eruptions, next work should take the size and the density of particles into account and study whether the size and density influence the signal attenuation.

References

Aranzulla, M., Cannavo, F., Scollo, S., et al. (2013). Volcanic ash detection by GPS signal. *GPS Solutions, 17*(4), 485–497.

Fee, D., Steffke, A., & Garces, M. (2010). Characterization of the 2008 Kasatochi and Okmok eruptions using remote infrasound arrays. *Journal of Geophysical Research Atmospheres, 115*(D2).

Fournier, N., & Jolly, A. D. (2014). Detecting complex eruption sequence and directionality from high-rate geodetic observations: The August 6, 2012 Te Maari eruption, Tongariro, New Zealand. *Journal of Volcanology and Geothermal Research, 286,* 387–396.

Gouhier, M., Harris, A., Calvari, S., et al. (2012). Erratum to: Lava discharge during Etna's January 2011 fire fountain tracked using MSG-SEVIRI. *Bulletin of Volcanology, 74*(5), 1261–1261.

Grapenthin, R., Freymueller, J. T., & Kaufman, A. M. (2013). Geodetic observations during the 2009 eruption of Redoubt Volcano, Alaska. *Journal of Volcanology and Geothermal Research, 259*(6), 115–132.

Houlie, N., Briole, P., Nercessian, A., et al. (2005). Sounding the plume of the 18 August 2000 eruption of Miyakejima volcano (Japan) using GPS. *Nature, 414*(5), 583–583.

Jin, S. G., & Komjathy, A. (2010). GNSS reflectometry and remote sensing: New objectives and results. *Advances in Space Research, 46*(2), 111–117. https://doi.org/10.1016/j.asr.2010.01.014.

Jin, S. G., & Najibi, N. (2014). Sensing snow height and surface temperature variations in Greenland from GPS reflected signals. *Advances in Space Research, 53*(11), 1623–1633.

Jin, S. G., Cardellach, E., & Xie, F. (2014). *GNSS remote sensing: Theory, methods and applications* (pp. 276). Springer, Netherlands, ISBN: 978-94-007-7481-0.

Jin, S. G., Qian, X. D., & Kutoglu, H. (2016). Snow depth variations estimated from GPS-reflectometry: A case study in Alaska from L2P SNR data. *Remote Sensing, 8*(1), 63.

Jin, S. G., Qian, X. D., & Wu, X. (2017). Sea level change from BeiDou navigation satellite system-reflectometry (BDS-R): First results and evaluation, global planet. *Change, 149,* 20–25. https://doi.org/10.1016/j.gloplacha.2016.12.010.

Jin, S. G., Feng, G. P., & Gleason, S. (2011). Remote sensing using GNSS signals: Current status and future directions. *Advances in Space Research, 47*(10), 1645–1653. https://doi.org/10.1016/j.asr.2011.01.036.

Larson, K. M., Braun, J. J., Small, E. E., et al. (2010). GPS multipath and its relation to near-surface soil moisture content. *IEEE Journal of Selected Topics in Applied Earth Observations & Remote Sensing, 3*(1), 91–99.

Larson, K. M. (2013). A new way to detect volcanic plumes. *Geophysical Research Letters, 40*(11), 2657–2660.

Ohta, Y., & Iguchi, M. (2015). Advective diffusion of volcanic plume captured by dense GNSS network around Sakurajima volcano: A case study of the vulcanian eruption on July 24, 2012. *Earth, Planets and Space, 67*(1), 157.

Scollo, S., Prestifilippo, M., Spata, G., et al. (2009). Monitoring and forecasting Etna volcanic plumes. *Natural Hazards & Earth System Sciences, 9*(5), 11993.

Scollo, S., Prestifilippo, M., Pecora, E., et al. (2014). Height estimation of the 2011–2013 Etna lava fountains. *Annals of geophysics = Annali di geofisica, 57*(2), 1–25.

Zhang, Q. Y. (2018). Detection of volcano plume using GNSS signal-to-noise ratio (SNR) and radio occultation data, Master Thesis, Shanghai University.

Part IV
Summary

Chapter 15
Summary and Prospective

15.1 Current Status

The worldwide seismometers could estimate rapidly the location, the magnitude, and the focal mechanism of earthquakes, but the detailed rupture process and seismic anomalies are still difficult to be monitored precisely and understood well due to limited observations or the lack of dense near-field observations. The horizontal extent of the source plays a fundamental role in the risk assessment of earthquake and its secondary disasters, such as tsunami and landslide warning. Global Navigation Satellite System (GNSS) is a powerful tool not only for the crust displacement but also the seismic atmospheric/ionospheric variation, especially for regions with dense GNSS continuous operating stations. GNSS seismic ionospheric disturbances may give us a new perspective to monitor and understand the earthquake. First, precise GNSS TEC time series are obtained, and the high frequency noise and the TEC background trend caused by the temporal and spatial variations are removed by the filtering with a certain window. For earthquakes and volcanoes, the filtering window length is set as 3–10 min for mixed disturbances (including signal related to different modes), i.e., 1.7–5.6 mHz, just covering the dominant frequencies of acoustic resonance modes between the ground surface and the lower thermosphere (Matsumura et al. 2011). Usually, Four-order zero-phase Butterworth filter is used because of its maximally flat magnitude. Moving average method could also be used to remove the background TEC variation (Tsugawa et al. 2011). Using TEC residual series, seismo-ionospheric anomalies can be detected following the earthquakes (Jin et al. 2015).

Since the first time the TEC anomalous signal was observed by Calais and Minster (1995) after the Northridge earthquake in California on January 17, 1994, numerous studies have been intended to study pre-/co-/post-seismic ionospheric disturbances related to earthquakes (Liu et al. 2004; Heki and Enomoto 2013). These preliminary results implicated that the energy released during the rupture of the fault was not only propagating in the solid Earth but also leaking out to the Earth's atmosphere and even to the ionosphere. The vertical displacement, at teleseismic distance, induced by the

S. Jin et al., *GNSS Atmospheric Seismology*,
https://doi.org/10.1007/978-981-10-3178-6_15

Rayleigh wave produces an acoustic wave that propagates vertically until the ionosphere by dynamic coupling. During the upward propagation, the double effects of the conservation of kinetic energy and the exponential decreasing of the atmospheric density, strongly amplify the acoustic wave that arrives in the ionosphere and creates strong variations in the plasma density detected by GNSS and plasma velocity detected by Doppler sounder. Particular emphasis was related to the Doppler sounder detection of the ionospheric signature induced by the Rayleigh wave propagation. Consequently, a new perspective of earthquake monitoring is provided by the analysis of seismic ionospheric disturbances estimated through observing the ionosphere. However, the seismic ionospheric disturbance amplitude and propagation characteristics are very complicated and still not understood comprehensively following the earthquake with different magnitudes, focal mechanism, and complex space weather. The mechanism of solid Earth/ocean–atmosphere–ionosphere coupling still needs to be improved for different cases and be verified with more GNSS and other measurements (Fig. 15.1). The seismic ionospheric disturbance amplitude increases proportionally with the earthquake magnitude generally. Comparing to strike-slip earthquake, normal/reverse fault earthquakes are usually followed with much more intense seismic ionospheric disturbances. The vertical displacement in the earthquake zone plays a more important role in the formation of seismic ionospheric disturbances.

This book presents the GNSS atmospheric seismology, including theory, methods, results, characteristics, and coupling between solid Earth and the fluid envelopes, nominally the ocean, atmosphere and ionosphere. Case studies on earthquakes, volcano and tsunami are presented, including the 2005 Northern California offshore earthquake, 2008 Wenchuan earthquake, 2011 Tohoku earthquake, 2012 Haida Gwaii earthquake, 2018 Alaska earthquake, etc. using the GPS data from the Crustal Movement Observation Network of China (CMONOC) and GPS Earth Observation Network of Japan (GEONET) as well as Plate Boundary Observatory (PBO), etc.

15.2 Seismo-Atmospheric Disturbances

The seismo-ionospheric disturbances following the 2008 Wenchuan earthquake are obtained from 28 continuous GPS sites of the CMONOC (Fig. 15.2). Dramatic disturbance appears several minutes after the main shock. The disturbance is presented as an N-shaped pulse. Its amplitude decreases rapidly with the increase of the time and the distance of IPP's location and the epicenter. In order to further confirm our results, 3 days of 28 continuous GPS stations data from the Crustal Movement Observation Network of China (CMONOC) are used during May 11–13, 2008. We filter the TEC time series dI(t) at 28 continuous GPS stations through removing the trend and regular changes with a window width of 3–10 min to obtain seismo-ionospheric disturbances characteristics (Jin et al. 2010). Moreover, an accumulated mean TEC disturbance is calculated for all GPS stations. Significant seismo-ionospheric disturbances are found over GPS stations within about 500 km far from the epicenter

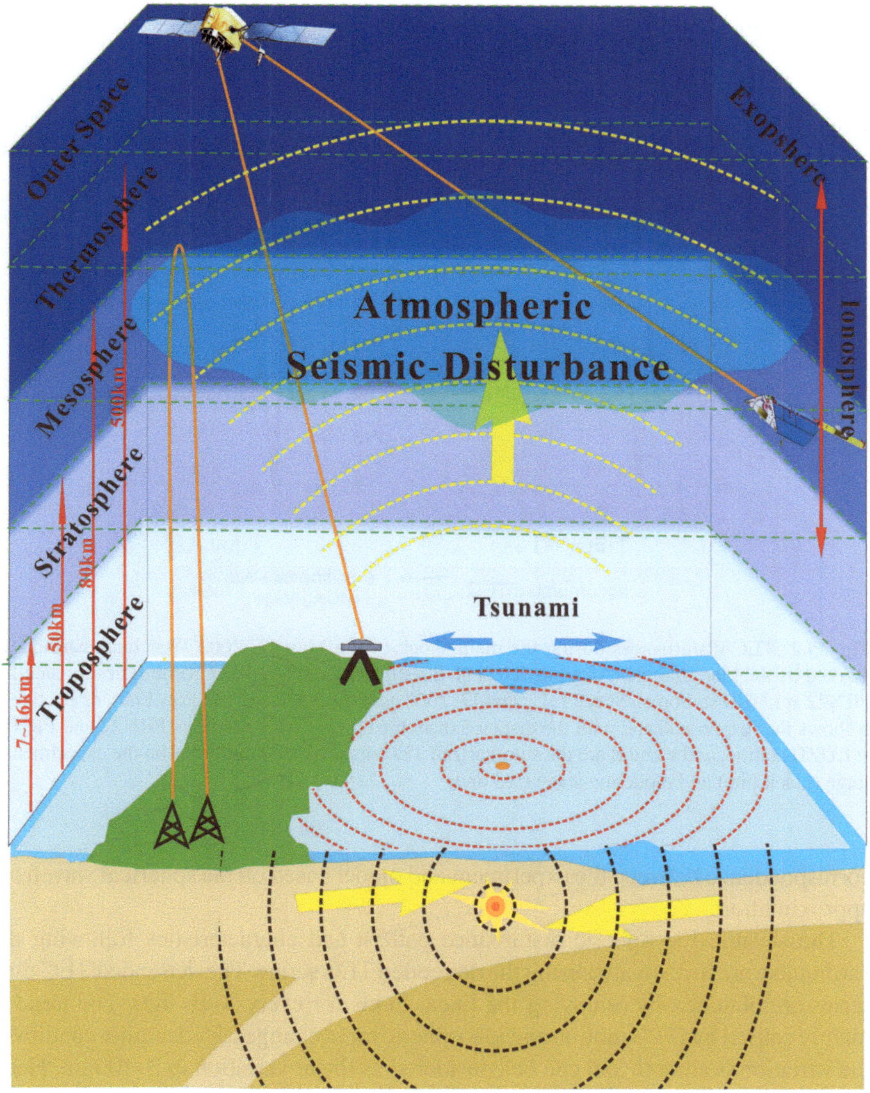

Fig. 15.1 Seismic ionospheric disturbances from GNSS observations

during the main shock on May 12, 2008, while there are no apparent disturbances on the day before or after the main shock. Furthermore, a strong acoustic wave of N-shape shock is found with a plane waveform and a half-period of about 200 s propagating northeastward at an average speed of 600 m/s. The disturbance wavefront is nearly parallel with the rupture direction (Jin et al. 2010). The directional divergence of the seismic ionospheric disturbance may be the reason for the insufficient

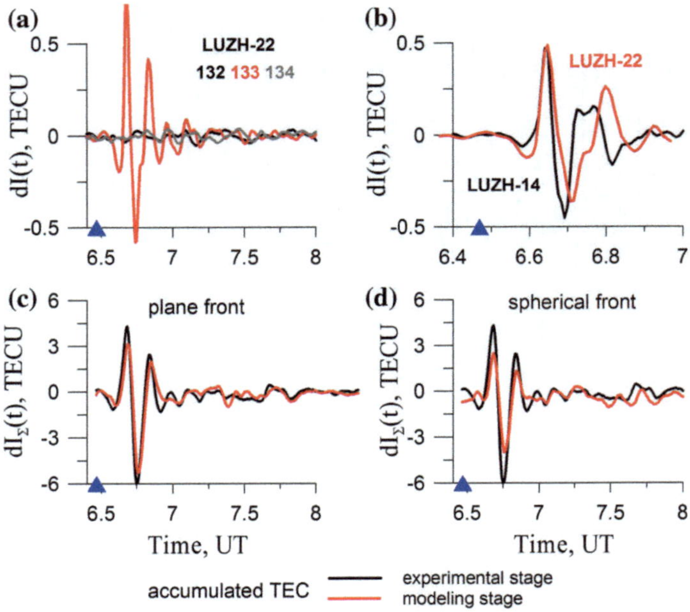

Fig. 15.2 TEC disturbances during the main shock of the May 12, 2008 Wenchuan earthquake marked by the blue triangle (Afraimovich et al. 2010). **a** is the filtered TEC time series for satellite PRN22 at LUZH station on May 11, 12, and 13, 2008 with the black, red and gray lines, respectively, **b** shows the response shape with the distance from the epicenter for satellites PRN22 and PRN14 at LUZH station, and **c** and **d** are the summarized TEC disturbance time series on the experimental stage (black line) and modeling stage (red line)

correspondence between the experiment and model based on the spherical wavefront approximation.

The seismic ionospheric disturbance pattern and characteristics following the earthquake are investigated using the detrended TEC series. The detrended TEC time series are obtained by removing the linear items for every 3–10 min. The trend is mainly caused by IPP's motion and ionospheric rapid changes. Under quiet condition, the variation is smooth and can be considered as linear variation in 3–10 min. Here, we do not run an average for the detrended TEC series as Tsugawa (2011), since the TEC time series derived from carrier phase observations have high precision and are not affected by observation noise greatly. Furthermore, we can keep higher frequencies signals in the detrended TEC time series and also avoid the disturbance amplitude attenuation due to the average process. Some recent studies comparing the unperturbed TEC with the perturbed TEC are found an enhancement before the event or a hole (decreasing) just after and before the main shockwave (Heki 2011; Heki and Enomoto 2013). The doubt about the pre-seismic TEC enhancement is that the post-seismic decrease will pull down the fit lines and the pre-seismic enhancement probably is a fake signal. Some recent works were trying to dynamically explain the hole formation as a consequence of the neutral-plasma coupling during the wave propagation (Kamogawa et al. 2012). More difficult issue is to support the hypothesis of the enhancement that involves electromagnetic behavior of the fault before the rupture.

More other pre-/co-/post-seismic ionospheric anomalies are obtained and analyzed. For example, the seismic ionospheric disturbances following the Tohoku earthquake occurred on March 11, 2011 are investigated from dense GPS network observations in Japan. The pre-seismic ionospheric disturbance is found about 4 h before the main shock. This phenomenon may be the precursors of this strong earthquake. Comparing to pre-seismic effects, co-/post-seismic ionospheric disturbances are more apparent. The first peak appears 10 min after the main shock. Disturbances are attenuating with the time, and sudden increase of anomalous arc occurs about 80 and 150 min after onset. Also, three types of disturbances are found after the strong earthquake happened on March 11, 2011. These three peaks should be related to the different types of atmospheric waves, such as acoustic wave in near field, acoustic wave generated by the Rayleigh wave propagation, tsunami-generated wave, and aftershocks effect (Jin et al. 2014). The ionospheric disturbance spreads out from the epicenter as a quasi-circular propagation pattern with the time, reproducing the seismic radiation diagram and the formation of Rayleigh wave (Rolland et al. 2011). Furthermore, the strongest disturbances mainly appear in the near-field. The seismic ionospheric disturbance amplitudes decrease dramatically with the SIPs' location far away from the epicenter, especially in the near-field (Jin et al. 2014). For the amplitudes variation near the epicenter (0–500 km), three peaks are visible in the amplitude series. The disturbances related to the Rayleigh wave and acoustic wave spread out to 500 km away in about 10 minutes, indicating that the disturbance effects induced by the first two types of waves are slight after 06:06UT, considering the time delay of the disturbance propagating from ground to the ionospheric height. So the two secondary peaks should be related to the two stronger aftershocks. For 500–1500 km, the average amplitudes are attenuated much more slowly. In addition, obvious disturbances in filtered TEC series with multi-frequencies are found with several minutes after the main shock. The multi-frequencies disturbances from GPS ionospheric TEC can detect the acoustic wave and the Rayleigh wave (3–7 mHz) and tsunami-generated gravity wave (<3 mHz) (Matsumura et al. 2011; Occhipinti et al. 2013). The mean seismic ionospheric disturbance with 1–3 mHz has larger amplitude than the one with 3–7 mHz. With the increase of the spreading time, the high frequency disturbance attenuates faster.

Figure 15.3 shows the relationship between time offset and IPP distance from the epicenter of the 2011 Tohoku earthquake. The stripes' slopes, called disturbance propagation velocities here, mainly ranging from 2000 to 3200 m/s, 500 to 1200 m/s, and 120 to 270 m/s displayed in Fig. 15.3. The first and second propagation modes begin at about 10 min after main shock epoch, while the third propagation mode begins several minutes later. The source mechanism is the acoustic-gravity wave generated by the vertical displacement of the shock that reaches the ionosphere in 8–10 min (Occhipinti et al. 2013) and propagates until the first 1000 km (Liu et al. 2011), a pure acoustic wave coupled with the Rayleigh wave that propagates with an horizontal speed of 3.5 km/s and later a pure gravity wave coupled with the tsunami

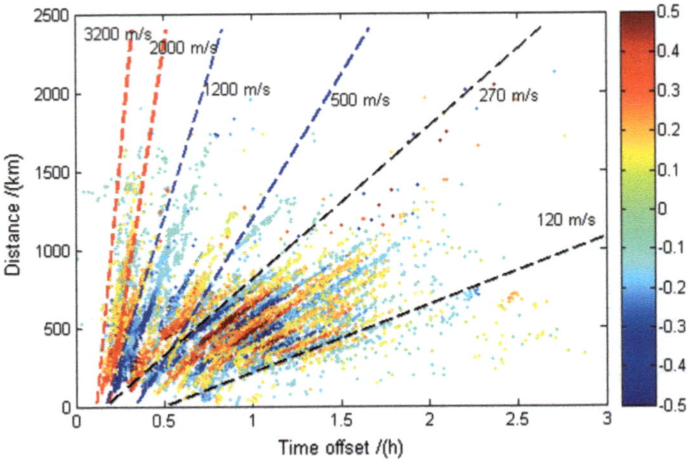

Fig. 15.3 Time offset referring to the epoch of main shock versus IPP distance from the epicenter. Origin of the axes stands for the main shock epoch and the epicenter location of the 2011 Tohoku earthquake. Each dot is corresponding to the peak TEC residual of one TEC residual series. Here residual values of larger than 0.5 TECU are set as 0.5 and residual values of smaller than −0.5 TEC are set as −0.5 TECU

propagation that reaches the ionosphere in 40–50 min and visible in the ionosphere at the epicentral distance of around 500 km (Occhipinti et al. 2013).

The ionospheric disturbances related to the Rayleigh wave, acoustic wave, and tsunami-generated gravity wave are probably produced following the giant earthquake (Artru et al. 2004; Matsumura et al. 2011; Occhipinti et al. 2013). The Rayleigh and acoustic wave propagate with the speed of about 2–3 km/s and around 1 km/s, respectively (Jin et al. 2014). From the in situ ocean Bottom Pressure Records (BPR), two obvious disturbances are found in three ocean bottom pressure residual series (Jin et al. 2017). The first disturbance with a speed of 3071 m/s is related to the seismic Rayleigh wave, while the second one with a speed of 234 m/s is related to the tsunami propagation. From the TEC disturbance frequency, the tsunami-generated gravity wave is less than 3 mHz, while the Brünt-Vaïsalla frequency is in order of 3 mHz. Due to the time shifts corresponding to peaks of average 3–7 mHz TEC disturbances, the waves spread out from the near-field to the far-field with speeds of several kilometers per second, which are more consistent with the Rayleigh wave. In addition, it has been found that not only the main shock but also the giant aftershocks seismic ionospheric with acoustic mode could be detected by GPS measurement even though their intervals are no longer than one hour (Jin et al. 2014).

Furthermore, the strong correlation is found between the CIDs and the vertical ground motion recorded by seismometers nearby the epicenter of the 2012 Haida Gwaii earthquake. The total electron content (TEC) series with lower elevation angles have higher perturbation amplitudes. Furthermore, the simulated ionospheric disturbance following a vertical Gauss pulse on the ground based on the Finite Difference

Time Domain (FDTD) method confirms the ionospheric Rayleigh wave signature in the near field and the vertical ground motion dependence theoretically. The vertical ground motion is the dominant source of the ionospheric Rayleigh wave and affects the CIDs waveform directly (Jin et al. 2017).

15.3 Tsunami-Generated Gravity Wave Coupling

The detection of the tsunami signal in the ionosphere slowly follows observations of the seismic signature in the ionosphere related to the source displacement or to the propagation of the Rayleigh waves. Indeed, the pure gravity wave generated by the tsunami in the overlying atmosphere/ionosphere creates a similar signature of the usually detected Traveling Ionospheric Disturbances (TID) (Artru et al. 2005). Therefore, the first undoubtful observation of the tsunami signal in the ionosphere followed the 2014 Mw = 9.1 Sumatra earthquake. The tsunami wave propagation in the Indian Ocean was broadly detected by ionospheric sounding with different techniques, mainly the TEC from GPS and Altimeters (Occhipinti et al. 2006). The observations were clearly supported by 3D numerical modeling (Occhipinti et al. 2006).

The tsunami is an oceanic Internal Gravity Wave (IGW) induced by the large volume displacement of ocean water, which is a long period gravity wave with the frequency smaller than the atmospheric Brunt-Vaisalla frequency (Lognonné et al. 1998). Based on the linear analysis, an internal gravity wave, coupled with the tsunami is also generated in the overlying atmosphere (Hines 1972; Occhipinti et al. 2006). To understand the tsunami-IGW propagation and mechanism of the tsunami–atmosphere coupling, a Boussinesq approximation is used (Spiegel and Veronis 1960). On the basis of the Navier–Stokes equation, Occhipinti et al. (2006) have developed a pseudo-spectral propagator $\frac{dV}{dz} = A \cdot V$ with the vector V and the matrix A as

$$V = \begin{pmatrix} \tilde{u}_z^* \\ P^* \end{pmatrix} \tag{15.1}$$

$$A = \begin{pmatrix} -\frac{1}{\Omega}\left(k_x \frac{du_{x0}}{dz} + k_y \frac{du_{y0}}{dz}\right) \frac{1}{2}\frac{d\ln\rho_0}{dz} & -\frac{i\left(k_x^2+k_y^2\right)}{\Omega} \\ i\left(\Omega + \frac{g}{\Omega}\frac{d\ln\rho_0}{dz}\right) & -\frac{1}{2}\frac{d\ln\rho_0}{dz} \end{pmatrix} \tag{15.2}$$

where $\tilde{u}_z^* = \sqrt{\rho_0}\tilde{u}_z$ and $P^* = \frac{\tilde{P}}{\sqrt{\rho_0}}$ are the normalized vertical velocity $\tilde{u}_z$ and pressure $\tilde{P}$ under the *omega-k* domain, respectively, $\Omega = \omega - u_{x0}k_x - u_{y0}k_y$ is the frequency induced by the wind (Nappo 2012; Sun et al. 2007), ρ_0 is the unperturbed atmospheric density, u_{x0} and u_{y0} are the meridional and zonal background wind, respectively. Following Occhipinti et al. (2013), for a realistic atmosphere, the vertical k-number k_z and the frequency ω can be expressed as

$$k_z = \sqrt{k_h^2\left(\frac{N^2}{\omega^2} - 1\right) - \left(\frac{N^2}{2g}\right)^2}$$
$$\omega^2 = \frac{k_h^2 N^2}{k_z^2 + k_h^2 + \left(\frac{N^2}{2g}\right)^2} \tag{15.3}$$

Therefore, the horizontal and vertical group velocity v_g^h and v_g^z can be estimated as

$$v_g^h = \frac{\partial \omega}{\partial k_h} = \frac{k_h N^2 (D - k_h^2)}{\omega D^2}$$
$$v_g^z = \frac{\partial \omega}{\partial k_z} = \frac{k_z k_h^2 N^2}{\omega D^2} \tag{15.4}$$

where $D = k_z^2 + k_h^2 + \left(\frac{N^2}{2g}\right)^2$ is the denominator of the associated dispersion equation.

Figure 15.4 shows the vertical (a) and horizontal (b) IGW group velocities with the ocean depth with ocean–tsunamis–atmosphere coupling. The vertical group velocity is very important, while the horizontal group velocity is also useful to estimate the epicenter distance where the IGWs begin interacting with the ionosphere.

In addition, when IGWs reach and interact with the ionospheric plasma, it will result in a perturbation in the plasma density and velocity due to collisions between the ions and neutral molecules. Recently, several works have been performed and

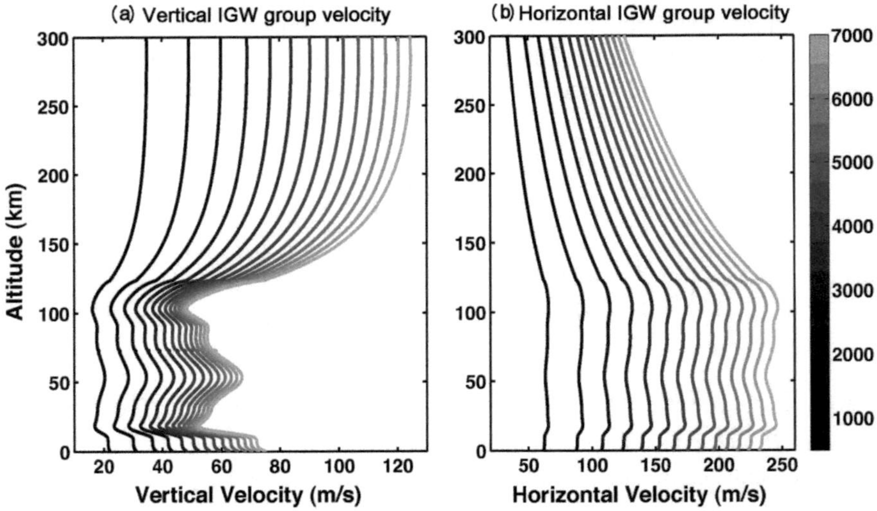

Fig. 15.4 The vertical (**a**) and horizontal (**b**) IGW group velocities with the ocean depth with ocean–tsunamis–atmosphere coupling

validated in detecting the tsunami by GNSS ionospheric TEC, e.g., tsunamis following the great 2004 Mw = 9.1 Sumatra earthquake, 2010 Chile earthquake and 2011 Mw = 9.0 Tohoku Earthquake (Galvan et al. 2012; Rolland et al. 2010; Occhipinti et al. 2013). Furthermore, numerical modeling of IGW reproduces the main features observed in the airglow images, which show interesting likenesses between the model and data, and explain the nature of the airglow observation and the role of the bathymetry in the ionospheric observation (Occhipinti et al. 2011). In the near future, a more complex three-dimensional (3D) modeling on the tsunami–atmosphere–ionosphere coupling should be done with the viscosity and thermal conduction as already tested in a 1D case by Hickey et al. (2009).

15.4 Volcanic Atmospheric Disturbances

Large volcanic eruptions can excite atmosphere infrasonic acoustic and gravity waves and cause oscillations in the atmosphere. The oscillations might have resulted from the lower atmospheric trapped waves excited by the eruption (Nakashima et al. 2015). Some part of the energy leaks upward out to the ionosphere and triggers the ionosphere disturbance. Such ionospheric disturbances can be detected by total electron contents (TEC), which was thought to be the results of mechanical coupling between the volcanic explosions caused pressure and the solid Earth. GNSS is a powerful tool to investigate the atmospheric or ionospheric response to the eruption of volcanoes, which gives us a new perspective to make a better understanding of the volcanic effect with the disturbance of ionosphere. For example, Heki (2006) used GPS TEC to study the Asama Volcano, Japan in 2014 and found a clear N-shaped disturbance signal about 11–13 min after the eruption. Dautermann et al. (2009) studied the July 13, 2003 explosion of the Soufriere Hills Volcano (SHV), Montserrat, and also showed different types of atmospheric disturbances, namely 1.4 mHz internal gravity wave and ~4 mHz atmospheric eigenfrequency. Since the energy released by the volcanic eruptions propagates both in the solid Earth and in the atmosphere, seismic data and GPS TEC data are often combined to study the volcano. For example, volumetric borehole strainmeter data and GPS TEC data were jointly analyzed after the explosion of SHV (Dautermann et al. 2009). Their spectral analysis showed an amplitude peak at 4 mHz for both data sets, with similar waveforms and signal duration. Detailed spatial distribution and time evolution of ionospheric disturbances and seismic data recorded by seismometer were combined to investigate the Kelud volcano on February 2014. Both spectra had peaks at 3.7 and 4.8 mHz (Nakashima et al. 2015).

Atmospheric and ionospheric anomalies following other volcanoes were further investigated using GPS, e.g., the Calbuco volcanic eruption on April 23, 2015. Significant TEC disturbances with the amplitude of 0.1–0.4 TECU are found about one hour after the eruptions and the disturbance lasts more than one hour for the first eruption and about half hour for the second eruption. The propagation velocities estimated from the travel time diagram are around 800 m/s and 900 m/s for the two eruptions, respectively. Meanwhile, the temperature anomaly from COSMIC vertical profiles over the 2015 Calbuco volcano is also detected, which is further confirmed by the spatial interpolation of the 200 hPa pressure level. The volume of the temperature anomaly is almost similar to the real-time image by the NASA VIIRS. The difference between anomalous and non-anomalous zones is about 8 °C, which may be the post-volcanic plumes perturbations. The SO_2 also shows anomalous enhancement on the April 23, 2015, immediately after the eruption day. All these results suggest the influence of volcanic eruption on the lower atmosphere and ionosphere. Although a number of cases have been studied the ionospheric disturbance to the volcanic eruptions, it is still difficult to understand the coupling of atmospheric or ionospheric variations and the volcanic eruptions owing to the complicated environment of the atmosphere around the Earth and the volcano. It still needs more works to study the mechanism of the atmospheric/ionospheric disturbances and the coupling between the volcano and the atmosphere.

15.5 Prospective

In this book, theory, methods, and progresses of GNSS atmospheric seismology are introduced and results are presented. The seismic ionospheric disturbances are mainly driven by the lithosphere–atmosphere–ionosphere coupled airwaves from the vertical motion of seismic waves. The seismo-ionospheric effects are related to not only the main shock but also the strongest aftershocks. The ionospheric perturbations related to tsunami and/or Rayleigh wave propagation are detectable at teleseismic distances. The ionospheric effects following the earthquake are attenuated with the increase of the time and distance between the IPP and the epicenter, especially in the near field. In general, the post-seismic ionospheric disturbance spread out with three patterns, which are related to the source displacement, to the Rayleigh wave, and the tsunami. As more and more GNSS continuous operating stations and Multi GNSS constellations, it is possible to estimate the source extent with seismic ionospheric effect based on the theory of solid Earth/ocean–atmosphere coupling. The basic parameters of earthquake, such as epicenter location and main shock time, could be estimated rapidly with the seismic ionospheric propagation if there are enough GNSS measurements around the epicenter. With this work, we wish to motivate future works in GNSS ionospheric seismology to estimate the source extent and the magnitude in order to support the classic techniques for earthquake and tsunami warning system.

In addition, solar and geomagnetic activities also have a significant effect on the amplitude of seismic ionospheric disturbance. The disturbance amplitude is higher relatively in disturbed geomagnetic days, which show that the filtered TEC time series are affected by not only earthquake generated disturbances but also the geomagnetic variations. As we are known, the variation of Earth's ionosphere is complex because of its high temporal dynamic and spatial discrepancy under the influence of various factors, such as space weather, natural event, anthropogenic activities, etc. It is still a challenge to extract clear and exact the earthquake generated ionospheric disturbance for the worldwide earthquakes occurred at different time with different external conditions (space weather, time of day, location, etc.). The relationship between the earthquake and the ionospheric variation is not estimated well and understood comprehensively up to now. The exact amplitude and mechanism of seismic ionospheric disturbances following earthquakes with different magnitudes, focal mechanism, depth and external conditions still need more studies in the future.

Furthermore, some possible pre-seismic ionospheric anomalies were also observed in the TEC, NmF2 (F2 peak electron density) or HmF2 (F2 peak height) time series, but the mechanism of the lithosphere–atmosphere–ionosphere coupling is still unclear and the debate still exists. Ionospheric anomalies with several hours to several days before the main shock onset were reported. Kamogawa (2006) proposed the possible mechanisms for energy-transport from the lithosphere to the atmosphere-ionosphere. Generally, the atmospheric electric field generated on or near the ground surface during the seismic period is assumed to cause the ionospheric anomalies. However, whether these pre-seismic ionospheric disturbances are related to the earthquake or not is still controversial. It is difficult to extract the clear background variation of TEC. Up to now, there is no explicit definition for pre-seismic ionospheric disturbances. One of effects is the filtering or the detrended method, especially the disturbance amplitude. Furthermore, some pre-seismic ionospheric anomaly is probably a fake signal induced by the data processing rather than the earthquake precursor (Kamogawa and Kakinami 2013), but maybe possible to extract pre-seismic anomalies with more and denser accurate observations in the future.

In addition, Pulinets and Ouzounov (2011) examined the characteristics of Radon emission in seismically active regions and found that the increase of the surface temperature may be due to vertical electric fields, ultimately causing thermal anomalies. Freund (2000) showed that rock fractures could activate mobile positive holes, which eventually leads to thermal anomalies that make bulk clouds. The spreading of these positive holes often ionizes the atmosphere, which may absorb infrared radiation. The light ions and inert gases are observed in high concentration over the seismogenic region, but the real source of these gas particles is still controversial. The light ions and water particles move under the influence of lithosphere–atmosphere coupling. The current studies showed that the earthquake-induced electric field travels along the seismic breeding zone. The gases and water particles injected in this electric field are directed towards the horizontal and vertical plane. This vertical motion is the source of the dynamo force, which is capable of producing magnetic fluctuations before and after the earthquake. A number of different theories and models were proposed to explain thermal anomalies and some simulation results agreed that thermal anomalies before the impending earthquake are due to the release of light ions from the rock's surface, which is subjected to stress (Freund 2000). The effect of other geophysical phenomenon is also assumed to participate in the origination of thermal anomalies. Most of the constituent of these light ions are O_2, N_2, CO, and water particles. These air components perturbations may be due to the accumulation of high

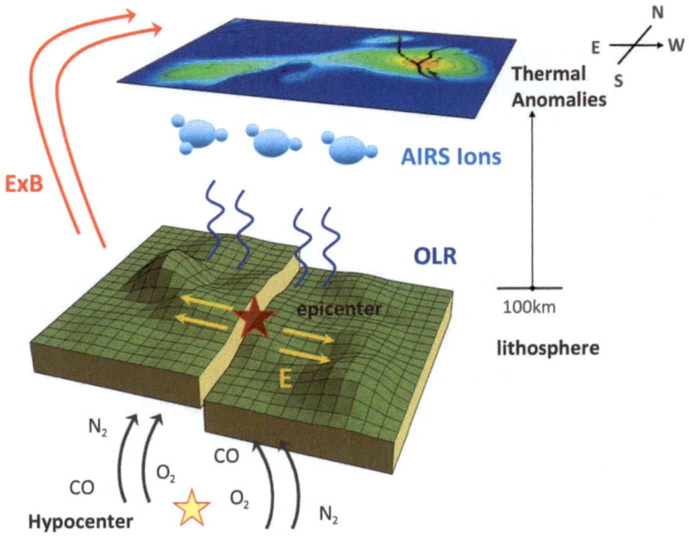

Fig. 15.5 Interaction of atmospheric components due to lithosphere–atmosphere coupling (LAC), including thermal anomalies, OLR radiation, AIRS ions, E × B drifts and earthquakes

energy in the system. It needs more studies on the gas chemistry variations in the atmosphere to gain a better interpretation of thermal anomalies related to earthquakes in the future.

Finally, there is a complex reaction between the geochemical gases and light ions in the upper atmosphere. In the upper atmosphere, the mutual repulsive force between the ions can cause the eastward motion of thermal anomalies. The eastward motion of thermal anomalies exemplifies lithosphere–atmosphere coupling (LAC), which causes E × B drifts and moves the free ions from the lithosphere to the atmosphere along plasma tubes and propagates thermal perturbations at specific altitudes. The phenomenon of plasma drift and its association in LAC coupling is shown in Fig. 15.5 (Shah 2017). Thermal anomalies from MODIS and OLR measurements may be due to this mechanism but no explicit definition is revealed. A complete understanding of the mechanism of lithosphere–atmosphere coupling is still needed more works.

With the development of the next generation of multi-frequency and multisystem GNSS constellations, including the U.S.'s modernized GPS-IIF and planned GPS-III, Russia's restored GLONASS, and the coming European Union's GALILEO system and China's Beidou Navigation Satellite System as well as a number of Space-Based Augmentation Systems, such as Japan's Quasi-Zenith Satellite System (QZSS) and India's Regional Navigation Satellite Systems (IRNSS), more GNSS observations will be available on ground and space-borne platform in the near future (Jin et al. 2011). Furthermore, more and more permanent GNSS stations are being established around the globe, including more than 600 permanent worldwide IGS stations and thousands of regional continuous GNSS network stations, e.g., ~1000 continuous

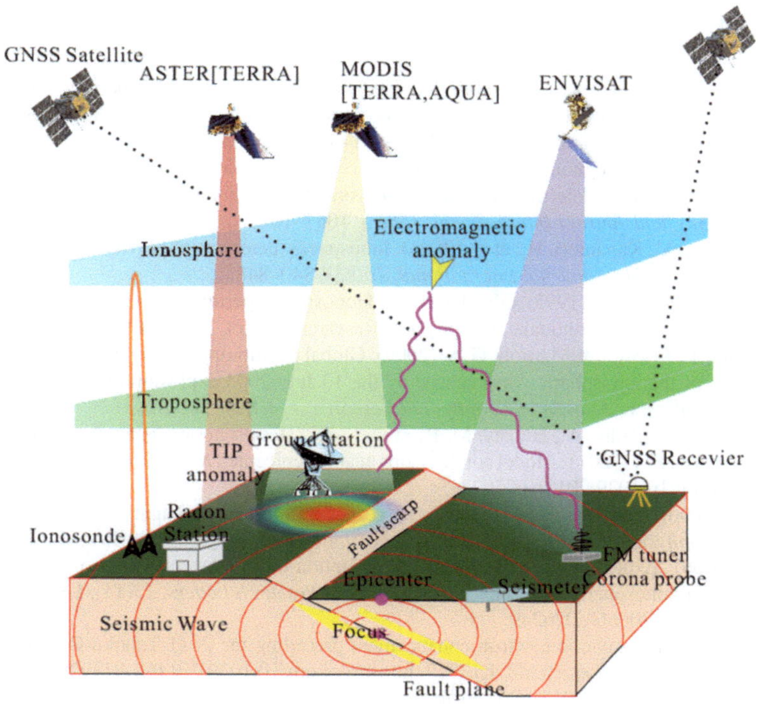

Fig. 15.6 Lithospheric-atmospheric-ionospheric coupling (LAIC) and monitoring from GNSS and other remote sensing

GPS stations of the GPS Earth observation network in Japan and more than 1000 continuous GPS stations in China. These dense ground-based GNSS network can receive more line-of-sight (LOS) signals from multi-frequency GNSS satellite constellations (over 100) through the atmosphere and ionosphere. These new signals will promote great improvements in the observation precision and the temporal-spatial resolution coverage of the atmosphere (Fig. 15.6).

Furthermore, more and more new techniques and other remote sensing satellites are available (Fig. 15.6), including MODIS, Advanced Space-borne Thermal Emission and Reflection Radiometer (ASTER) on NASA's Terra satellite, Environmental Satellite (ENVISAT), etc. together with dense ground in situ measurements, e.g., electro-magnetometer and seismometer, more detailed near-real-time micro-changes, processes and evolutions of the seismic tropospheric and ionospheric disturbances will be monitored and investigated. In theory, the possible mechanism of lithospheric–atmospheric–ionospheric coupling (LAIC) will be modeled and understood late, particularly, pre-seismic anomalies. These expected new advances in measurements and theories may provide pre-/co-earthquake/volcano signals for possible prediction of geohazards in the future.

References

Afraimovich, E. L., Ding, F., Kiryushkin, V., Astafyeva, E., Jin, S. G., & Sankov, V. (2010). TEC response to the 2008 Wenchuan earthquake in comparison with other strong earthquakes. *International Journal of Remote Sensing, 31*(13), 3601–3613.

Artru, J., Farges, T., & Lognonne, P. (2004). Acoustic waves generated from seismic surface waves: propagation properties determined from Doppler sounding observations and normal-mode modeling. *Geophysical Journal International, 158*(3), 1067–1077.

Artru, J., Ducic, V., Kanamori, H., et al. (2005). Ionospheric detection of gravity waves induced by tsunamis. *Geophysical Journal International, 160*(3), 840–848.

Calais, E., & Minster, J. B. (1995). GPS detection of ionospheric perturbations following the January 17, 1994, Northridge earthquake. Geophysical Research Letters, *22*(9), 1045–1048.

Dautermann, T., Eric, C., & Mattioli, G. S. (2009). Global positioning system detection and energy estimation of the ionospheric wave caused by the 13 July 2003 explosion of the soufrière hills volcano, montserrat. *Journal of Geophysical Research, 114*(B2), 311–320.

Galvan, D. A., Komjathy, A., Hickey, M. P., et al. (2012). Ionospheric signatures of Tohoku-Oki tsunami of March 11, 2011: Model comparisons near the epicenter. *Radio Science, 47*(4), RS4003. https://doi.org/10.1029/2012rs005023.

Freund, F. (2000). Time-resolved study of charge generation and propagation in igneous rocks. *Journal Geophysical Research, 105,* 11001–11019. https://doi.org/10.1029/1999JB900423.

Heki, K. (2006). Explosion energy of the 2004 eruption of the Asama volcano, central Japan, inferred from ionospheric disturbances. *Geophysical Research Letters, 33*(14), L34303. https://doi.org/10.1029/2006GL026249.

Heki, K. (2011). Ionospheric electron enhancement preceding the 2011 Tohoku-Oki earthquake. *Geophysical Reseach Letters, 38,* L17312. https://doi.org/10.1029/2011GL047908.

Heki, K., & Enomoto, Y. (2013). Preseismic ionospheric electron enhancements revisited. *Journal Geophysical Research: Space Physics, 118,* 6618–6626. https://doi.org/10.1002/jgra.50578.

Hickey, M. P., Schubert, G., & Walterscheid, R. L. (2009). Propagation of tsunami-driven gravity waves into the thermosphere and ionosphere. *Journal of Geophysical Research: Space Physics, 114*(A8), A08304.

Hines, C. O. (1972). Gravity waves in the atmosphere. *Nature, 239,* 73–78.

Jin, S. G., Zhu, W. Y., & Afraimovich, E. (2010). Co-seismic ionospheric and deformation signals on the 2008 magnitude 8.0 Wenchuan Earthquake from GPS observations, *International Journal of Remote Sensing., 31*(13), 3535–3543. https://doi.org/10.1080/01431161003727739.

Jin, S. G., Feng, G. P., & Gleason, S. (2011). Remote sensing using GNSS signals: current status and future directions. *Advances in Space Research, 47*(10), 1645–1653. https://doi.org/10.1016/j.asr.2011.01.036.

Jin, S. G., Jin, R., & Li, J. H. (2014). Pattern and evolution of seismo-ionospheric disturbances following the 2011 Tohoku earthquakes from GPS observations. *Journal of Geophysical Research: Space Physics, 119*(9), 7914–7927. https://doi.org/10.1002/2014JA019825.

Jin, S. G., Occhipinti, G., & Jin, R. (2015). GNSS ionospheric seismology: Recent observation evidences and characteristics. *Earth Science Reviews, 147,* 54–64. https://doi.org/10.1016/j.earscirev.2015.05.003.

Jin, S. G., Jin, R., & Li, D. (2017). GPS detection of ionospheric Rayleigh wave and its source following the 2012 Haida Gwaii earthquake. *Journal of Geophysical Research: Space Physics, 122*(1), 1360–1372. https://doi.org/10.1002/2016JA023727.

Kamogawa, M. (2006). Preseismic lithosphere-atmosphere-ionosphere coupling. *Eos, Transactions American Geophysical Union, 87*(40), 417–424.

Kamogawa, M., Kakinami, Y., Watanabe, S., Liu, J.-Y., & Watanabe, Y. (2012). Seismo-tsunamigenic ionospheric hole triggered by M 9.0 2011 off the Pacific coast of Tohoku earthquake. *Terrestrial, Atmospheric and Oceanic Sciences, 23,* 327–331.

Kamogawa, M., & Kakinami, Y. (2013). Is an ionospheric electron enhancement preceding the 2011 Tohoku-Oki earthquake a precursor? *Journal of Geophysical Research: Space Physics, 118*, 1751–1754. https://doi.org/10.1002/jgra.50118

Liu, J.-Y., Chuo, Y. J., Shan, S. J., Tsai, Y. B., Chen, Y. I., Pulinets, S. A., et al. (2004). Pre-earthquake ionospheric anomalies registered by continuous GPS TEC measurements. *Annales Geophysicae, 22*(5), 1585–1593.

Liu, J.-Y., Chen, C.-H., Lin, C.-H., Tsai, H.-F., Chen, C.-H., Kamogawa, M. (2011). Ionospheric disturbances triggered by the 11 March 2011 M9. 0 Tohoku earthquake. *Journal of Geophysical Research: Space Physics, 116*(A6), A06319. https://doi.org/10.1029/2011ja016761.

Lognonné, P., Clévédé, E., & Kanamori, H. (1998). Computation of seismograms and atmospheric oscillations by normal-mode summation for a spherical earth model with realistic atmosphere. *Geophysical Journal International, 135*(2), 388–406.

Matsumura, M., Saito, A., Iyemori, T., Shinagawa, H., Tsugawa, T., Otsuka, Y., et al. (2011). Numerical simulations of atmospheric waves excited by the 2011 off the Pacific coast of Tohoku Earthquake. *Earth, Planets and Space, 63*(7), 885–889.

Nakashima, Y., Heki, K., Takeo, A., Cahyadi, M. N., Aditiya, A., & Yoshizawa, K. (2015). Atmospheric resonant oscillations by the 2014 eruption of the kelud volcano, indonesia, observed with the ionospheric total electron contents and seismic signals. *Earth & Planetary Science Letters, 434,* 112–116.

Nappo, C. J. (2012). *An introduction to atmospheric gravity waves* (Vol. 102). Academic Press.

Occhipinti, G., Lognonné, P., Kherani, E. A., & Hebert, H. (2006). Three-dimensional waveform modeling of ionospheric signature induced by the 2004 Sumatra tsunami. *Geophysical Research Letters, 33*(20), L20104. https://doi.org/10.1029/2006GL026865.

Occhipinti, G., Rolland, L., Lognonne, P., & Watada, S. (2013). From Sumatra 2004 to Tohoku-Oki 2011: The systematic GPS detection of the ionospheric signature induced by tsunamigenic earthquakes. *Journal of Geophysical Research: Space Physics, 118*(6), 3626–3636.

Occhipinti, G., Coisson, P., Makela, J. J., Allgeyer, S., Kherani, A., Hebert, H., et al. (2011). Three-dimensional numerical modeling of tsunami-related internal gravity waves in the Hawaiian atmosphere. *Earth, Planets and Space, 63*(7), 847–851.

Pulinets, S., Ouzounov, D. (2011). Lithosphere–Atmosphere–Ionosphere Coupling (LAIC) model–An unified concept for earthquake precursors validation. *Journal of Asian Earth Sciences, 41*(4), 371–382.

Rolland, L., Occhipinti, G., Lognonné, P., Loevenbruck, A. (2010). The 29 September 2009 Samoan tsunami in the ionosphere detected offshore Hawaii. *Geophysical Research Letters, 37*, L17191.

Rolland, L. M., Lognonne, P., Astafyeva, E., Kherani, E. A., Kobayashi, N., Mann, M., et al. (2011). The resonant response of the ionosphere imaged after the 2011 off the Pacific coast of Tohoku Earthquake. *Earth, Planets and Space, 63*(7), 853–857.

Shah, M. (2017). *Statistical analysis and modelling of seismic ionospheric anomalies from GNSS and other satellite observations*, Ph.D. Thesis, University of Chinese Academy of Sciences.

Spiegel, E. A., & Veronis, G. (1960). On the Boussinesq approximation for a compressible fluid. *The Astrophysical Journal, 131,* 442.

Sun, L., Wan, W., Ding, F., & Mao, T. (2007). Gravity wave propagation in the realistic atmosphere based on a three-dimensional transfer function model. *Annales Geophysicae, 25*(9), 1979–1986.

Tsugawa, T., Saito, A., Otsuka, Y., Nishioka, M., Maruyama, T., Kato, H., et al. (2011). Ionospheric disturbances detected by GPS total electron content observation after the 2011 off the Pacific coast of Tohoku Earthquake. *Earth, Planets and Space, 63*(7), 875–879.

Correction to: Volcanic Plumes Detection from GNSS SNR

Correction to:
Chapter 14 in: S. Jin et al.,
***GNSS Atmospheric Seismology*,**
https://doi.org/10.1007/978-981-10-3178-6_14

In the original version of the book, the following belated correction has been incorporated:

The reference citation 'Larson et al. 2010' in Chapter 14 was changed to 'Larson, 2013'.
The text "..1227.60 MHz and wavelength λ 1 equal to 19.032 cm" was changed to "..1227.60 MHz and wavelength λ 2 equal to 24.42 cm".
The correction chapter has been updated with the change.

The updated version of this chapter can be found at
https://doi.org/10.1007/978-981-10-3178-6_14

S. Jin et al., *GNSS Atmospheric Seismology*,
https://doi.org/10.1007/978-981-10-3178-6_16

Index

S. Jin et al., *GNSS Atmospheric Seismology*,
https://doi.org/10.1007/978-981-10-3178-6

MIX
Papier aus verantwortungsvollen Quellen
Paper from responsible sources
FSC® C105338

If you have any concerns about our products,
you can contact us on
ProductSafety@springernature.com

In case Publisher is established outside the EU,
the EU authorized representative is:
Springer Nature Customer Service Center GmbH
Europaplatz 3, 69115 Heidelberg, Germany

Printed by Libri Plureos GmbH
in Hamburg, Germany